Technical Writing

Principles, Strategies, and Readings

Diana C. Reep

The University of Akron

Allyn and Bacon

Boston London Toronto Sydney Tokyo Singapore

For Faye and Sid Dambrot

A Division of Simon & Schuster, Inc.
160 Gould Street
Needham Heights, Massachusetts 02194

Executive Editor: Joseph Opiela
Series Editorial Assistant: Amy Capute
Editorial-Production Administrator: Rowena Dores
Editorial-Production Service: Editorial Inc.
Text Designer: Pat Torelli
Cover Administrator: Linda Dickinson
Composition and Manufacturing Buyer: Louise Richardson

Library of Congress Cataloging-in-Publication Data

Reep, Diana C.
Technical writing: principles, strategies, and readings / Diana C. Reep.
p. cm.
Includes bibliographical references and index.
ISBN 0-205-12937-4
1. English Language—Technical English. 2. English Language—Rhetoric.
3. Technical writing. I. Title.
P. E1475.R44 1991
808-.0666—dc20 90-26094
CIP

Credits

Pages 50–53. "Education: Our Best Vaccine Against AIDS," *Crib Sheet* 26(4) (April 1988). Reprinted with permission of Children's Hospital Medical Center of Akron, Akron, Ohio.

Pages 54–59. Schiff, B., "Intercooling," from *AOPA Pilot* 28 (May 1985). Reprinted with permission of the author.

Credits continued on page 619, which constitutes an extension of the copyright page.

Printed in the United States of America

10 9 8 7 6 5 4 3 2 95 94 93 92

Contents

Preface *xv*
Thematic Table of Contents for Part Two Readings *xi*

PART ONE **Technical Writing: Ways of Writing** **1**

1 **TECHNICAL WRITING ON THE JOB** **3**

Writing in Organizations 4
Reader/Purpose/Situation 4 Diversity of Technical Writing 5
Writing as a Process 6
One Writer's Process 6 Stages of Writing 8
Writing with Others 16
Writing with a Partner 16 Writing on a Team 17 Handling Conflicts among Writers 19 Writing for Management Review 19 Writing with a Technical Editor 21
Chapter Summary 22
Supplemental Readings in Part Two 22
Models 24
Exercises 34

2 **AUDIENCE** **35**

Analyzing Readers 36
Subject Knowledge 36 Position in the Organization 38 Personal Attitudes 38 Reading Style 39 Multiple Readers 40
Finding out about Readers 42
Informal Checking 42 Interviewing 43

Testing Reader-Oriented Documents 45
Readers Answering Questions 45 *Readers Performing a Task 46*
Readers Thinking Aloud 47
Chapter Summary 47
Supplemental Readings in Part Two 48
Models 49
Exercises 61

3 **ORGANIZATION** **65**

Sorting Information 66
Select Major Topics 66 *Identify Subtopics 67*
Constructing Outlines 67
Informal Outlines 67 *Formal Outlines 69* *Topic Outlines 70*
Sentence Outlines 70
Developing Effective Paragraphs 70
Unity 72 *Coherence 73* *Development 75* *Patterns for Presenting Information 76*
Chapter Summary 82
Supplemental Readings in Part Two 82
Models 83
Exercises 92

4 **DOCUMENT DESIGN** **95**

Understanding Design Features 96
Purpose of Design Features 96 *Design Principles 97*
Creating Graphic Aids 98
Purpose of Graphic Aids 99 *Tables 100* *Figures 101*
Using Format Elements 112
Written Cues 113 *White Space 117* *Typographic Devices 118*
Chapter Summary 121
Supplemental Readings in Part Two 121
Models 122
Exercises 131

5 **REVISION** **133**

Creating a Final Draft 134
Making Global Revisions 135
Content 135 *Organization 136*

Making Fine-Tuning Revisions 137
Overloaded Sentences 138 Clipped Sentences 139 Lengthy Sentences 140 Passive Voice 140 Jargon 142 Sexist Language 143 Concrete versus Abstract Words 145 Gobbledygood 146 Fad Words 147 Wordiness 148
Chapter Summary 150
Supplemental Readings in Part Two 150
Models 151
Exercises 157

6 **DEFINITION** **160**

Understanding Definitions 161
Writing Informal Definitions 162
Writing Formal Sentence Definitions 162
Writing Expanded Definitions 165
Cause and Effect 166 Classification 166 Comparison or Contrast 167 Description 167 Development 167 Etymology 168 Examples 168 Method of Operation or Process 169 Negation 169 Partition 169
Placing Definitions in Documents 170
Within the Text 170 In an Appendix 170 In Footnotes 171 In a Glossary 171
Chapter Summary 171
Supplemental Readings in Part Two 172
Models 173
Exercises 181

7 **DESCRIPTION** **183**

Understanding Description 184
Planning Descriptions 185
Subjects 185 Kinds of Information 186 Detail and Language 187 Graphic Aids 189
Writing Descriptions 190
Organization 190 Sections of a Technical Description 191 Sample Technical Descriptions 193
Chapter Summary 197
Supplemental Readings in Part Two 198

Models 199
Exercises 209

8 **INSTRUCTIONS, PROCEDURES, AND PROCESS EXPLANATIONS 212**

Understanding Instructions, Procedures, and Process Explanations 213
Writing Instructions 214
Readers 214 Organization 215 Details and Language 220 Troubleshooting Instructions 222
Writing Procedures 222
Procedures for One Employee 223 Procedures for Several Employees 224 Procedures for Handling Equipment and Systems 224
Writing Process Explanations 226
Readers 226 Organization 227 Details and Language 229
Chapter Summary 231
Supplemental Readings in Part Two 232
Models 233
Exercises 249

9 **LETTERS AND MEMOS 252**

Understanding Letters and Memos 253
Letters 253
Memos 253
Developing Effective Tone 253
Natural Language 254 Positive Language 255 You-Attitude 256
Organizing Letters and Memos 256
Direct Organization 257 Indirect Organization 259 Persuasive Organization 261
Writing Memos as a Manager 265
Selecting Letter Format 267
Date Line 267 Inside Address 267 Salutation 267 Body 269 Close 269 Signature Block 269 Notations 269 Second Page 270
Selecting Memo Format 270
Writing Job-Application Letters 270
Writing Résumés 271
Chapter Summary 275
Supplemental Readings in Part Two 275

Models 276
Exercises 294

10 **SHORT AND LONG REPORTS** **298**

Understanding Reports 299
Developing Short Reports 300
Opening-Summary Organization 300 Delayed-Summary Organization 302
Developing Long Reports 305
Planning Long Reports 305 Gathering Information 306 Taking Notes 310 Interpreting Data 312 Drafting Long Reports 313
Chapter Summary 315
Supplemental Readings in Part Two 316
Models 317
Exercises 338

11 **FORMAL REPORT ELEMENTS** **340**

Selecting Formal Report Elements 341
Writing Front Matter 341
Title Page 341 Transmittal Letter or Memo 342 Table of Contents 342 List of Figures 343 List of Tables 343 Abstract and Executive Summary 343
Writing Back Matter 346
References 346 Glossary/List of Symbols 346 Appendixes 347
Documenting Sources 347
APA System: Citations in the Text 349 APA System: List of References 350 Number-Reference System 355
Chapter Summary 356
Supplemental Readings in Part Two 356
Models 357
Exercises 375

12 **TYPES OF REPORTS** **377**

Understanding Conventional Report Types 378
Writing a Feasibility Study 378
Purpose 378 Organization 379
Writing an Incident Report 381
Purpose 381 Organization 381

Writing an Investigative Report 383
Purpose 383 Organization 384
Writing a Progress Report 387
Purpose 387 Organization 387
Writing a Trip Report 390
Purpose 390 Organization 390
Writing a Proposal 392
Purpose 392 Organization 393
Chapter Summary 397
Supplemental Readings in Part Two 397
Models 398
Exercises 425

13 **ORAL PRESENTATIONS** **428**

Understanding Oral Presentations 429
Purpose 429 Advantages 429 Disadvantages 430 Types of Oral Presentations 430
Organizing Oral Presentations 431
Preparing for Oral Presentations 434
Delivering Oral Presentations 436
Nerves 436 Rehearsal 436 Voice 436 Professional Image 436 Gestures 437 Eye Contact 437 Notes and Outlines 437 Visual Equipment 437 Questions 438
Joining a Team Presentation 438
Chapter Summary 439
Supplemental Readings in Part Two 439
Exercises 440

Appendix **GUIDELINES FOR GRAMMAR, PUNCTUATION, AND MECHANICS** **441**

Grammar 441
Dangling Modifiers 441 Misplaced Modifiers 442 Squinting Modifiers 442 Parallel Construction 443 Pronoun Agreement 443 Pronoun Reference 444 Reflexive Pronouns 444 Sentence Faults 445 Subject/Verb Agreement 446
Punctuation 447
Apostrophe 447 Colon 448 Comma 448 Dash 449 Exclamation Point 449 Hyphen 449 Parentheses 450

Quotation Marks 450 *Question Mark* 451 *Semicolon* 451 *Slash* 451

Mechanics 452

Acronyms and Initialisms 452 *Brackets* 452 *Capital Letters* 453 *Ellipsis* 454 *Italics* 455 *Measurements* 455 *Numbers* 456 *Symbols and Equations* 457

PART TWO **Technical Writing: Advice from the Workplace** **459**

BAGIN, C. B., and VAN DOREN, J. "How to Avoid Costly Proofreading Errors" 460

BENSON, P. J. "Writing Visually: Design Considerations in Technical Publications" 463

BERRY, E. "How to Get Users to Follow Procedures" 474

BLANK, S. J. "Greater Concern for Ethics and the 'Bigger Backyard' " 482

BOE, E. "The Art of the Interview" 485

BOWMAN, J. P., and OKUDA, T. "Japanese-American Communication: Mysteries, Enigmas, and Possibilities" 489

BROWN, A. "Is Ethics Good Business?" 495

BUEHLER, M. F. "Defining Terms in Technical Editing: The Levels of Edit as a Model" 501

CALABRESE, R. "Designing and Delivering Presentations and Workshops" 511

DAVIDSON, J. P. "Astute DP Professionals Pay Attention to Business Etiquette" 524

"Eliminating Gender Bias in Language" 528

ELSEA, J. G. "Strategies for Effective Presentations" 530

ERDLEN, J. D. "A Good Résumé Counts Most" 535

HOLCOMBE, M. W., and STEIN, J. K. "How to Deliver Dynamic Presentations: Use Visuals for Impact" 537

HORTON, W. "Writing Online Documentation" 543

"Is It Worth a Thousand Words?" 547

KIRTZ, M. K., and REEP, D. C. "A Survey of the Frequency, Types and Importance of Writing Tasks in Four Career Areas" 550

KLIEM, R. L. "Writing Technical Procedures" 553

MARRA, J. L. "For Writers: Understanding the Art of Layout" 557

MARTEL, M. "Combating Speech Anxiety" 563

PERRYMAN, P. "Technical Writing for Computer Software" 566

PLUNKA, G. A. "The Editor's Nightmare: Formatting Lists Within the Text" 573

"The Practical Writer" 580

REDISH, J. "Creating Computer Menus that People Understand Easily" 585

RIDGWAY, L. "The Writer as Market Researcher" 590

"Six Graphic Guidelines" 596

"When You're Not the Whole Staff" 602

WICCLAIR, M. R. and FARKAS, D. K." "Ethical Reasoning in Technical Communication: A Practical Framework" 605

Thematic Contents for Part Two Readings

Audience: Testing Reader Use of Documents

RIDGWAY, L. "The Writer as Market Researcher" 590

Business Etiquette

DAVIDSON, J. P. "Astute DP Professionals Pay Attention to Business Etiquette" 524

Computer Documentation

HORTON, W. "Writing Online Documentation" 543
PERRYMAN, P. "Technical Writing for Computer Software" 566
REDISH, J. "Creating Computer Menus that People Understand Easily" 585

Document Design

BENSON, P. J. "Writing Visually: Design Considerations in Technical Publications" 463
BERRY, E. "How to Get Users to Follow Procedures" 474
"Is It Worth a Thousand Words?" 547
KLIEM, R. L. "Writing Technical Procedures" 553
MARRA, J. L. "For Writers: Understanding the Art of Layout" 557
"Six Graphic Guidelines" 596

Ethics

BLANK, S. J. "Greater Concern for Ethics and the 'Bigger Backyard' " 482
BROWN, A. "Is Ethics Good Business?" 495
WICCLAIR, M. R., and FARKAS, D. K. "Ethical Reasoning in Technical Communication: A Practical Framework" 605

Graphics: Oral Presentations

HOLCOMBE, M. W., and STEIN, J. K. "How to Deliver Dynamic Presentations: Use Visuals for Impact" 537

Graphics: Written Communication

BENSON, P. J. "Writing Visually: Design Considerations in Technical Publications" 463
"Is It Worth a Thousand Words?" 547
"Six Graphic Guidelines" 596

International Communication

BOWMAN, J. P., and OKUDA, T. "Japanese-American Communication: Mysteries, Enigmas, and Possibilities" 489

Job Search

BOE, E. "The Art of the Interview" 485
DAVIDSON, J. P. "Astute DP Professionals Pay Attention to Business Etiquette" 524
ERLDEN, J. D. "A Good Résumé Counts Most" 535
KIRTZ, M. K., and REEP, D. C. "A Survey of the Frequency, Types and Importance of Writing Tasks in Four Career Areas" 550

Lists

PLUNKA, G. A. "The Editor's Nightmare: Formatting Lists Within the Text" 573

Manuals

MARRA, J. L. "For Writers: Understanding the Art of Layout" 557
PERRYMAN, P. "Technical Writing for Computer Software" 566
RIDGWAY, L. "The Writer as Market Researcher" 590

Oral Presentations: Audience

CALABRESE R. "Designing and Delivering Presentations and Workshops" 511
ELSEA, J. G. "Strategies for Effective Presentations" 530

Oral Presentations: Delivery

CALABRESE, R. "Designing and Delivering Presentations and Workshops" 511
ELSEA, J. G. "Strategies for Effective Presentations" 530
MARTEL, M. "Combating Speech Anxiety" 563

Oral Presentations: Graphic Aids

HOLCOMBE, M. W., and STEIN, J. K. "How to Deliver Dynamic Presentations: Use Visuals for Impact" 537

Oral Presentations: Organization

CALABRESE, R. "Designing and Delivering Presentations and Workshops" 511

Oral Presentations: Conferences and Workshops

CALABRESE, R. "Designing and Delivering Presentations and Workshops" 511

Policies and Procedures

BERRY, E. "How to Get Users to Follow Procedures" 474
KLIEM, R. L. "Writing Technical Procedures" 553

Proofreading

BAGIN, C. B., and VAN DOREN, J. "How to Avoid Costly Proofreading Errors" 460

Revising and Editing

BUEHLER, M. F. "Defining Terms in Technical Editing: The Levels of Edit as a Model" 501
"Eliminating Gender Bias in Language" 528
"The Practical Writer" 580

Style

"Eliminating Gender Bias in Language" 528
"The Practical Writer" 580

Writing Process

"When You're Not the Whole Staff" 602

Preface

Technical Writing: Principles, Strategies and Readings is designed for students who study technical writing as part of their career preparation in science, technology, social services, and business. The book is a unique combination of instructional chapters, readings from the real world, models from various disciplines, and exercises that offer students in-class practice of principles and forms as well as well as out-of-class assignments that call for research and development of a complete technical document.

Instructional Chapters

Part One contains thirteen chapters of instruction. Five chapters discuss the writing process on the job, including team writing; analyzing the audience; organizing documents; designing documents; and revising and editing. The other eight chapters discuss specific document types and communication strategies, including definitions; descriptions; instructions; procedures; process explanations; letters and memos; oral presentations; formal reports and types of reports, such as feasibility studies, progress reports, and proposals. Instructors can use these instructional chapters in the order presented here, or they can concentrate on specific topics, following their own pedagogical approach.

These concise instructional chapters include (1) checklists to guide students in planning, drafting, and revising documents; (2) sample outlines for document types to provide students with starting points in drafting their own documents; (3) models of technical documents; and (4) exercises for practice and for developing complete technical documents.

Models

The models that accompany each chapter represent a variety of documents typically written on the job: information brochures, newsletter articles, short and long reports, company procedures bulletins, police accident reports, memos and letters, social service agency court reports, company product descriptions, operating manuals, and scientific research reports.

Some models, such as those in Chapter 1, represent successive drafts of a document, so students can analyze changes writers make during the drafting process. Student models are also included; some are in response to exercises in the chapter.

Most models include commentary that explains to students how writers use specific writing strategies and make context decisions. Models without detailed commentary allow instructors to guide students as they analyze the organization and writing strategies used in the model. The models are not presented as "perfect specimens." Students are encouraged to critique and revise them. Discussion questions to encourage student analysis accompany the models. Many models also include writing exercises specific to the model or calling for a document similar to the model. Most of these exercises can be used for in-class practice or homework.

Exercises

The classroom-tested exercises that accompany the models or appear at the end of each chapter include both in-class and out-of-class writing tasks. Exercises vary in difficulty and in professional subject and may require students to (1) revise a sample of poor writing, (2) develop a full original document, (3) read one or more articles in Part Two and use them in a writing task, (4) prepare an oral presentation connected to a written document, (5) collaborate with other students in drafting a document or making an oral presentation.

Appendix

The appendix offers a convenient reference to the fundamentals of grammar, punctuation, and mechanics of style for such items as measurements, acronyms, and equations. Instructors may assign portions of the appendix, or students may use it to check their writing for correctness.

Readings

Part Two contains 28 short "how-to" articles by people who write on the job and by instructors who teach technical communication. All the articles are concerned with technical communication topics. Some offer further information on issues covered in the chapters, such as document design and oral presentations; others discuss supplementary topics the instructor may wish to include in the course, such as ethics and international communication. The thematic table of contents for these readings provides easy access to readings that discuss technical communication topics of interest. Also, each instructional chapter includes a list of the readings in Part Two that complement the topics covered in the chapter.

Instructor's Manual

The instructor's manual available with this book offers suggestions to those instructors who wish to try new approaches with familiar topics and to new instructors who are teaching technical writing for the first time. The manual includes suggestions for course policies, sample course outlines, a directory of document types appearing in the book, chapter-by-chapter teaching suggestions, suggestions for using the readings, a list of further readings, and a list of academic journals in professional communication.

Acknowledgments

Many people contributed in various ways to the development of this book. I am grateful for the assistance of professional people who provided technical writing models and for information about on-the-job writing: Marjorie R. Kohls, Midwesco, Inc., Niles, Illinois; Jacquelyn Biel, Kompas/Biel and Associates, Milwaukee, Wisconsin; William H. Lambert, Jr., Diebold, Inc., North Canton, Ohio; Sidney Dambrot, Ford Motor Company, Brook Park, Ohio; Keith Price, The Timken Company, Canton, Ohio. I would also like to thank the reviewers who helped in the development of the text: Jane Allen, New Mexico State University; Nancy Blyler, Iowa State University; Virginia Book, University of Nebraska: Davida Charney, Pennsylvania State University; David Farkas, University of Washington, Seattle; Deborah Journet, University of Louisville; Peter McGuire, Georgia Institute of Technology; Robert Mehaffey, American River College; Charles Sides, Northeastern University; Katherine Staples, Austin Community College; Thomas Warren, Oklahoma University; and Don Zimmerman, Colorado State University. I am indebted to The University of Akron and especially Dean Claibourne Griffin for a Faculty Improvement Leave, allowing me to spend one semester in full-time work on this book. Those friends who sustained me with extraordinary support and encouragement include Dawn Trouard, Gerald Alred, Martin McKoski, Thomas Dukes, Eric Birdsall, Ellen Heib, Emily Ann Allen, Michael

Mikolajczak, Janet Marting, Kay Whitford, Penny Ryan, and Nick Ranson. For encouragement and assistance above and beyond the call of friendship, I thank Faye Dambrot.

At Allyn and Bacon, Joseph Opiela led this project from the initial idea to the completed manuscript; Amy Capute and Rebecca Dudley provided essential assistance in many matters. My thanks to research assistants Roberta Ungar and Diana Loucks for library work. Finally, my immense gratitude to Sonia Dial who cheerfully typed and retyped multiple drafts of every page of this book.

Part One

Technical Writing: Ways of Writing

CHAPTER 1

Technical Writing on the Job

Writing in Organizations
- Reader/Purpose/Situation
- Diversity in Technical Writing

Writing as a Process
- One Writer's Process
- Stages of Writing
  - Planning
  - Multiple Drafting
  - Revising and Editing

Writing with Others
- Writing with a Partner
- Writing on a Team
  - Team Planning
  - Team Drafting
  - Team Revising and Editing
- Handling Conflict among Writers
- Writing for Management Review
- Writing with a Technical Editor

Chapter Summary

Supplemental Readings in Part Two

Endnotes

Models

Exercises

Writing in Organizations

No matter what your job is, writing will be important to your work because you will have to communicate your technical knowledge to others, both inside and outside the organization. Consider these situations. An engineer writes an article for a professional engineering journal, describing a project to restructure a city's sewer pipelines. A police officer writes a report for every arrest and incident that occurs on a shift. An artist writes a grant application asking for funds to create a large, postmodern, steel sculpture. A dietitian writes a brochure about choosing foods low in cholesterol for distribution to participants in a weight-control seminar. Anyone who writes about job-related information prepares technical documents that supply information to readers who need it for a specific purpose.

Surveys of people on the job indicate that writing consumes a significant portion of their working time. One study reported that engineers spend 25% of their job-related time writing technical documents and 23% reading technical and business materials.[1] Another study reported that 62.8% of the upper-level managers, 39.3% of the midlevel managers, and 43.5% of the low-level managers surveyed said that written communication was "extremely important" as a management tool.[2] And a survey of people in four different career areas reported that 39.4% in management positions, 22.2% in technical fields, 20.8% in clerical positions, and 31% in social-service occupations said they spend 21% to 40% of their job-related time writing.[3] These responses clearly indicate that writing is an important element in many careers.

Reader/Purpose/Situation

Three elements to consider in writing any technical documents are reader, purpose, and writing situation. The reader of a technical document seeks information for a specific purpose, and the writer's goal is to design a document that will serve the reader's needs and help the reader understand and use the information efficiently. The writing situation consists of both reader and purpose, as well as such factors as the sponsoring organization's size, budget, ethics, deadlines, policies, competition, and priorities. Consider this example of a writing situation.

Lori Vereen, an occupational therapist, must write a short article about the new Toddler Therapy Program, which she directs, at Children's Hospital. Her article will be in the hospital's monthly newsletter, which is sent to people who have donated funds in the past ten years. Newsletter readers are interested in learning about hospital programs and new medical technology, and many of the readers will donate funds for specific programs. Reading the articles in the newsletter helps them decide how to allocate their donations.

Lori understands her readers' purpose in using the newsletter. She also knows that the hospital's management wants to encourage readers to donate to specific programs (writer's purpose). Although she could include in her article much specialized information about new therapy techniques for children with cerebral palsy, she decides that her readers will be more interested in learning how the children progress through the program and develop the ability to catch a ball or draw a circle. Scientific details about techniques to enhance motor skills will not interest these readers as much, or inspire as many donations, as stories about children needing help will. Lori's writing situation is restricted further by hospital policy (she cannot use patients' names), space (she is limited to 700 words), and time (she has one day to write her article).

Like Lori, you should consider reader, purpose, and situation for every on-the-job writing task you have. These three elements will influence all your decisions about the document's content, organization, and format.

Diversity in Technical Writing

Science and engineering were once considered the only subjects of technical writing, but that limitation no longer applies. All professional fields require technical documents that help readers perform tasks and understand specific tasks. These sentences are from technical documents that provide readers with guidelines for properly completing a specific procedure:

- "Retired employees who permanently change their state of residence must file Form 176-B with the Human Resources Office within 90 days of moving to transfer medical coverage to the regional plant serving their area." (This sentence is from a manufacturing company's employee handbook. Readers are employees who need directions for keeping their company benefit records up to date.)
- "Be sure your coffeemaker is turned OFF and unplugged from the electrical outlet." (This sentence is from a set of instructions packed with a new product. Readers are consumers who want to know how to use and care for their new coffeemaker properly.)
- "The diver must perform a reverse three-and-a-half somersault tuck with a 3.4 degree of difficulty from a 33-ft platform." (This sentence is from the official regulations for a college diving competition. The readers are coaches who need to train their divers to perform the specific dives required for the meet.)
- "For recover installations, pressure-relieving metal vents should be installed at the rate of 1 for every 900 square feet of roof area." (This sen-

tence is from a catalog for roofing products and systems. The readers are architects and engineers who may use these products in a construction project. If they do, they will include the manufacturer's specifications for the products in their own construction design documents.)

Although these sentences involve very different topics, they all represent technical writing because they provide specific information to clearly identified readers who will use the information for a specific purpose.

Writing as a Process

The process of writing a technical document includes three general stages—planning, multiple drafting, and revising—but remember, this process is not strictly linear. If you are like most writers, you will revise your decisions about the document many times as you write. You may get an entirely new idea about format or organization while you are drafting, or you may change your mind about appropriate content while you are revising. You probably will also develop a personal writing process that suits your working conditions and preferences. Some writers compose a full draft before tailoring the document to their readers' specific needs; others analyze their readers carefully before gathering any information for the document. No one writing system is appropriate for all writers, and even experienced writers continue to develop new habits and ways of thinking about writing.

One Writer's Process

Margo Keaton is a mechanical engineer who works for a large construction firm in Chicago. Presently, her major project is constructing a retirement community in a Chicago suburb. As the project manager, she meets frequently with the general contractor, subcontractors, and construction foremen. At one meeting with the general contractor and six subcontractors, confusion arises concerning each subcontractor's responsibility for specific construction jobs, including wiring on control motors and fire-safing the wall and floor openings. Because no one appears willing to take the responsibility for the jobs or to accept statements at the meeting as binding, Margo realizes that, as project manager, she will have to determine each subcontractor's precise responsibility.

When she returns to the office she shares with five other engineers, she considers the problem. Responsibility became confused because some subcontractors were having trouble keeping their costs within their initial bids for the construction project. Eliminating some work would ease their financial burden. Then too, no one, including the general contractor, is certain about who has definite responsibilities for the wiring system and for fire-safing the openings. Margo's task is to sort out the information and write a report that will delegate responsibility to the appropriate subcontractors. As Margo thinks about the problem, she realizes that her audience for this report consists of several readers who will each use the report differently. The general contractor needs the information to understand the chain of responsibility, and the subcontractors need the report as an assignment of duties. All the readers will have to accept Margo's report as the final word on the subject.

While analyzing the situation, Margo makes notes on the problem, the objectives of her report, and the information that should be included. She finds that she needs the original specifications for the project, previous reports, and the original bids. Looking at her list, she decides that she also needs to look up correspondence with the subcontractors to check for a record of any changes in the original agreements. She then lists the people she regards as her readers for this report.

The next morning Margo collects the relevant specifications, originally prepared by the design engineer, and locates the correspondence with the subcontractors. She finds several progress reports and the minutes of past meetings, but they do not mention the specific responsibilities under dispute. After she checks all her material, she decides to enlarge her list of readers to include the design engineer, whose specifications she intends to quote, and the owner's representative, who attended all the construction meetings and heard the arguments over responsibilities. Although the design engineer and the owner's representative are not directly involved in the argument over construction responsibilities, Margo knows they want to keep up to date on all project matters.

Margo writes her notes on sticky notepad paper so that she can spread the sheets out on the side table next to her desk. She finds it helpful to move the notes around while she thinks about ways to organize her report.

Margo makes her first decision about organization because she knows that some of her readers will resent any conclusions in her report that assign them more responsibility. The opening summary she usually uses in short reports will not suit this situation. Some of her readers might become angry and stop reading before they reach the explanation based on the specifications. Instead, Margo decides to structure the report so that the documentation and explanation come first, followed by the assignment of responsibilities. She plans to quote extensively from the specifications.

In her opening she will review the misunderstanding and remind her readers why this report is needed. Because the general contractor ultimately has the task of enforcing assignment of responsibilities, she decides to address the report to him and send copies of it to all other parties.

Moving her notes around on the table into a rough outline, Margo groups her information according to specific tasks, such as wiring the control motors. After that, she moves her notes again so that each group begins with a definition of the specific task and ends with the name of the subcontractor responsible for it. Because the original design specification indicated which subcontractor was to do what and yet everyone seems confused, she decides that she should not only quote from the documents, but also paraphrase the quotations to ensure that her readers understand. By the end of the report, she will have assigned responsibility for each task in the project.

As she writes her draft, she also chooses words carefully because her readers will resent any tone that implies they have been trying to avoid their responsibilities or that they are not intelligent enough to figure out the chain of responsibility for themselves. She writes somewhat quickly because she has already made her major organizational decisions. The report turns out to be longer than she expected because of the need to quote and paraphrase so extensively, but she is convinced that such explanations are crucial to her purpose and her readers' understanding. Looking at her first draft again, she changes her mind about the order of the sections and decides to discuss the least controversial task first to keep her readers as calm as possible for as long as possible.

She gives the draft to Kimberly, the secretary, to put on the word processor and tells Kimberly to mark any sections that seem unclear. Margo always asks the secretary to check sentence clarity for her because she believes that if Kimberly cannot understand her meaning, her readers may not understand either. When Kimberly returns the report, she has marked one sentence that seems to be missing something.

Margo's revision at this stage focuses on three major considerations: (1) Is all the quoted material adequately explained? (2) Are there enough quotations to cover every issue and clearly establish who is to do what? (3) Is her language neutral so that no one will think she is biased or dictatorial? She decides that she does have enough quoted material to document each task and that, aside from the one garbled sentence Kimberly noticed, her explanations are clear. Finally, she reads the report carefully for language and makes a few more word changes.

As a check, she takes the report to her supervisor and asks him whether he thinks it will settle the issue of job responsibilities. When Margo writes a bid for a construction project, her supervisor always reads the document carefully to be sure that special conditions and costs applying to the project are covered thoroughly. For this report, however, he assumes that Margo, as project manager, has included all the pertinent information. After a quick reading, he says the report is fine.

Margo then edits her report for surface accuracy—clear sentence structure and correct grammar and punctuation. Then she reads the report aloud so that she can hear any sentences that may still be unclear. After she is satisfied, she returns the report to Kimberly, who makes the changes, prints the final copy, and distributes it.

Stages of Writing

Margo Keaton's personal writing process enables her to control her writing tasks even while working at a busy construction company where interruptions occur every few minutes. She relies on notes and lists to keep track of both information and her decisions about a document because she often has to leave the office in the middle of a writing task. Her notes and lists enable her to pick up where she left off when she returns. She uses a table to arrange her notes because she finds it easier to think about a document's content if she sees all the information spread out before her.

Aside from using personal devices for handling the writing process, most writers go through the same three general stages as they develop a document. Remember that all writers do not go through these stages in exactly the same order, and writers often repeat stages as they make new decisions about the content and format of a document.

Planning

In the planning stage, a writer analyzes reader, purpose, and the writing situation; gathers information; and tentatively organizes the document. All these activities may recur many times during the writing.

Analyzing Readers. No two readers are exactly alike. They differ in knowledge, needs, abilities, attitudes, relation to the situation, their purpose in using the document. All readers are alike, however, in that they need documents that provide information they can understand and use. As a writer, your task is to create a document that will fit the precise needs of your readers. Margo Keaton's readers were all technical people, so she was free to use technical terms without defining them. Because she knew some readers would be upset by her report, she made her first organizational decision—she would not use an opening summary. Detailed strategies for thinking about readers are discussed in Chapter 2. In general, however, consider these questions whenever you analyze readers and your readers' needs:

- Who are my specific readers?
- Why do they need this document?
- How will they use it?
- Do they have a hostile, friendly, or neutral attitude toward the subject?
- What is the level of their technical knowledge about the subject?
- How much do they already know about the subject?
- Do they have preferences for some elements, such as tables, headings, or summaries?

Analyzing Purpose. A document should accomplish something. Remember that a document actually has two purposes: (1) what the writer wants the reader to know or do and (2) what the reader wants to know or do. The writer of instructions, for example, wants to explain a procedure so that readers can perform it. Readers of instructions want to follow the steps to achieve a specific result. The two purposes obviously are closely related, but they are not identical. Margo Keaton wrote her report to assign responsibilities (writer's purpose), as well as to enable her readers to understand the situation and plan their actions accordingly (reader's purpose). A writer should consider both in planning a document. Furthermore, different readers may have different purposes in using the same document, as Margo Keaton's readers did. In analyzing your document's purpose, you may find it helpful to think first of these general purposes:

To instruct—The writer tells the reader how to do a task and why it should be done. Documents that primarily instruct include training and operator manuals, policy and procedure statements, and consumer instructions. Such documents deal with

- The purpose of the procedure
- Steps in performing the procedure
- Special conditions that affect the procedure

To record—The writer sets down the details of an action, decision, plan, or agreement. The primary purpose of minutes, file reports, and laboratory reports is to record events both for those currently interested and for others who may be interested in the future. Such documents deal with

- Tests or research performed and results
- Decisions made and responsibilities assigned
- Actions and their consequences

To inform (for decision making)—The writer supplies information and analyzes data to enable the reader to make a decision. For decision making a reader may use progress reports, performance evaluations, or investigative reports. Such documents deal with

- Specific facts that materially affect the situation
- The influence the facts have on the organization and its goals
- Significant parts of the overall situation

To inform (without decision making)—The writer provides information to readers who need to understand data but do not intend to take any action or make a decision. Technical writing that informs without expectation of action by the reader includes information bulletins, literature reviews, product descriptions, and process explanations. Such documents deal with

- The specific who, what, where, when, why, and how of the subject
- A sequence of events showing cause and effect
- The relationship of the information to the company's interests

To recommend—The writer presents information and suggests a specific action. Documents with recommendation as their purpose include simple proposals, feasibility studies, and recommendation reports. Such documents deal with

- Reasons for the recommendation
- Expected benefits
- Why the recommendation is preferable to another alternative

To persuade—The writer urges the reader to take a specific action or to reach a specific conclusion about an issue. To persuade, the writer must convince the reader that the situation requires action and that the information in the report is relevant and adequate for effective decision making. A report recommending purchase of a specific piece of equipment, for instance, may present a simple cost comparison between two models. However, a report that argues the need to close a plant in one state and open a new one in another state must persuade readers about the practicality of such a move. The writer will have to (1) explain why the facts are relevant to the problem, (2) describe how they were obtained, and (3) answer potential objections to the plan. Documents with a strong persuasive purpose include construction bids, grant applications, technical advertisements, technical news releases, and reports dealing with sensitive topics, such as production changes to reduce acid rain. Such documents emphasize

- The importance or urgency of the situation
- The consequences to the reader or others if a specific action is not taken or a specific position is not supported
- The benefits to the reader and others if a specific action is taken or a specific position is supported

To interest—The writer describes information to satisfy a reader's intellectual curiosity. Although all technical writing should satisfy readers' curiosity, writing that has interest as its main purpose includes science articles in popular magazines, brochures, and pamphlets. Such documents deal with

- How the subject affects daily life
- Amusing, startling, or significant events connected to the subject
- Complex information in simplified form for general readers

The general purpose of Margo Keaton's report was to inform her readers about construction responsibilities. Her specific purpose was to delegate responsibilities so that the subcontractors could work efficiently. The readers' specific purpose was to understand their duties. Remember that technical documents have both a general purpose and a specific purpose relative to the writing situation. Remember also that documents generally have multiple purposes because of the specific needs of the readers. For instance, a report that recommends purchasing a particular computer model also must provide enough information about capability and costs so that the reader can make a decision. Such a report also may include information about the equipment's design to interest the reader and may act as a record of costs and capability as of a specific date.

Strategies for analyzing readers' purpose are included in Chapter 2. Consider these questions in determining purpose:

- What action (or decision) do I want my reader to take (or make)?
- How does the reader intend to use this document?
- What effect will this document have on the reader's work?
- Is the reader's primary use of this document to be decision making, performing a task, or understanding information?
- If there are multiple readers, do they all have the same goals? Will they all use the document in the same way?
- Do my purpose and my reader's purpose conflict in any way?

Analyzing the Writing Situation. No writer on the job works completely alone or with complete freedom. The organization's environment may help or hinder your writing and certainly will influence both your document and your writing process. The organizational environment in which you write includes (1) the roles and authority both you and your readers have in the organization and in the writing situation, (2) the communication atmosphere, that is, whether information is readily available to employees or only to a few top-level managers, (3) preferences for specific documents, formats, or types of information, (4) the organization's relationship with the community, customers, competitors, unions, and government agencies, (5) government regulations controlling both actions and communication about those actions, and (6) trade or professional associations with standards or ethical codes the organization follows. You can fully understand an organization's environment only by working in it, because each is a unique combination of individuals, systems, relationships, goals, and values.

When Margo Keaton analyzed her writing situation, she realized that (1) her readers were hostile to the information she was providing, (2) work on the retirement community could not continue until the subcontractors understood and accepted her information about construction responsibilities, and (3) the delay caused by the dispute among subcontractors jeopardized her own position because she was responsible for finishing the project on time and within budget. Margo's writing situation, therefore, included pressures from readers' attitudes and time constraints. In analyzing your writing situation, consider these questions:

- Is this subject controversial within the organization?
- What authority do my readers have relative to this subject?

- What events created the need for this document?
- What continuing events depend on this document?
- Given the deadline for this document, how much information can be included?
- What influence will this document have on company operations or goals?
- Is this subject under the control of a government agency or specific regulations?
- What external groups are involved in this subject, and why?
- Does custom indicate a specific document for this subject or a particular organization and format for this kind of document?

Gathering Information. Generally, you will have some information when you begin a writing project. Some writers prefer to analyze reader, purpose, and situation and then gather information; others prefer to gather as much information as possible early in the writing process and then decide which items are needed for the readers and purpose. Information in documents should be (1) accurate, (2) relevant to the readers and purpose, and (3) up to date or timely. Margo Keaton's information gathering was focused on existing internal documents because she had to verify past decisions. Many sources of information are available beyond your own knowledge and company documents. Chapter 10 discusses how to find information from outside sources.

Organizing the Information. As you gather information, you will probably think about how best to organize your document so that your readers can use the information efficiently. As Margo Keaton gathered information for her report, she organized the material by posting her notes on her side table and moving them around as she added new ones. When she finished her note taking, she grouped the information according to specific tasks and then put her notes in order for each task. General strategies for organizing documents are discussed in Chapter 3. In organizing information, a writer begins with two major considerations: (1) how to group the information into specific topics and (2) how to arrange the information within each topic.

Grouping Information into Topics. Arranging information into groups requires looking at the subject as a whole and recognizing its parts. Sometimes you know the main topics from the outset because the subject of the document is usually organized in a specific way. In a report of a research experiment, for example, the information would probably group easily into topics, such as research purpose, procedure, specific results, and conclusions. Always consider what groupings will help the reader use the information most effectively. In a report comparing two pieces of equipment, you might group information by such topics as cost, capability, and repair costs to help your reader decide which model to purchase. If you were writing the report for technicians who will maintain the equipment, you might group information by such topics as safety factors, downtime, typical repairs, and maintenance schedules. Consider these questions when grouping information:

- Does the subject matter have obvious segments? For example, a process explanation usually describes a series of distinct stages.
- Do some pieces of information share one major focus? For example, data about equipment purchase price, installation fees, and repair costs might be grouped under the major topic cost, with subtopics covering initial cost, installation costs, and maintenance costs.
- Does the reader prefer that the same topics appear in a specific type of document? For example, some readers may want "benefits" as a separate section in any report involving recommendations.

Arranging Information within a Topic. After grouping the information into major topics and subtopics, organize it effectively within each group. Consider these questions:

- Which order will enable the reader to understand the material easily? For example, product descriptions often describe a product from top to bottom or from bottom to top so that readers can visualize the connecting parts.
- Which order will enable the reader to use this document? For example, instructions should present the steps in the order in which the reader will perform them. Many managers involved in decision making want information in descending order of importance so that they can concentrate on major issues first.
- Which order will help the reader accept this document? For example, when Margo Keaton organized her sections, she held back information her readers were not eager to have until the end of the report.

By making a master list of your topics in order or by moving your notes physically, as Margo Keaton did, you will be able to visualize the document's structure. The master list of facts in order is an *outline.* Although few writers on the job take the time to make a formal outline with Roman numerals or decimal numbering, they usually make some kind of informal outline because the information involved in most documents is too lengthy or complicated to be organized coherently without using a guide. Some writers use lists of the major topics; others prefer more detailed outlines and list every major and minor item. Outlines do not represent final organization decisions. You may reorganize as you write a draft or as you revise, but an outline will help you control a writing task in the midst of on-the-job interruptions. After a break in the writing process, you can continue writing more easily if you have an outline.

Multiple Drafting

Once you have tentatively planned your document, a rough draft is the next step. At this stage, focus on thoroughly developing the information you have gathered, and do not worry about grammar, punctuation, spelling, and fine points of style. Thinking about such matters during drafting will interfere with your decisions about content and organization. Follow your initial plan of organization, and write quickly. When you have a completed draft, then think about revision strategies. A long, complicated document may require many drafts, and most documents except the simplest usually require several drafts. Margo Keaton wrote her first rough draft from beginning to end using the notes she had organized; other writers may compose sections of the document out of order and then put them in order for a full draft. Keep your reader and purpose firmly in mind while you write, because ideas for new information to include or new ways to organize often occur during drafting.

Revising and Editing

Revision takes place throughout the writing process, but particularly after you have begun drafting. Read your draft and rethink these elements:

- *Content*—Do you need more facts? Are your facts relevant for the readers and purpose?
- *Organization*—Have you grouped the information into topics appropriate for your readers? Have you put the details in an order that your readers will find easy to understand and use? Can your readers find the data easily?
- *Headings*—Have you written descriptive headings that will guide your readers to specific information?
- *Openings and closings*—Does your opening establish the document's purpose and introduce the readers to the main topic? Does your closing provide a summary, offer recommendations, or suggest actions appropriate to your readers and purpose?
- *Graphic aids*—Do you have enough graphic aids to help your readers understand the data? Are the graphic aids appropriate for the technical knowledge of your readers?
- *Language*—Have you used language appropriate for your readers? Do you have too much technical jargon? Have you defined terms your readers may not know?
- *Reader usability*—Can your readers understand and use the information effectively? Does the document format help your readers find specific information?

After you are satisfied that you have revised sufficiently for your reader and purpose, edit the document for correct grammar, punctuation, spelling, and company editorial style. For Margo Keaton, revision centered on checking the report's content to be sure that every detail relative to assigning construction responsibilities was included. She also asked her supervisor and the secretary to read the report to check that the information was clear and that the tone was appropriate for the sensitive issue. Her final editing focused on grammar and punctuation. Remember that no one writes a perfect first draft. Revising and editing are essential for producing effective technical documents.

Writing with Others

Writing on the job usually means collaborating with others. A survey of members of six professional associations, such as the American Institute of Chemists and the American Consulting Engineers Council, found that 87% of those responding said they sometimes wrote in collaboration with others, and 98% said that effective writing was "very important" or "important" to successfully doing their work.[4]

You may informally ask others for advice when you write, or you may write a document with a partner or a team of writers. When writers work together to produce a document, problems often arise beyond the usual questions about content, organization, style, and clarity. Overly harsh criticism of another's work can result in a damaged ego; moreover, frustration can develop among writers when their writing styles and paces clash. In addition to other writers, you may also collaborate with a manager who must approve the final document, or you may work with a technical editor who must prepare the document for publication.

Writing with a Partner

Writing a document with a partner often occurs when two people report research or laboratory tests they have conducted jointly. In other situations, two people may be assigned to a feasibility study or field investigation, and both must write the final report. Sometimes one partner will write the document and the other will read, approve, and sign it. More often, however, the partners will collaborate on the written document, contributing separate sections or writing the sections together. Here is a checklist for writing with a partner:

- Plan the document together so that you both clearly understand the document's reader and purpose.
- Divide the task of gathering information so that neither partner feels overworked.

- Decide who will draft which sections of the document. One partner may be strong in certain topics and may prefer to write those sections.
- Consult informally to clarify points or to change organization and content during the drafting stage.
- Draft the full document together so that you both have the opportunity to suggest major organizational and content changes.
- Edit individually for correctness of grammar, punctuation, usage, and spelling, and then combine results for the final draft. Or if one partner is particularly strong in these matters, that person may handle the final editing alone.

Writing on a Team

In addition to writing alone or with a partner, writers frequently work on technical documents with a group. An advantage to writing with a team is that several people, who may be specialists in different areas, are working on the same problem, and the final document will reflect their combined knowledge and creativity—several heads are better than one. A disadvantage of writing with a group is the conflicts that arise among ideas, styles, and working methods—too many cooks in the kitchen. An effective writing team must become a cohesive problem-solving group in which conflict is used productively.

Team Planning

The initial meetings of the writing team are important for creating commitment to the project as well as mutual support. All team writers should feel equally involved in planning the document. Here is a checklist for team planning:

- Select someone to act as a supervisor of the team, call meetings, and organize the agenda. The supervisor also can act as a discussion leader, seeking consensus and making sure that all questions are covered.
- At meetings, select someone to take notes on decisions and assignments. These notes should be distributed to team members as soon as possible.
- Clarify the writing problem so that each team member shares the same understanding of reader and purpose.
- Generate ideas about strategy, format, and content without evaluating them until all possible ideas seem to be on the table. This brainstorming should be freewheeling and noncritical so that members feel confident about offering ideas.

- Discuss and evaluate the various ideas; then narrow the suggestions to those which seem most suitable.
- Arrive at a group agreement on overall strategy, format, and content for the document.
- Organize a tentative outline for the major sections of the document so that individual writers understand the shape and boundaries of the sections they will write.
- Divide the tasks of collecting information and drafting sections among individual writers.
- Schedule meetings for checking progress, discussing content, and evaluating rough drafts. Frequent meetings help team members stay committed to the goals of the project and the efforts of each individual.
- Schedule deadlines for completing rough drafts.

Team Drafting

Meetings to work on the full document generally focus on the drafts from individual writers. Team members can critique the drafts and offer revision suggestions. At these meetings, the group also may decide to revise the outline, add or omit content, and further refine the format. Here is a checklist for team drafting:

- Be open to suggestions for changes in any area. No previous decision is carved in stone; the drafting meetings should be sessions that accommodate revision.
- Encourage each team member to offer revision suggestions at these meetings. Major changes in organization and content are easier to incorporate in the drafting stage than in the final editing stage.
- Ask questions to clarify any points that seem murky.
- Assign someone (usually the supervisor) to coordinate the full draft and, if needed, send it for review to people not on the team, such as technicians, marketing planners, lawyers, and upper-level management.
- Schedule meetings for revising and editing the full draft.

Team Revising and Editing

The meetings for revising and editing should focus on whether the document achieves its purpose. This stage includes final changes in organization or content as well as editing for accuracy. Here is a checklist for revising and editing:

- Evaluate comments from outside reviewers and incorporate their suggestions or demands.

- Assign any needed revisions to individual writers.
- Assign one person to evaluate how well the format, organization, and content fit the document's reader and purpose.
- Check that the technical level of content and language is appropriate for the audience.
- Decide how to handle checking grammar, punctuation, and spelling in the final draft. One writer who is strong in these areas might volunteer to take on this editing.

Handling Conflict among Writers

Even when writers work well together, some conflict during a project is inevitable; however, conflict can stimulate effort and creativity. Remember these strategies for turning conflict into productive problem solving:

- Discuss all disagreements in person rather than through memos or third parties.
- Assume that others are acting in good faith and are as interested as you in the success of the project.
- Avoid feeling that all conflict is a personal challenge to your skills and judgment.
- Listen carefully to what others say; do not interrupt them before they make their points.
- Paraphrase the comments of other people to make sure you understand their meaning.
- Solicit ideas for resolving conflicts from all team members.
- Acknowledge points on which you can agree.
- Discuss issues as if a solution/compromise is possible instead of trying to prove other people wrong.

By keeping the communication channels open, you can eliminate conflict and develop a document that satisfies the goals of the writing team.

Writing for Management Review

The procedure called *management review*—in which a writer composes a document that must be approved and perhaps signed by a manager—holds special problems. First, management review almost inevitably involves criticism from a writer's superior. Second, the difference in company rank between manager and writer makes it difficult for the two to think of themselves as partners. Finally, the writer often feels responsible for the real work of producing the document, while the manager is free to cast slings and arrows at it without having struggled through the writing process. Management review works best when writers feel that they share in making the decisions instead of merely taking orders as they draft and redraft a document. Here are guidelines for both writers and managers who want a productive management review:

- Hold meetings in a quiet workplace where the manager will not be interrupted with administrative problems.
- Discuss the writing project from the beginning so that you both agree on purpose, reader, and expected content.
- Review the outline together before the first draft. Changes made at this point are less frustrating than changes made after sections are in draft.
- As the project progresses, meet periodically to review draft sections and discuss needed changes.
- Be certain that meetings do not end before you both agree on the goals of the next revision and the strategies for changes.

The manager involved in a management review of a document is ultimately responsible for guiding the project. In some cases, the manager is the person who decides what to write and to whom. In other situations, the manager's primary aim is to direct the writer in developing the final document's content, organization, and format. Here are guidelines for a manager who wants a successful management review:

- Criticize the problem, not the person.
- Avoid criticism without specifics. To say, "This report isn't any good," without pinpointing specific problems will not help the writer revise.
- Criticize the project in private meetings with the writer; do not discuss the review process publicly.
- Analyze the situation with the writer rather than dictate solutions or do all the critiquing yourself.
- Project an objective, calm frame of mind for review meetings.
- Concentrate on problem solving rather than on the writer's "failure."
- Have clear-cut reasons for requesting changes, and explain them to the writer. Avoid changes made for the sake of change.
- If the standard company format does not fit the specific document very well, encourage the writer to explore new formats.
- Seek outside reviewers for technical content or document design before a final draft.

- Listen to the writer's ideas and be flexible in discussing writing problems.
- Begin the review meetings by going over the successful parts of the document.

Most managers dislike having to give criticism, but subordinates dislike even more having to accept it. Dwelling on the criticism itself will interfere with the writing process. Here are guidelines for a writer involved in a management review:

- Concentrate on the issues that need to be solved, not on your personal attachment to the document.
- Accept the manager's criticism as guidance toward creating an improved document rather than as an attack on you.
- Remember that every document can be improved; keep your ego out of the review process.
- Ask questions to clarify your understanding of the criticism. Until you know exactly what is needed, you cannot adequately revise.

Writing with a Technical Editor

A writer of technical documents may also work with a technical editor during the review. Documents with a large audience, such as policy manuals or consumer booklets, usually require a technical editor. In a typical situation, the editor begins work on the document after it has been through several drafts.

The writer should have a draft that is accurate and free of errors in grammar, punctuation, and spelling. The editor will review several areas. First, the editor will judge the document's structure and check for consistency throughout the sections. Second, the editor will mark the draft for clarity, revising sentences that seem overly long, garbled, ambiguous, or awkward. Third, the editor will check punctuation, grammar, mechanics, and spelling. Last, the editor will mark for standard company style. Both the editor and the writer should concentrate on the issues rather than on personal preferences. Here are guidelines for writers and editors working together:

- Discuss how editorial changes clarify the document's meaning or serve the reader's purpose rather than focusing on personal preferences in style or format.
- Do not try to cover all editorial matters for a long document in one editorial meeting. Schedule separate meetings for format, grammar and mechanics, organization, and consistency.

- Check to ensure that editorial changes do not alter the document's meaning or appropriate tone.

After the editorial meetings, the writer incorporates the changes and produces a final draft. The editor will mark this draft for printing and check for any typing mistakes. The editor is responsible for the publication procedure. Remember that the editor's purpose is not to interfere with creativity, but to help produce a document that serves readers efficiently.

CHAPTER SUMMARY

This chapter discusses the importance of writing on the job, the writing process, and writing with others. Remember:

- Writing is important in most jobs and often takes a significant portion of job-related time.
- Technical writing covers many subjects in diverse fields, and every writing task involves analyzing reader, purpose, and writing situation.
- Each writer develops a personal writing process that includes the general stages of planning, multiple drafting, and revising and editing.
- The planning stage of the writing process includes analyzing reader, purpose, and writing situation; gathering information; and organizing the document.
- The multiple drafting stage of the writing process involves developing a full document and redrafting as writers rethink their original planning.
- The revising and editing stage of the writing process includes revising the document's content and organization and editing for grammar, punctuation, spelling, and company style.
- Writing on the job often requires collaborating with others by writing with a partner, writing on a team, writing for management review, or writing with a technical editor.
- Conflict among writers working together can be productive if writers concentrate on the project and exchange ideas freely.

SUPPLEMENTAL READINGS IN PART TWO

"When You're Not the Whole Staff," *Simply Stated.*

ENDNOTES

1. Charlene M. Spretnak, "A Survey of the Frequency and Importance of Technical Communication in an Engineering Career," *The Technical Writing Teacher* 9.3(Spring 1982):133–136.

2. Marie E. Flatley, "A Comparative Analysis of the Written Communication of Managers at Various Organizational Levels in the Private Business Sector," *Journal of Business Communication* 19.3(Summer 1982):35–49.

3. Mary K. Kirtz and Diana C. Reep, "The Writing Seminar: The View from the Participant's Chair," paper presented at the Association for Business Communication International Meeting, Los Angeles, November 1986.

4. Andrea Lunsford and Lisa Ede, "Why Write . . . Together: A Research Update," *Rhetoric Review* 5.1(Fall 1986):71–81.

Model 1-1 Commentary

This model illustrates one writer's first activity in planning an administrative bulletin at a manufacturing company. Because the company has defense contracts, many written and visual materials include classified information that must be coded according to Department of Defense regulations. The company administrative bulletin covers the correct procedures for marking classified materials. Readers are the six clerks who work in the Security Department, and they will mark the materials correctly. To ensure consistency in security procedures, the clerks are accustomed to following company guidelines in administrative bulletins.

The writer begins planning the document by scanning the Department of Defense regulations and noticing those which apply to his company. He then lists these items as a preliminary outline.

Discussion

Assume you are the writer in this situation. Discuss the kinds of information you would want to have about your readers and how they intend to use this document before you plan the bulletin further.

MARKING CLASSIFIED INFORMATION
- BOOKS, PAMPHLETS, BOUND DOCS
- CORRESPONDENCE + NONBOUND DOCS
- ARTWORK
- PHOTOS, FILM, MICROFILM
- LETTERS OF TRANSMITTAL
- SOUND TAPES
- MESSAGES
- UNMARKABLE MATERIAL
- CHARTS, TRACINGS, DRAWINGS

AUTHORITY FOR CLASSIFYING
- CONTRACTING AGENCY

CLASSIFIED — ALL INFO CONNECTED TO PROJECT

OTHER INFO — COMPANY MAY CLASSIFY — REGIONAL OFFICE REVIEWS

MARKING PARAGRAPHS FOR DIFFERENT CLASSIFICATIONS
- PREFERRED
- OR STATEMENT ON FRONT
- OR ATTACH CLASSIFICATION GUIDE FOR CONTENT

NOTE: MATERIAL IN PRODUCTION
- EMPLOYEES NOTIFIED OF CLASSIFICATION

Model 1-2 Commentary

After talking to several Security Department clerks about the kinds of classified materials they handle and their preferences for written guidelines in the company bulletins, the writer expands and rearranges his list of items.

He identifies those with the authority to classify materials, names the document that lists the specifications for classification, and states the circumstances under which the company may decide to classify materials. He also groups the types of materials in his list into categories—printed, graphic, and sound. Based on his discussions with the Security Department clerks, he now lists the most common classified materials first and the least common last. He also records two unusual circumstances that he must explain in the bulletin—materials that cannot be marked in the usual manner and documents that include paragraphs classified at different levels.

Discussion

In his revised list, the writer includes more detail under each item. Discuss why he might choose to expand and rearrange his first list at this stage rather than write a full draft.

AUTHORITY TO CLASSIFY - CONTRACTING AGENCY
CLASSIFIED INFO - ACCORDING TO FORM 264
INFO NOT IN CONTRACT - COMPANY CAN CLASSIFY
IF NEEDED - REGIONAL OFFICE SHOULD REVIEW
MARKING CLASSIFIED INFO - NO TYPING - USE
DATE - CLASSIFICATION - NAME & ADDRESS OF
FACILITY
① BOUND DOCS - BOOKS, PAMPHLETS: TOP & BOTTOM,
COVERS, TITLE, FIRST & LAST PAGES
② CORRESPONDENCE & NONBOUND - TOP & BOTTOM,
IF PARTS DIFFER, HIGHEST CLASSIFICATION PRE-
VAILS
③ LETTERS OF TRANSMITTAL - FIRST PAGE
④ CHARTS, TRACINGS & DRAWINGS - UNDER LEGEND,
TITLE BLOCK OR SCALE, TOP & BOTTOM
⑤ ARTWORK - TOP & BOTTOM OF BOARD & PAGE
⑥ PHOTOS, FILMS, MICROFILMS - OUTSIDE OF CON-
TAINER, BEGINNINGS & ENDS OF ROLLS - TITLE BLOCK
⑦ SOUND TAPES - ON CONTAINERS + BEGINNING &
END OF RECORDING
⑧ MESSAGES - TOP & BOTTOM, FIRST & LAST WORDS
OF TRANSMITTED ORAL MESSAGE
⑨ UNMARKABLE MATERIAL - TAGGED - PRODUCTION
EMPLOYEES NOTIFIED
⑩ PARAGRAPHS - IF VARYING CLASSIFICATIONS IN
FORCE - EACH MARKED FOR DEGREE
- OR STATEMENT ON FRONT OR IN TEXT
- OR ATTACH GUIDE FOR EACH PART
** MARK EACH PARAGRAPH IF POSSIBLE

Model 1-3 Commentary

The writer's first draft of the bulletin is based on his expanded outline. In this draft, the writer follows company style requirements by numbering the bulletin and titling it. He also capitalizes "company" wherever it appears. He develops each point in his revised outline into full, detailed sentences, including instructions for each item and special considerations for specific items.

Discussion

In his discussions with the clerks, the writer learned that they prefer to divide their work so that each handles only specific types of classified materials. Each clerk, therefore, will use the guidelines in the bulletin that apply to those materials. Discuss how the writer tries to accommodate these work habits in his first draft by organization and sentence structure.

First Draft

Military Security Guide—Bulletin 62A

The contracting federal agency shall have the authority for classifying any information generated by the Company. All information developed or generated by the Company while performing a classified contract will be classified in accordance with the specifications on the "Contract Security Classification Specification," Form 264. Information generated by the Company shall not be classified unless it is related to work on classified contracts; however, the facility management can classify any information if it is believed necessary to safeguard that information in the national interest. Moreover, the information classified by Company management should be immediately reviewed by the regional security office.

MARKING CLASSIFIED INFORMATION AND MATERIAL

All classified information and material must be marked (not typed) with the proper classification, date of origin, and the name and address of the facility responsible for its preparation.

1. Bound Documents, Books, and Pamphlets shall be marked with the assigned classification at the top and bottom on the front and back covers, the title page, and first and last pages.

2. Correspondence and Documents Not Bound shall be marked on the top and bottom of each page. When the separate components of a document, such as sections, etc., have different classifications, the overall classification is the highest one for any section. Mark sections individually also.

3. Letters of Transmittal shall be marked on the first page according to the highest classification of any component. A notation may be made that, upon removal of classified material, the letter of transmittal may be downgraded or declassified.

4. Charts, Drawings and Tracings shall be marked under the legend, title block, or scale and at the top and bottom of each page.

5. Artwork shall be marked on the top and bottom margins of the mounting board and on all overlays and cover sheets.

6. Photographs, Films, and Microfilms shall be marked on the outside of the container. In addition, motion picture films shall be marked at the beginning and end of each roll and in the title block.

7. Sound Tapes shall be marked on their containers and an announcement made at the beginning and end of the recording.

8. Messages, such as telegrams, sent electronically shall have the classification marking at the top and bottom of each page. In addition, the first and last word of the message shall be the classification.

9. Classified Material that cannot be marked shall be tagged with the classification and other markings. Material still in production that cannot be tagged requires that all employees be notified of the proper classification.

10. Paragraphs in documents, bound or nonbound, which are of different classifications, shall be marked to show the degree of classification, if any, of the information contained therein. Or a statement on the front of the document or in the text shall identify the parts of the document that are classified and to what degree. Or an appropriate classification guide shall be attached to cover the classified contents of the document. Marking paragraphs individually is the preferred method. Neither of the other two alternatives may be used until it is determined that paragraph marking is not possible.

Model 1-4 Commentary

Before redrafting, the writer asks the senior clerk to read his first draft and make suggestions. He then writes a second draft, reorganizing according to her comments.

In the second draft, the writer uses a headline that clarifies the purpose of the bulletin and includes headings in the document to separate groups of information. In addition to numbering the classified materials, the writer now underlines the key words that identify each item.

Item 10 in the first draft is a separate section in the second draft because the senior clerk noticed that the information about marking paragraphs applies to all written materials.

Discussion

1. On the basis of what you know about the readers' intended use of this bulletin and work habits, discuss further changes in organization that might make the bulletin easier to use. For instance, what organization changes might help the clerk who works only with items 1, 2, and 3?

2. Compare the first draft with the second draft. Assume you are a clerk who must mark a 20-page booklet in which the sections have different classifications. Discuss how the organization changes from first draft to second draft will help you find the information you need.

Second Draft

Military Security Guide—Bulletin 62A

MARKING CLASSIFIED INFORMATION AND MATERIALS

Authority

The contracting federal agency shall have the authority for classifying any information generated by the Company. All information developed or generated by the Company while performing a classified contract will be classified in accordance with the specifications on the "Contract Security Classification Specification," Form 264. Information generated by the Company shall not be classified unless it is related to work on classified contracts; however, management can classify any information that may affect the national interest. The information classified by Company management should be immediately reviewed by the regional security office.

Required Marking

All classified information and material must be marked (not typed) with (1) the proper classification, (2) date of origin, and (3) the name and address of the facility responsible for its preparation.

Types of Information and Materials

1. Bound Documents, Books, and Pamphlets shall be marked with the assigned classification at the top and bottom on the front and back covers, the title page, and first and last pages.

2. Correspondence and Documents Not Bound shall be marked on the top and bottom of each page. If the components of a document, such as sections or chapters, have different classifications, the overall classification is the highest one for any section. Individual sections shall be marked also.

3. Letters of Transmittal shall be marked on the first page according to the highest classification of any component. A notation may be made that, upon removal of classified material, the letter of transmittal may be downgraded or declassified.

4. Charts, Drawings and Tracings shall be marked under the legend, title block, or scale and at the top and bottom of each page.

5. Artwork shall be marked on the top and bottom margins of the mounting board and on all overlays and cover sheets.

6. Photographs, Films, and Microfilms shall be marked on the outside of the container. In addition, motion picture films shall be marked at the beginning and end of each roll and in the title block.

7. Sound Tapes shall be marked on their containers and an announcement made at the beginning and end of the recording.

8. Electronic Messages, such as telegrams, shall have the classification marking at the top and bottom of each page. In addition, the first and last word of the message shall be the classification.

9. Classified Material that cannot be marked shall be tagged with the classification and other markings. All employees shall be notified of the proper classification for material still in production that cannot be tagged.

Individual Paragraphs

Paragraphs of documents, bound or nonbound, which are of different classifications, shall be marked to show the degree of classification. Material shall be marked in one of three ways:

- Individual paragraphs shall be marked separately.
- A statement on the front of the document or in the text shall identify the parts of the document that are classified and to what degree.
- An appropriate classification guide shall be attached to provide the classifications of each part of the document.

Marking paragraphs individually is the preferred method. Neither of the other two alternatives may be used until facility management determines that paragraph marking is not possible.

CHAPTER 1 Exercises

1. Interview someone you know who has a job that requires some writing. Ask the person about his or her personal writing process. What kinds of documents and for whom does the person write? How do company requirements affect the writing process? Is team writing or management review usually involved? What activity in the writing process does the person find most difficult? How much rewriting does the person do for a typical writing task? What part of the person's writing process would he or she like to change or strengthen? Write a memo to your instructor describing how this person handles a typical writing task. Use the memo format shown in Chapter 9.

2. Select a business letter, memo, or report addressed to you. Write a short, informative memo to your instructor describing the purpose of the document from the writer's perspective and your own. Discuss how the specific content, organization, and language in the document help you use the document. Point out any problems that interfere with your using the document effectively. Attach a copy of the piece of writing to your memo. Use the memo format shown in Chapter 9. *Or* make enough copies of the piece of writing for your classmates and present a brief oral critique of the writing. Read the article by Martel entitled, "Combating Speech Anxiety," in Part II, before giving your oral report.

3. Your task is to write a description of another student's job—either a current job or a recent vacation job. The description you write will be included in *Ways to Earn Money for College*, a booklet for incoming freshmen who want to work part time during the school year or full time during vacations. In class, discuss what kinds of information would be most helpful to these freshmen. Then interview a student and determine his or her job duties, salary, hours, and so on. Next, organize the job information you collected by topics and put the details in the order you decide is appropriate under each topic. Then write a draft of the job description.

Next, in pairs in class, each assume the role of a manager who will review the other's writing. Read each other's drafts and make suggestions for revision. Then revise your rough draft according to the suggestions you received from your "manager." Submit to your instructor (a) your first draft with the notes from your "management review" and (b) your revised draft.

CHAPTER 2

Audience

Analyzing Readers
- Subject Knowledge
- Position in the Organization
- Personal Attitudes
- Reading Style
- Multiple Readers
  - Primary Readers
  - Secondary Readers

Finding out about Readers
- Informal Checking
- Interviewing
  - Preparing for the Interview
  - Conducting the Interview

Testing Reader-Oriented Documents
- Readers Answering Questions
- Readers Performing a Task
- Readers Thinking Aloud

Chapter Summary

Supplemental Readings in Part Two

Models

Exercises

Analyzing Readers

Each reader represents a unique combination of characteristics and purpose that will affect your decisions about document content and format. To prepare an effective technical document, therefore, analyze your readers during the planning stage of the writing process. Consider your readers in terms of these questions:

- How much technical knowledge about the subject do they already have?
- What positions do they have in the organization?
- What are their attitudes about the subject or the writing situation?
- How will they read the document?
- What purpose do they have in using the document?

If you have multiple readers for a document, you also need to consider the differences among your readers.

Subject Knowledge

Consider how much information your readers already have about the main subject and subtopics in the document. In general, you can think of readers as having one of these levels of knowledge about any subject:

- *Expert level*—Readers with expert knowledge of a subject understand the theory and practical applications as well as most of the specialized terms related to the subject. Expert knowledge implies years of experience and/or advanced training in the subject. A scientist involved in research to find a cure for emphysema will read another scientist's report on that subject as an expert, understanding the testing procedures and the discussion of results. A marketing manager may be an expert reader for a report explaining possible strategies for selling a home appliance in selected regions of the country.

  Expert readers generally need fewer explanations of principles and fewer definitions than other readers, but the amount of appropriate detail in a document for an expert reader depends on purpose. The expert reader who wants to duplicate a new genetic test, for instance, will want precise information about every step in the test. The expert reader who is interested primarily in the results obtained will need only a summary of the test procedure.

- *Semiexpert level*—Readers with semiexpert knowledge of a subject may vary a great deal in how much they know and why they want information. A manager may understand some engineering principles in a report but probably is more interested in information about how the project affects company planning and budgets, subjects in which the manager is an expert. An equipment operator may know a little about the scientific basis of a piece of machinery but is more interested in information about handling the equipment properly. Other readers with semiexpert knowledge may be in similar fields with overlapping knowledge. A financial analyst may specialize in utility stocks but also has semiexpert knowledge of other financial areas. Semiexpert readers, then, may be expert in some topics covered in a document and semiexpert in other topics. To effectively use all the information, the semiexpert reader needs more definitions and explanations of general principles than the expert reader does.
- *Nonexpert level*—Nonexpert readers have no specialized training or experience in a subject. These people read because (1) they want to use new technology or perform new tasks or (2) they are interested in learning about a new subject. Nonexperts using technology for the first time or beginning a new activity are such readers as the person using a stationary exercise bike, learning how to play golf, or installing a heat lamp in the bathroom. These readers need information that will help them use equipment or perform an action. They are less interested in the theory of the subject than in its practical application. Nonexperts who read to learn more about a subject, however, are often interested in some theory. For instance, someone who reads an article in a general science magazine about the disappearance of the dinosaurs from the Earth will probably want information about scientific theories on the cause of the dinosaurs' disappearance. If the reader becomes highly interested in the topic and reads widely, he or she then becomes a semiexpert in the subject, familiar with technical terms, theories, and the physical qualities of dinosaurs.

All readers, whatever their knowledge level, have a specific reason for using a technical document, and they need information tailored to their level of knowledge. Remember that one person may have different knowledge levels and objectives for different documents. A physician may read (1) a report on heart surgery as an expert, seeking more information about research, (2) a report recommending new equipment for a clinic as a semiexpert who must decide whether to purchase the equipment, (3) an article about space travel as a nonexpert who enjoys learning about space, and (4) an owner's manual for a new camera as a consumer who needs instructions for operating the camera.

Position in the Organization

Your reader's hierarchical position in the company and relationship with you are also important characteristics to consider. Readers are either external (outside the company) or internal (inside the company). Those outside the company include customers, vendors, stockholders, employees of government agencies or industry associations, competitors, and the general public. The interests of all these groups center on how a document relates to their own activities. Within all companies, the hierarchy of authority creates three groups of readers:

- *Superiors*—Readers who rank higher in authority than the writer are superiors. They may be executives who make decisions based on information in a document. Superiors may be experts in some aspects of a subject, such as how cost projections will affect company operations, and they may be semiexperts in the production systems. If you are writing a report to superiors about a new company computer system, your readers would be interested in overall costs, the effect of the system on company operations, expected benefits companywide, and projections of future computer uses and needs.
- *Subordinates*—Readers who rank lower in authority than the writer are subordinates. They may be interested primarily in how a document affects their own jobs, but they also may be involved in some decision making, especially for their own units. If your report on the new computer system is for subordinates, you will probably emphasize information about specific models and programs, locations for the new computers, how these computers support specific tasks and systems, and how the readers will use the computers in their jobs.
- *Peers*—Those readers on the same authority level as the writer are peers, although they may not be in the same technical field as the writer. Their interests could involve decision making, coordinating related projects, following procedures, or keeping current with company activities. Your report on the new computer system, if written for peers, might focus on how the system will link departments and functions, change current procedures, and support company or department goals.

Personal Attitudes

Readers' personal responses to a document or a writing situation often influence the document's design. As you analyze your writing situation, assess these considerations for your readers:

- *Emotions*—Readers can have positive, negative, or neutral feelings about the subject, the purpose of the document, or the writer. Even when readers try to read objectively, these emotions can interfere. If your readers have a negative attitude about a subject, organizing the information to move from generally accepted data to less accepted data or starting with shared goals may help them accept the information.
- *Motivation*—Readers may be eager for information and eager to act. On the other hand, they may be reluctant to act. Make it easy for your readers to use the information by including items that will help them. Use lists, tables, headings, indexes, and other design features to make the text more useful. Most important, tell readers why they should act in this situation.
- *Preferences*—Readers sometimes have strong personal preferences about documents they must use. Readers may demand features such as lists or charts, or they may refuse to read a document that exceeds a specified length or does not follow a set format. If you discover such preferences in your readers, adjust your documents to suit them.

Reading Style

Technical documents usually are not read, nor are they meant to be read, from beginning to end like a mystery novel. Readers read documents in various ways, determined partly by their need for specific information and partly by personal habit. These reading styles reflect specific readers' needs:

- *Readers use only the summary or abstract.* Some readers want only general information about the subject, and the abstract or opening summary will serve this purpose. An executive who is not directly involved in a production change may read only the abstract of a report that describes the change in detail. This executive needs up-to-date information about the change, but only in general form. A psychologist who is looking for research studies about abused children may read the abstract of an article in order to decide whether or not the article contains the kind of information needed. Relying on the abstract for an overview or using the abstract to decide whether to read the full document saves time for busy readers.
- *Readers check for specific sections of information.* A reader may be interested only in some topics covered in a lengthy document. A machine operator may read a technical manual to find the correct operating procedures for a piece of equipment, but a design engineer may look for a description of the machine and a service technician will turn to the section on maintenance. Long reports and manuals often have multiple readers who are interested only in the information relevant to their jobs

or departments. Use descriptive headlines to direct these readers to the information they need.

- *Readers scan the document, pausing at key words and phrases.* Sometimes readers quickly read a document for a survey of the subject, but they concentrate only on information directly pertaining to topics that affect them. The manager of an insurance company annuity department may scan a forecasting report to learn about company planning and expected insurance trends over the next ten years. The same manager, however, will read carefully all information that in any way affects the annuity department. Such information may be scattered throughout the report so that the manager will look for key words. To help such a reader, use consistent terms throughout.
- *Readers study the document from beginning to end.* Readers who need all the information in order to make a decision are likely to read carefully from beginning to end. A reader who is trying to decide what automobile to buy may read the manufacturers' brochures from beginning to end, looking for information to aid in the decision. Someone who needs to change an automobile tire will read the instructions carefully from beginning to end in order to perform the steps correctly. For these readers, highlight particularly important information in lists or boxes, direct their attention to information through headings, and summarize main points to refresh their memories.
- *Readers evaluate the document critically.* Someone who opposes a project or the writer's participation in it may read a document looking for information that can be used as negative evidence. The multiple readers of a report recommending a company merger, for instance, may include those opposed to such a plan. Such readers will focus on specific facts they believe are inadequate to support the recommendation. If you know that some of your readers are opposed to the general purpose of your report, anticipate criticisms of the plan or of your data and include information that will respond to these criticisms.

Multiple Readers

Technical documents usually have multiple readers. Sometimes the person who receives a report will pass it on informally to others. Manuals may be used by dozens of employees, and documents outlining policies and procedures may apply to employees in all departments of an organization. Readers inside and outside a company study annual reports. Consumer instruction manuals are read by millions. This diversity among readers presents extra problems in document design. In addition to analyzing readers based on technical knowledge, positions in the organization, attitudes, and reading styles, you should consider whether they are primary or secondary.

Primary Readers

Primary readers will take action or make decisions based on the document. An executive who decides whether to accept a recommendation to change suppliers and purchase equipment is a primary reader. A consumer who buys a blender and follows the enclosed instructions for making drinks is a primary reader. The technical knowledge, relationship to the organization and to the writer, and personal preferences of these two primary readers are very different, but each is the person who will use the document most directly.

Secondary Readers

Secondary readers do not make decisions or take direct action because of the document, but they are affected or influenced by it. A technician may be asked to read a report about changing suppliers and to offer an opinion. If the change takes place, the technician will have to set up maintenance procedures for the new equipment. A report suggesting new promotion policies will have one or more primary readers who are authorized to make decisions. All the employees who have access to the report, however, will be secondary readers with a keen interest in the effect of any changes on their own chances for promotion. Remember that secondary readers are not necessarily secondary in their interest in a subject, but only in their power or authority to act on the information.

After deciding which readers are primary and which are secondary, plan adjustments in the document according to the readers' characteristics discussed earlier. Your primary readers are the first priority, but adjustments for secondary readers are necessary too.

Content. Include the technical detail that is appropriate for the primary audience. If the secondary readers are more technically knowledgeable than the primary ones, you can add statistical appendixes, detailed illustrations, or supplementary technical materials for them. If the secondary readers are less technically knowledgeable than the primary ones, you can add appendixes that illustrate the information more simply, include glossaries of technical terms, or provide summaries that highlight important points.

Structure. Divide the document into sections that will serve both primary and secondary audiences. Headings, such as "Recommendations," "Objectives," or "Cost Analysis," will guide specific readers to the information they need. Use graphics and appendixes to clarify information for readers with different technical knowledge and interests. You can also develop specific parts of the document for different readers. Include an abstract for nonexpert readers or a summary of main points for decision makers.

Style. Use language appropriate for your primary readers. If they are nonexperts in the subject, include definitions of technical terms and provide checklists of important points. If your primary readers are experts, use the appropriate technical terms and include glossaries for secondary readers, if needed.

If your document has multiple primary readers, plan for those who need the most help in understanding the subject and the technical language. For instance, when preparing a bulletin for dozens of regional sales representatives, some of whom have been with the company many years and some of whom are newly hired, include the amount of detail and the language most appropriate for the least experienced in the group. New sales representatives need full explanations, but experienced people can skim through the bulletin and read the sections they have a specific interest in if you include descriptive headings.

In some situations, the differences among primary readers or between primary and secondary readers may be so great that you decide to write separate documents. For example, one report of a medical research program cannot serve both the general public and experts. The two groups of readers have entirely different interests and abilities for dealing with the specialized medical information. Even within a company, if the amount of technical information needed by one group will shut out other readers, separate documents may be the easiest way to serve the whole audience.

Finding out about Readers

Sometimes you will write for readers whom you know well. But how do you find out about readers whom you do not know well or whom you will never meet? Gathering information about your readers and how they plan to use a document can be time-consuming, but it is essential at the planning stage of the writing process. Use both informal inquiries and formal interviews to identify readers' characteristics and purposes.

Informal Checking

When you are writing for people inside your organization, you often can find out about them and their purposes by checking informally in these ways:

- Talk in person or by telephone with the readers themselves or with those who know them, such as the project director, the person who assigned you the writing task, and people who have written similar documents. If you are writing to a high-level executive, you may not feel comfortable calling the executive, but you can talk with others who are familiar with the project, the purpose of the document, and the reader's characteristics.

- Check the readers' reactions to your drafts. You can find out how your readers will use the information in your document by sharing drafts during the writing process. If you are writing a procedures manual, sharing drafts with the employees who will have to follow the procedures will help you clarify how they intend to use the document and what their preferences are.
- Analyze your organization's chain of responsibility relative to the document you are writing. For example, the marketing director may supervise 7 regional managers, who, in turn, supervise 21 district managers, who, in turn, supervise the sales representatives. Perhaps all are interested in some aspects of your document. By analyzing such organizational networks, you can often identify secondary readers and adjust your document accordingly.
- Brainstorm to identify readers' characteristics for groups. If you are writing a document for a group of readers, such as all registered nurses at St. Luke's Hospital, list the characteristics that you know the readers have in common. You may know, for example, that the average nurse at St. Luke's has been with the hospital for 4.5 years, has a B.S. degree, grew up in the area, and attends an average of two professional training seminars a year. Knowing these facts about your readers will help you tailor your document to their needs.

Interviewing

Writers on the job use interviews as part of the writing process in two ways: (1) to gather information about the subject from experts and technicians and (2) to find out readers' purposes and intended uses of a document. Whether you are interviewing experts for information about the subject or potential readers to decide how to design the document, a few guiding principles will help you control the interview and use the time effectively.

Preparing for the Interview

Interviewers who are thoroughly prepared get the most useful information while using the least amount of time.

- *Make an appointment for the interview.* Making an appointment shows consideration for the other person's schedule and indicates that the interview is a business task, not a casual chat. Be on time, and keep the interview within the estimated length.
- *Do your background research before the interview.* Interview time should be as productive as possible. Do not try to discuss the document before having at least a general sense of the kinds of information and overall

structure required. You are then in a position to ask more specific questions about content or ask for and use opinions about document design.

- *Prepare your questions ahead of time.* Write out specific questions you intend to ask. Ask the expert for details that you need to include in the document. Project yourself into your readers' minds and imagine how the document might serve their purposes; then frame your questions to pinpoint exactly how your readers will respond to specific sections, headings, graphic aids, and other document features.
- *Draft questions that require more than yes or no answers.* Interview time will be more productive if each question requires a detailed answer. Notice the following ineffective questions and their more effective revisions:

  Ineffective: Do you expect to use the manual daily for repair procedures?

  Effective: For which repair and maintenance procedures do you need to consult the manual?

  This revision will get more specific answers than the first version.

  Ineffective: What do you think operators need in a manual like this one?

  Effective: Which of the following features are essential in an operator's manual for this equipment? (List all that you can think of and then ask if there are others.)

  The first version may get a specific answer, but it will not necessarily be comprehensive. Check all possibilities during an interview to avoid repeating questions later.

Conducting the Interview

Think of the interview as a meeting that you are leading. You should guide the flow of questioning, keep the discussion on the subjects, and cover all necessary topics.

- *Explain the purpose of the interview.* Tell the person why you want to ask questions and how you think the interview will affect the final document.
- *Break the ice if necessary.* If you have never met the person before, spend a few minutes discussing the subject in general terms so that the two of you can feel comfortable with each other.
- *Listen attentively.* Avoid concentrating on your questions so much that you miss what the other person is saying. Sometimes in the rush to cover all the topics, an interviewer is more preoccupied with checking off the questions than with listening carefully to answers.
- *Take notes.* Take written notes so that you can remember what the reader told you about special preferences or needs. If you want to tape

the interview, ask permission before arriving. Remember, however, that some people are intimidated by recording devices.

- *Group your questions by topic.* Do not ask questions in random order. Thoroughly cover each topic, such as financial information, before moving on to other subjects.
- *Ask follow-up questions.* Do not stick to your list of questions so rigidly that you miss asking obvious follow-up questions. Be flexible, and follow a new angle if it arises.
- *Ask for clarification.* If you do not understand an answer, ask the person for more detail immediately. It is easier to clarify meaning on the spot than days later.
- *Maintain a sense of teamwork.* Although you need to control the interview, think of the meeting as a partnership to solve specific questions of content and format.
- *Keep your opinions out of the interview.* What you think about the project or the purpose of the document is not as relevant during an interview as what the other person thinks about these topics. Concentrate on gathering information rather than on debating issues.
- *Ask permission to follow up.* End the interview by requesting a chance to call with follow-up questions if needed.
- *Thank the interviewee.* Express your appreciation for the person's help and cooperation.

Immediately after the interview, write up your notes in a rough draft so that you can remember the details.

Testing Reader-Oriented Documents

Along with analyzing and interviewing readers, writers sometimes evaluate audience needs through user tests of a document while it is still in draft. Manuals and instructions are particularly good candidates for user tests because their primary purpose is to guide readers in performing certain functions and because they are meant for groups of readers. When potential readers test a document's usefulness before the final draft stage, a writer can make needed changes in design and content based on actual experience with the document. There are three types of user tests.

Readers Answering Questions

Asking readers to answer questions after reading a document helps a writer decide whether the content will be easily understood by the audience. A utility company may want to distribute a brochure that explains to consumers how electricity use is computed and how they can reduce their energy consumption. Through a user test, the writer can determine if the brochure answers consumers' questions. Several versions of the same brochure also can be tested to determine which version readers find most useful. The test involves these steps:

1. A group of typical consumers reads the draft or drafts of the brochure.
2. The consumers then answer a series of questions on the content and perhaps try to compute an energy-consumption problem based on the information in the brochure.
3. The consumers also complete a demographic questionnaire, which asks for ages, sex, job titles, education level, length of time it took to complete the questions, and their opinions as to the readability of the brochure.
4. The writer then analyzes the results to see how correctly and how fast the consumers answered the questions and what the demographic information reveals about ease of use for different consumer groups.
5. Finally, the writer revises any portions of the brochure that proved difficult for readers and then retests the document.

Readers Performing a Task

In another situation, a writer may be most concerned about whether readers will be able to follow instructions and perform a task correctly. A manual may be tested in draft by operators and technicians who will use the final version. This user test follows the same general pattern as the preceding one, but the emphasis is on performing a task.

1. Selected operators follow the written instructions to perform the steps of a procedure.
2. The length of time to complete the task is recorded, as well as how smoothly the operators proceeded through the steps of the procedure.
3. The operators next complete a demographic questionnaire, comment on how effective they found the instructions, and point out areas that were unclear.
4. Analysis of the results focuses on correctness in completing the task, the length of time required, and the portions of the document that received most criticism from the readers.
5. The writer revises the sections that were not clear to the readers and retests the document.

Readers Thinking Aloud

Another type of user test requires readers to think aloud as they read through a document. This method (called *protocol analysis*) allows the writer or design-review team to record readers' responses everywhere in the document, analyze readers' comments, and revise content, structure, and style accordingly. Readers may be puzzled by the terminology, the sequence of items, too much or too little detail, the graphics, or sentence clarity. The writer can then revise these areas before the final draft.

User tests have become increasingly important for audience analysis, especially in large corporations. The Document Design Center of the American Institutes for Research in Washington, D.C., established a Usability Test Laboratory in 1985 to assist businesses in testing documents and evaluating readers' responses to written materials.

CHAPTER SUMMARY

This chapter discusses analyzing readers and testing reader use of documents. Remember:

- Writers analyze readers according to the level of their readers' technical knowledge, positions in the organization, attitudes, reading styles, and positions as primary or secondary readers.
- Readers generally have one of three levels of technical knowledge—expert, semiexpert, and nonexpert.
- A reader's position in the organization may be as the writer's superior, subordinate, or peer.
- A reader's response to a document may be influenced by emotions, motivation, and preferences.
- Readers may differ in their reading styles. Some read only the abstract or introduction; others look for specific sections or topics; others read from beginning to end; and some may read in order to criticize the information or project.
- When a document has multiple readers, the writer must develop content, organization, and style to serve both primary and secondary readers.
- Writers find out about their readers through informal checking and formal interviewing.
- Writers test document effectiveness by asking readers to (1) read the document and answer questions about content, (2) read the document and then perform the task explained in the document, and (3) talk about their reactions to the document.

SUPPLEMENTAL READINGS IN PART TWO

Ridgway, L. "The Writer as Market Researcher," *Technical Communication.*

Model 2-1 Commentary

This article about AIDS is from a newsletter (April 1988) distributed by Children's Hospital Medical Center of Akron, Ohio, to all employees. The multiple readers for this newsletter include both medical and nonmedical personnel. Many employees also take the newsletter home to their families. The articles are written primarily for the nonmedical readers who are interested in news about the hospital and medical treatments.

The headline "Fact vs. Fiction" calls attention to the article's purpose of informing readers about this disease. To support the "facts" idea, the writer quotes a nurse as an authority in explaining how people can and cannot get AIDS.

Statistics about the disease are presented in easy-to-read graphics, and a simple definition chart explains the terms included in the name of the disease.

The article ends optimistically, telling readers that medical experts are working on finding a cure.

Discussion

Discuss why the kinds of information in this article are suited to most of the newsletter readers.

Children's Hospital Medical Center of Akron **April, 1988; Vol. 26 No. 4**

crib sheet

Fact vs. Fiction

Education: Our best vaccine against AIDS

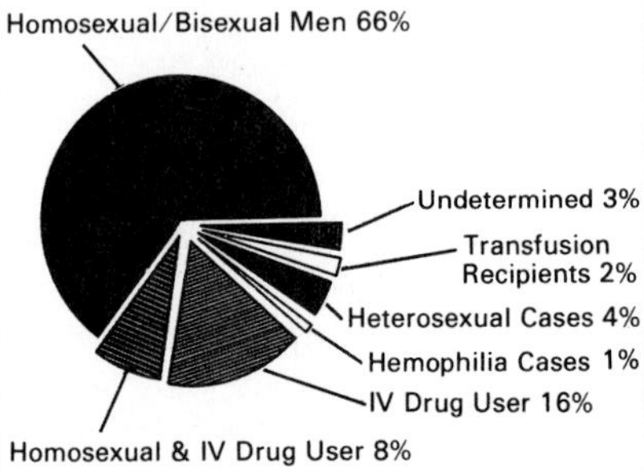

It seems that every generation must meet the challenges presented by a new, often deadly disease. In the early 1900s, it was influenza. Midway through the century it was polio. Today, it's Acquired Immune Deficiency Syndrome, or AIDS.

AIDS was first identified in 1981; the human immunodeficiency virus (HIV) which causes AIDS was just identified in 1984. Thus physicians, scientists and other health care professionals are just beginning to unravel the complexities of the disease. A cure is still several years away.

In the meantime we can all do our part in winning the battle against AIDS by learning just what this disease is—and what it is not. As Infection Control Nurse Lorie Lerner-DurJava noted, "Education is our best vaccine against AIDS."

AIDS is caused by HIV, which cripples the body's immune system. Individuals whose immune systems are functioning normally are able to overcome most illnesses. However, those same illnesses, or "opportunistic infections," can prove fatal for people with AIDS because of their body's weakened ability to fight them off.

It is important to remember that everyone infected with the HIV virus may not develop AIDS. However, these individuals are infectious and can still pass the virus on to others.

Some of these people may never develop symptoms, while others will develop symptoms ranging from mild to severe. AIDS is the final stage of an HIV infection—that is, when the immune (or infection fighting) system is so weakened that opportunistic infections can occur.

Although AIDS is considered a contagious disease, the virus is not easily transmitted. It is spread through sexual contact, blood to blood transmissions (for example, sharing intravenous drug needles) and from mothers to their unborn babies.

REPORTED CASES OF ADULT AIDS
1981 - AUGUST 31, 1987
N = 40,795

Lerner-DurJava noted that there are many misconceptions about how a person can become infected with the AIDS virus.

"You cannot get AIDS from toilet seats, touching or holding someone with AIDS, sharing eating utensils or household items, donating blood or going to school or work with someone who has AIDS," she said.

She also dispelled the notion that the virus could be spread by mosquitoes. "If mosquitoes transmitted AIDS, every child in Africa (where the disease is believed to have originated) would have it," she said, adding, "It's a very fragile virus and cannot live outside in the environment for very long."

At highest risk for AIDS are IV drug abusers and homosexual/bisexual men—and anyone who has sexual contact or shares needles with those in the high-risk groups. Children born to women in the high-risk categories also have a good chance of becoming infected with the virus.

As of January 1988, 52,000 cases of AIDS had been reported to the Centers for Disease Control (CDC); 29,000 of those have died. It's estimated that by 1991 the number of reported cases will be 300,000. In that year alone, the CDC expects to receive reports of 74,000 new cases.

continued on page 4

Page 4 Crib Sheet April, 1988

AIDS

continued from page 1

Those in the high-risk groups can reduce their chances of becoming infected with the HIV virus by taking such preventive measures as reducing the number of sexual partners they have, using condoms and not sharing needles.

The risk of transmission to health care professionals who work with AIDS patients and/or their blood and body substances is minimal. By following a few simple precautions, health care workers can effectively protect themselves against the virus. (See article on universal precautions on pages 5 - 6.)

As evidence of the fact that education can play a key role in combatting AIDS, Lerner-DurJava pointed out that the number of AIDS transmissions among homosexuals is now declining, thanks to an extensive education effort on the part of that population. Efforts to educate IV drug abusers, on the other hand, have not been as successful. The number of AIDS cases in that group continues to rise dramatically.

"Since the development of the AIDS antibody test to check the blood supply, the risk of transmitting the virus through blood transfusions has been significantly reduced," Lerner-DurJava said. "And we're doing a much better job of screening blood donors. Today the blood supply is 99.9 percent safe."

Despite the fact that the number of AIDS cases continues to grow, Lerner-DurJava maintains that the government and the medical profession are moving quickly to combat the spread of the disease. "We just identified the disease seven years ago," she said. "In 1987 alone, the federal government spent $502.1 million on AIDS research and education.

While researchers continue to search for a cure for AIDS, the rest of society—particularly those of us in the health care field—can help by easing the psychological and social stigma placed on persons with AIDS.

"Individuals with AIDS are losing their jobs and their homes. People run away from them. Some health care professionals refuse to treat them," said Lerner-DurJava.

"We must not refer to them as 'AIDS victims.' They're people with AIDS. We should treat them with the same respect and compassion as we treat those with cancer or other illnesses."

A wealth of information on AIDS is available to the public from the Centers for Disease Control in Atlanta, the American Red Cross and the AIDS hotline (1-800-332-AIDS). The Northeast Ohio Task Force on AIDS also offers a speakers bureau. For details, call 375-2960.

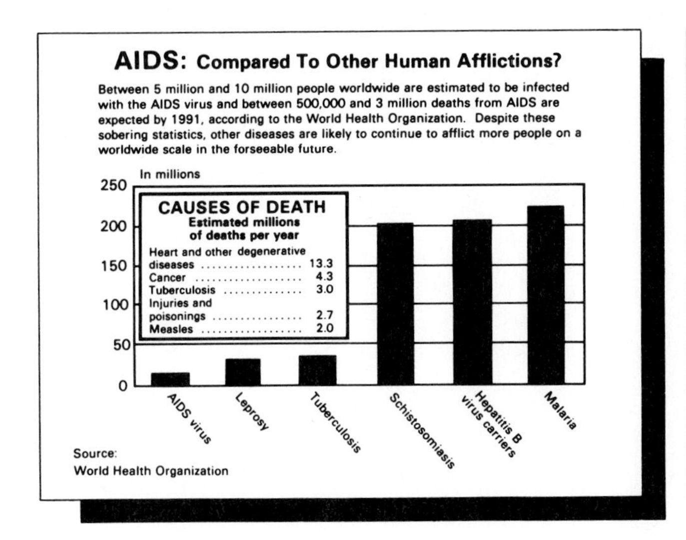

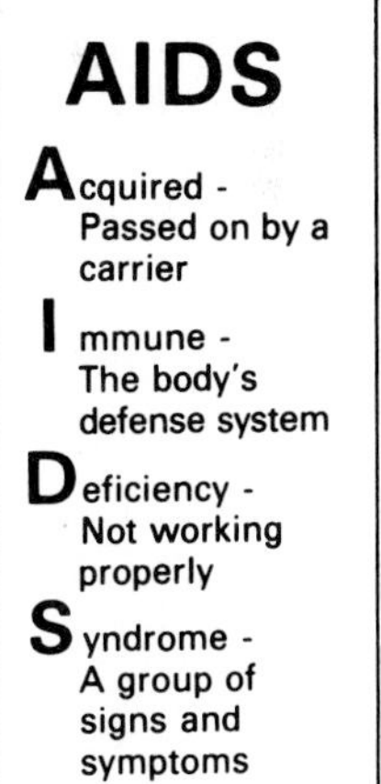

Model 2-2 Commentary

This article from *AOPA Pilot* (May 1985) discusses a piece of equipment that helps reduce high temperatures in airplane engines. The readers are members of the Aircraft Owners and Pilots Association (AOPA).

The writer first reviews the effect of engine temperatures on engine output and then introduces the intercooler, a device that will absorb heat and reduce engine temperatures. Included in the article is a technical description of the intercooler and comments by a pilot who has used it successfully.

The writer closes with information about prices and suppliers for readers who want to investigate the equipment further.

Discussion

1. This article has multiple readers. Although all the readers are probably members of the AOPA, they include pilots, owners of airplane rental agencies, mechanics, managers who sell aircraft, and people who provide airplane-related services, such as computer programs, purchasing consultants, insurers, and attorneys. A reader with general interest in flying could read the magazine at the library. Discuss which of these readers probably would be interested in the subject of the article, and identify how each reader would be likely to use the information.

2. In groups, discuss which kinds of information about the intercooler would be interesting to people attending an airshow at the lakefront in Chicago on Labor Day. Draft a 300- to 500-word discussion of the intercooler suitable for people at the airshow. Share your draft with the rest of the class.

Intercooling:

A Hassle-Free Means of Boosting Available Power and Efficiency

by Barry Schiff

The power output of a given piston engine is determined primarily by a combination of its propeller rpm, manifold pressure and fuel/air mixture. Atmospheric conditions also affect engine horsepower. Of these, one significant, frequently overlooked variable is engine inlet air temperature, which is approximately the same as ambient temperature in normally aspirated (nonturbocharged) engines.

Engines lose horsepower when the inlet temperature increases and becomes more powerful when it decreases. This is logical. After all, warm air is less dense than cool air and cannot support as much combustion.

A piston engine loses about one percent of its horsepower for each 10°F rise in engine inlet temperature. Conversely, power increases one percent for each 10°F decrease.

Consider, for instance, an aircraft departing an airport at an elevation of 2000 feet. The temperature is 112°F, which is 60°F warmer than the standard temperature (52°F) for this elevation. The engine will produce six percent less power than when departing the same airport on a standard day. (This is in addition to the aerodynamic performance losses that occur when operating at high density altitudes.)

The power loss associated with elevated inlet temperatures is much more significant, however, when operating a turbosupercharged (or turbocharged, for short) engine. The purpose of a turbocharger is to compress the ambient air and provide a boost of manifold pressure, thereby providing more horsepower at altitude than would otherwise be available. Turbocharging allows the engine of a Cessna T210, for instance, to produce as much manifold pressure and rpm at 15,000 feet as it does at sea level. Does this mean that the engine produces as much horsepower at 15,000 feet as it does at sea level? Unfortunately, no.

When air is compressed, its temperature rises. This explains why a tire gets warm when it is inflated. A turbocharger, however, compresses air much more than a tire pump. When a Cessna T210 is climbing with maximum-allowable power at 15,000 feet, for instance, the air being pumped into the engine is really hot—more than 200°F. At 25,000 feet, where the T210's turbocharger works harder to maintain a given manifold pressure, the compressor-discharge temperature increases to more than 250°F. Such heat can reduce engine power by almost 20 percent, even though engine rpm and manifold pressure have the same values as at sea level. In a manner of speaking, the heat generated by turbocharging is like operating an engine with the carburetor heat turned on.

(To partially compensate for the power-robbing effect of hot induction air, a few aircraft—such as Piper's Navajo—have density turbocharger controllers that automatically and steadily increase manifold pressure during climb with fixed throttle settings. This allows the engine(s) to produce nearly constant

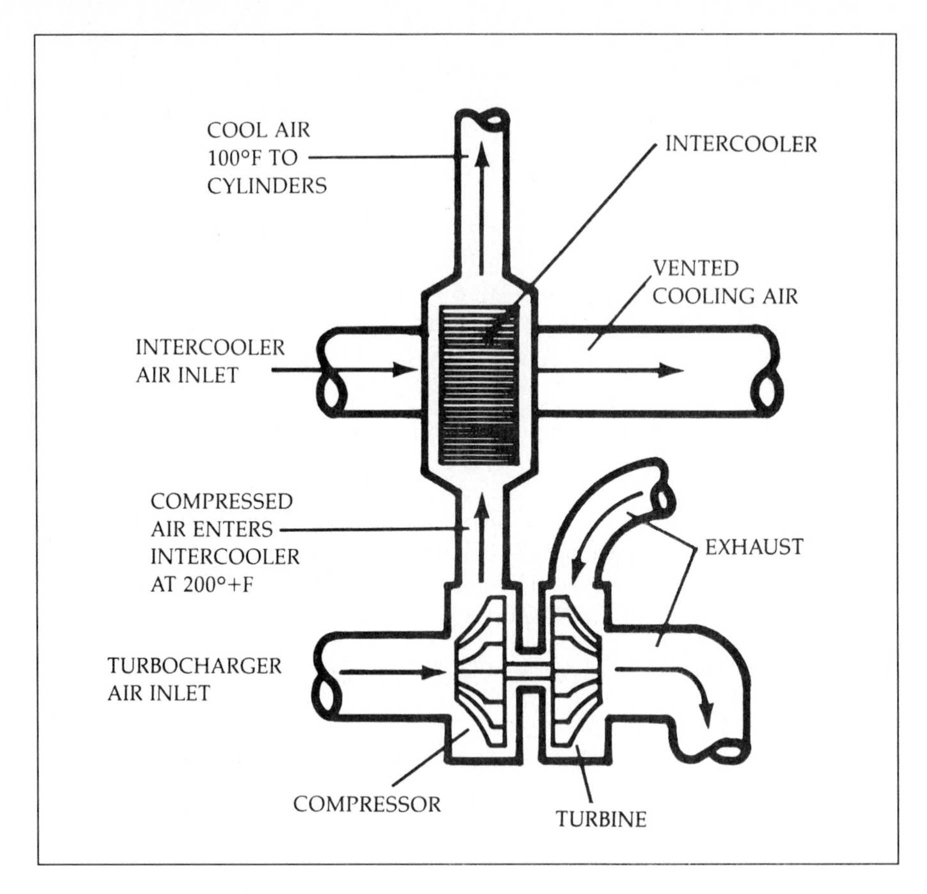

Compressed air from turbocharger enters intercooler, where heat is dissipated through radiation. Cooled compressed air then enters engine intake manifold.

horsepower until manifold pressure finally reaches the maximum allowable limit.)

It is obvious, therefore, that a turbocharged engine generally runs hottest at high altitude. Unfortunately, this makes it all the more difficult to cool the engine. This may seem incongruous because the standard temperature at 25,000 feet is −30°F. Although such ambient air is quite cold, it has very little density (which is why a turbocharger is needed in the first place) and cannot carry away heat as well as the denser air found at lower altitudes. The poor cooling property of thin air plus the effect of high induction temperatures are the reasons why turbocharged engines have such high operating temperatures at altitude.

High induction temperatures naturally result in high cylinder head temperatures (CHT). But since the metallic mass of an engine essentially is a heat sink, the temperature of the entire engine tends to increase as well. Often this is reflected in relatively warm oil temperatures.

Heat is one nemesis of a piston engine. Its cumulative effects can lead to piston, ring and cylinder-head failure as well as placing thermal stress on other operating components. (For instance, excessive CHT can lead to detonation, which can cause catastrophic engine failure.) Heat also is a primary reason why a turbocharged engine has a lower TBO (time between overhauls) than its normally aspirated counterparts. Another reason for its shorter life span is that turbocharged engines generally are operated at higher power settings a greater portion of the time.

Another problem associated with the heat of turbocharging is thermal shock, or shock cooling. This is caused by a rapid rate of engine cooling and can lead to cracked heads, warped valves and cracked cylinder barrels. When descending from an altitude where the engine is heat soaked, a pilot must be careful to keep the rate of cooling in check by making gradual power reductions and not increasing airspeed too much as the aircraft descends into the denser air of lower altitudes. Thermal shock most frequently occurs when attempting to lose altitude rapidly or when attempting to simultaneously lose altitude and reduce airspeed during a turbulence encounter or while trying to conform to an air traffic control clearance.

One way to reduce the problems caused by high induction temperatures is to install an intercooler between the compressor outlet of a turbocharger and the induction inlet of the engine. An intercooler is essentially a radiator, or heat exchanger, that allows ambient air to pass in close proximity to the compressed air. The ambient air then absorbs much of the heat of compression and carries it overboard. This results in cooler, denser induction air, thus reducing engine temperatures and increasing horsepower.

A convincing way to demonstrate an intercooler's effectiveness is to blow a hair dryer through an intercooler resting on a table. This air comes out the other side at a noticeably reduced temperature. And this is without the benefit of ambient air being ducted across the intercooler.

There is nothing new or magical about intercooling. Its first aviation application appears to have been on military aircraft where it was used to cool the air between two successive supercharger stages. (This explains how the intercooler got its name.) The intercooler is extraordinarily simple—it has no moving parts or controls and generally is the most reliable, maintenance-free accessory on an airplane. They are standard equipment on such aircraft as the Piper Malibu, the Cessna 400-series and some of the Beech Barons.

What is new, however, is that intercoolers are now available as retrofit items. Turboplus, Incorporated, of Gig Harbor, Washington, offers modification kits for the Piper Seneca II and III, Mooney 231, all 200-series Cessnas and the Piper Turbo Arrow. It is also developing intercooler systems for the Piper Turbo Lance and Saratoga. Riley International Corporation of Carlsbad, California, offers an intercooler for the Cessna P210 and expects Federal Aviation Administration approval for other 200-series Cessnas by the time this is published. A third entrant about to break into the intercooler aftermarket is Machen, Incorporated, of Spokane, Washington, which has developed intercooler expertise while converting Aerostars with 290-hp engines into Superstars with 350-hp engines. Its intercooler modification for a Piper Chieftain should be available this summer, followed by one for a Beech Duke. Both Turboplus and Machen are considering the development of new, more efficient intercooler systems for the Cessna 414 and 421.

Even some automobile manufacturers, such as Porsche, Saab and Volvo, are using intercoolers to improve performance and durability of their turbocharged models.

Although intercoolers can be used to recapture much of the power loss caused by the heat of turbocharger compression, most owners who have had them installed seem content to use them to improve operating economy and increase engine longevity.

The effect of an intercooler modification can be noticed immediately after engine start. The intercooler gauge (included with all intercooler modifications available thus far) digitally displays the temperature difference between the air entering and exiting the intercooler. A 15 to 20°F temperature drop is typical when the engine is idling. By flipping a three-way switch on the gauge, compressor discharge and engine inlet temperatures can be read directly.

During takeoff, induction air is cooled about 60°F, which can cause the engine to develop more than its rated power. Consequently, a pilot needs to compensate for the dense inlet air by reducing manifold pressure one inch for each 20°F of intercooling. Using this technique, 37 inches of manifold pressure results in as much horsepower as the same nonintercooled engine would with 40 inches.

Once the manifold pressure is reduced to compensate for intercooling, some pressure can be added back to compensate for ambient air that is warmer than standard. By increasing MP one inch for each 20°F that ambient temperature exceeds standard temperature (for a given elevation), the engine produces as much power as normally would be available at the same elevation on a standard day.

These manifold pressure corrections, however, are unnecessary when operating an engine that has an intercooler provided by the powerplant manufacturer as original equipment. This is because engine limits and power charts are established to automatically compensate for intercooling. (When Lycoming intercooled its TIO-540, for example, additional shaft horsepower became available with less manifold pressure, and time between overhauls increased from 1600 to 2000 hours because of reduced engine operating temperatures.)

Although manifold pressure reductions are required during takeoff for all of the Turboplus conversions, Riley claims that its modified Cessna P210 has been approved for takeoff and initial climb performance.

Pilots should be aware that piston-engine manufacturers are not required to prove engine reliability when ambient temperature exceeds 100°F. In other words, when an aircraft departs Palm Springs, California, where summer temperatures soar to 120°F, a pilot has no assurance that he can maintain engine temperatures within approved limits. An intercooler modification resolves this doubt.

Climb power also should be reduced one inch of manifold pressure per 20°F of intercooling. In this manner, the engine continues to develop about as much horsepower as it would without intercooling. Fuel flow and climb performance remain essentially unchanged. But be prepared for a pleasant surprise. The CHT needle will be comfortably in the green arc at a time when it normally snuggles near the red line. (During climb, a pilot has the option to use normal power settings to increase climb performance.)

Since engine temperatures are so cool, Martin Jacobson, who owns a Turboplus-modified Mooney 231, claims that he frequently climbs with the cowl flaps closed and the mixture leaned to 50°F on the rich side of peak exhaust gas temperature—something he says he would never consider before installing the intercooler.

After leveling off at altitude, the pilot chooses the percentage of power he would like to use. He then reduces manifold pressure to compensate for intercooling (a 150°F temperature drop across the intercooler is not unusual at altitude) and then makes a final adjustment for ambient temperature. It might seem strange to reduce manifold pressure by eight inches, but recall that induction air to the engine is much denser than it normally would be. Consequently, not as much manifold pressure is required to develop the same horsepower.

A glance at the instrument panel will reveal the CHT, oil temperature and EGT are noticeably lower; the entire engine is cooler.

The cylinder head temperatures of turbocharged engines often are kept cool by using a rich mixture that supplies an excess of fuel to the combustion chambers. This not only is an efficient use of fuel but also washes oil off the cylinder walls, thereby increasing ring and cylinder-wall wear. When an engine is modified with an intercooler, cylinder head temperatures are low enough that it is unnecessary to use as much fuel for cooling. This is one reason why the mixture of the Piper Malibu's intercooled engine can be leaned to 50° on the *lean* side of peak EGT instead of 25° or 50° on the rich side, for instance. Engines modified with intercoolers similarly can be operated with leaner mixtures (but not on the lean side of peak EGT, however). This reduces fuel flow by about 10 percent and improves range.

Those not interested in conserving fuel can operate their modified engines without reducing manifold pressure and use the additional horsepower created by intercooling to increase cruise speeds, cruise altitudes and service ceilings. The amount of additional performance depends, of course, on the type of aircraft involved. But at 23,000 feet, for instance, an intercooled Cessna T210 can cruise 28 knots faster than an unmodified T210 at the same power settings. The single-engine service ceiling of modified twins also is increased significantly.

Since an intercooled engine runs cooler than other engines, it does not present as much concern about rapid cooling during descent. This could be interpreted to mean that thermal shock is more easily avoided. Perhaps, but pilots must remember to consider the *rate* of cooling, not the total amount.

The negative aspects of intercooling include weight (8 to 10 pounds per engine) and some cooling drag (back pressure), which is overcome by a slight power increase. (The amount of back pressure depends on the efficiency of the design.) Prices for currently available modifications vary from a low of Riley's $4950 (installed) for a Cessna T210 to a high of $7995 (plus 46 hours of installation time and $400 to repaint the cowlings) for the Turboplus conversion of a Piper Seneca II or III. Prices for Machen's intercooler have yet to be announced.

Advocates claim that the price of admission to intercooling is worth the investment—it can be recaptured in terms of fuel savings, reduced engine maintenance, prolonged engine life, better performance and added safety.

Model 2-3 Commentary

This model shows the opening page of a report by research chemists for the U.S. Bureau of Mines. The report is an investigative study of the thermodynamics of certain chemical compounds. Readers are other research scientists interested in this experiment and the results that might influence their own research. The abstract is written for expert readers and does not provide a summary for nonexpert readers.

The report language is highly specialized. None of the technical terms or chemical symbols is defined because the writer expects only experts to read the report.

Discussion

Compare the language, kinds of information, and formats of Models 2–1, 2–2, and 2–3, and discuss how each document is appropriate for its readers.

Enthalpy of Formation of $2CdO \cdot CdSO_4$

by H. C. Ko[1] and R. R. Brown[1]

Abstract

The Bureau of Mines maintains an active program in thermochemistry to provide thermodynamic data for the advancement of mineral science and technology. As part of this effort, the standard enthalpy of formation at 298.15 K, $\Delta Hf°$, for $2CdO \cdot CdSO_4$ was determined by HCl acid solution calorimetry to be -345.69 ± 0.61 kcal/mol.

Introduction

Cadmium oxysulfate ($2CdO \cdot CdSO_4$) is known as an intermediate compound in the decomposition process of cadmium sulfate ($CdSO_4$). There is only one reported value (4)[2] for the enthalpy of formation of $2CdO \cdot CdSO_4$, and that value was determined by a static manometric method. The objective of this investigation was to establish the standard enthalpy of formation for this compound by HCl solution calorimetry, which is inherently a more accurate method. This investigation was conducted as part of the Bureau of Mines effort to provide thermodynamic data for the advancement of mineral technology.

Materials

Cadmium oxide: Baker[3] analyzed reagent-grade CdO was dried at 500°C for 2 hr. X-ray diffraction analysis indicated that the pattern matched the one given by PDF card 5–640 (2). Spectrographic analysis indicated 0.05 pct Pb to be present as the only significant impurity.

Cadmium oxysulfate: $2CdO \cdot CdSO_4$ was prepared by reacting stoichiometric quantities of anhydrous $CdSO_4$ and CdO in a sealed, evacuated Vycor tube according to the following procedure. The materials were blended and transferred to a Vycor reaction tube. The reaction tube was evacuated and heated to 330°C for 17 hr to remove traces of water that may have been introduced during blending and transferring. The tube was then sealed and heated to 600°C for 19 hr. The temperature was. . . .

[1] Research chemist, Albany Research Center, Bureau of Mines, Albany, Oreg.
[2] Underlined numbers in parentheses refer to items in the list of references at the end of this report.
[3] Reference to specific trade names does not imply endorsement by the Bureau of Mines.

Chapter 2 Exercises

1. Discuss these situations and decide what kinds of information the readers will need in order to make a decision.

a. During a plane flight, the flight attendant dropped the dinner tray on your lap, damaging a very expensive article of clothing. When the plane landed at its destination, airport personnel reimbursed you $7.50—the amount authorized by the airline. A dry cleaner later told you that the clothing could not be cleaned properly. You have decided to write to the president of the airline and ask for a full reimbursement for your ruined clothing. What kinds of information will you have to give the president of the airline before he or she can decide whether or not to reimburse you.

b. You are director of your community "Be a Mentor" program. Next month, selected high-school seniors will spend a full business day with an executive at a local company, observing typical business activities. The program has been a great success in other cities, and you want to get it off to a good start in your city. As director, you plan to write a letter to local business executives asking them to participate in the program by volunteering to advise a high-school senior about career choices and by spending a full business day with a student. What kinds of information does the business executive need to know in order to decide whether or not to participate in the program?

2. Find a research article on a current topic in your field. Read the article and identify the most important points. Then write a short summary of the important research information to be included in a newsletter to alumni of your school. The alumni are interested in hearing about new developments in the majors offered at your school, and each newsletter features one research summary such as the one you will write. Limit yourself to no more than 300 words. Give your article a title that will attract your readers' attention.

3. Find a technical description of a piece of equipment or system used in your field. Rewrite the description for a general reader who is interested in learning how the equipment or system works but who has no experience with it.

4. The following article on lawn care from *House Beautiful* (August 1988) was written for homeowners. Discuss in class which information would be most important to a reader who wants to have a beautiful green carpet of grass next summer. Using the information from the article, write a short guide to growing grass for that reader.

Garden Guide

by Ken Druse

A well-kept lawn is one of the most beautiful sights in nature, though it is hardly a natural occurrence. The idea of lawns originally came from pastures kept close-cropped by grazing animals. Today, *we* have to do the mowing and fertilizing.

Grass plants are grains, similar to wheat. Left untended, they grow tall, flower, set seed and dry. To keep them green, you trim, feed and water them—and before you know it, you're hooked to a cycle of perpetual care. But to me a verdant lawn is worth the work.

You can sow a pretty lawn easily on open ground in April, but to *keep* a lawn looking good—and to reduce maintenance the next season—late-summer care is crucial. August is a high-stress month for lawns. Some species, such as Zoisia grass, may brown out completely. Thorough, deep watering is paramount; shallow watering encourages surface roots that are damaged easily by sunscald and drought. If you're uncertain when your lawn needs watering, look for such signs as a silvery tinge or footprints that fail to spring back. If you're reseeding a patch or laying new sod, keep the soil damp for several days, until roots have become firmly anchored (you can tell by the new growth above). Feeding, too, must continue into fall.

Since you're constantly encouraging new growth, better keep the mower handy. The height of your grass depends on the variety. Bermuda grass may be clipped down to a quarter-inch, but most varieties should be kept about two inches tall. You'll find recommended heights printed on grass-seed packages. When your grass grows half an inch above the suggested height, mow it. At this time of year twice-weekly mowings may be necessary. Try changing the direction each time you mow, so the lawn doesn't look permanently striped.

If you suspect grub damage, which is evidenced by brown patches you can lift out of the lawn (often revealing grayish-white coiled Japanese beetle larvae), consider a biological solution. Milky spore disease, available at garden centers, is a host-specific organism that can be used to inoculate lawns. It takes a year or two to become effective, but it remains in the soil up to 20 years and will not harm children or animals. If you can enlist your neighbors in a mass inoculation, you will reduce the damage from adult beetles as well.

Thatch is another ongoing problem. It's simply the accumulation of dead rhizomes, matted roots and grass cuttings on the lawn surface that eventually suffocates the plants themselves. One good deterrent is to clear away the grass clippings after you mow the lawn. Sprinkle them around flowering plants as a mulch if you haven't used weed killer recently. The Ringer Corp. offers a cure that biodegrades the thatch material (for a free catalog, write or call: Ringer Corp., Dept. HB, 9959 Valley View Rd., Eden Prairie, MN 55344; 800–654–1047).

Now is the time to head for your garden center to check out fall lawn-care products that are beginning to line the shelves. Select items that will meet your specific needs, and don't put your mower away too soon. Long grass left to lie beneath a layer of snow can really smother a lawn.

If all this seems like a lot of hard work, keep in mind that late-summer lawn care will save time and energy and will give you a head start on the most beautiful spring lawn you've ever seen.

5. Find a user manual for a word-processing program. Rewrite one page of the introductory material for an eighth-grade student who is going to start using the program.

6. Assume that you are writing a career booklet for students at your school. The booklet will answer the most frequent questions students have about selecting a career, finding jobs, starting the application process, preparing for an interview, and other helpful information. Interview a fellow classmate about information he or she would want to have in such a booklet. Based on that interview, list the topics that should be included in the booklet. Then, in groups, compare your list with those of other students. As a group, prepare an outline that arranges the topics in an order you think appropriate for your readers. In class, present the group outline and discuss why the listed topics probably would appeal to an audience of new college graduates.

7. Individually, or in groups if your instructor prefers, develop a reader test as discussed in this chapter under "Testing Reader-Oriented Documents." Bring to class a short brochure intended for general readers and distributed by a utility company, a charitable organization, or a consumer-products company. Develop five questions designed to test a reader's understanding of the information in the brochure. Follow the testing procedure described under "Readers Answering Questions." Analyze the results and identify any sections of the brochure that could be improved. Rewrite one section of the brochure based on your reader test. Submit both the original brochure and your revision of one section to your instructor.

8. Read the article by Ridgway entitled "The Writer as Market Researcher" in Part Two. Bring to class a copy of the official directions issued on your campus to guide students in registering for classes each semester or quarter. In groups, develop a set of interview questions to test student understanding and use of these directions. Each group member will then interview three students outside the class, using these questions. The group members will pool their results and come to some conclusions about the usefulness of the registration directions. Report your group results to the rest of the class, and discuss how the registration directions might be revised to better serve the students on your campus.

CHAPTER 3

Organization

Sorting Information
- Select Major Topics
- Identify Subtopics

Constructing Outlines
- Informal Outlines
- Formal Outlines
- Topic Outlines
- Sentence Outlines

Developing Effective Paragraphs
- Unity
- Coherence
- Development
- Patterns for Presenting Information
  - Ascending or Descending Order of Importance Pattern
  - Cause and Effect Pattern
  - Chronological Pattern
  - Classification Pattern
  - Partition Pattern
  - Comparison and Contrast Pattern
  - Definition Pattern
  - Spatial Pattern

Chapter Summary

Supplemental Readings in Part Two

Endnotes

Models

Exercises

Sorting Information

In some writing situations, documents have predetermined organizational patterns. Social workers at a child welfare agency may be directed to present information in all case reports under the major topics of living conditions, parental attitudes, previous agency contact, and suspected child abuse. Because these major topics appear in all the agency case reports, readers who use them regularly, such as attorneys and judges, know where to look for the specific details they need. Such consistency is useful to readers if the documents have a standard format—only the details change from situation to situation.

If you are writing a document that does not have standard categories for information, however, your first step will be to organize the information into major topics and identify appropriate subtopics.

Select Major Topics

Begin organizing your document by sorting the information into major topics. Select the topics based on your analysis of your readers' interest in and need for information. If you are writing a report about columns or pillars, you should group your information into major topics that represent what your readers want to know about the subject. If your readers are interested in Greek and Roman column design, you can group your information according to classic column type—Doric, Ionic, and Corinthian. Readers interested in learning about column construction may prefer an organization based on construction materials, such as brick, marble, wood, stone, and metal. Readers interested in the architectural history of columns may prefer an organization based on time period, such as Ancient, Medieval, and Renaissance. Readers interested in the cultural differences in column design may prefer the information grouped according to geographic region, such as Asian, Mediterranean, and North American.

As you sort your information into major topics, remember that the topics should be similar in type and relatively equal in importance. You should not have one major topic based on column design and one based on time period. Such a mix of topics would confuse your readers. You should also be certain that the major topics you select are distinct. You would not want to use time periods that overlap, such as the Renaissance and the fifteenth century, or column designs that overlap, such as Doric and Tuscan. Consider carefully which major topics represent appropriate groups of information for your readers.

Identify Subtopics

After you determine your major topics, consider how to sort the information into appropriate subtopics. Do not simply lump facts together. Instead, think about what specific information you have that will support each major topic. If your major topics are based on classic column types, your readers may be interested in such subtopics as Greek design, Roman design, changes in design over time, and how the Greeks and Romans used the columns. These subtopics, or specific pieces of information, should appear under each of the major topics because your readers want to know how these details relate to each column type. When you have identified your major topics and have selected appropriate subtopics, you are ready to begin outlining.

Constructing Outlines

Think of an outline as a map that identifies the major topics and shows their location in a document. Outlines provide three advantages to a writer:

1. A writer can "see" the structure of a document before beginning a rough draft in the same way that a traveler can follow the route of a journey on a road map before getting into a car.
2. Having a tentative organizational plan helps a writer concentrate on presenting and explaining the information in the rough draft rather than on writing and organizing at the same time.
3. A writer can keep track of organizational decisions no matter how many interruptions occur in the writing process.

An outline usually begins as a short, informal list of topics and grows into a detailed list that includes all major topics and subtopics and groups them into chunks of information most appropriate for the reader. Consider this situation: Al Martinez's first project as a summer intern in the Personnel Department of Tri-Tech Chemical Company is to write a bulletin explaining the company rules on travel and business expenses. Al's readers will be both the managers who travel and the secretaries who handle the paperwork. At present, the pertinent information is scattered through more than a dozen memos written over the last three years. After reading all these memos and noting appropriate items, Al begins to organize his information. He begins with an informal outline.

Informal Outlines

An *informal outline* is a list of main topics in the order the writer expects to present them in the rough draft. Word processing has made outlining faster and easier than it used to be. With word processing, a writer can quickly (1) add or delete items and rearrange them without recopying sections that do not change and (2) keep the outline up to date by changing it to match revisions in rough drafts. In developing an informal outline, remember these guidelines:

1. List all the relevant topics in any order at first. Al listed the main topics for his bulletin in the order in which he found them discussed in previous memos in company files.
2. Identify the major groups of related information from your list. Looking at his list of items, Al saw immediately that most of his information centered around types of expenses and reimbursement and approval procedures. Al therefore clustered his information into groups representing these topics.
3. Arrange the information groups in an order that will best serve the readers' need to know. Al decided to present reimbursement procedures first because readers would need to know about those before traveling. He then organized groups representing types of expenses from most common to least common so that the majority of readers could find the information they needed as quickly as possible.

Figure 3-1 shows Al's informal outline for his bulletin. Writers usually use informal outlines to group and order information in the planning stage. Often, an informal outline is all that a writer needs. For some documents, such as a manual, however, a writer must prepare a formal outline that will be part of the document. Also, some writers prefer to work with formal outlines in the planning stage.

1. Travel Authorization and Reimbursement Rules
2. Meals
3. Hotels
4. Transportation
5. Daily Incidental Expenses
6. Spouse's Travel
7. Entertainment
8. Gifts
9. Loss or Damage
10. Nonreimbursed Items

Figure 3-1 Al's Informal Outline

Formal Outlines

A *formal outline* uses a special numbering system and includes subtopics under each major section. A formal outline is a more detailed map of the organization plan for a document. Writers usually use one of two numbering systems for outlines: (1) Roman and Arabic numerals or (2) decimals (also called the *military numbering system*). Choose whichever numbering system you prefer unless your outline will be part of the final document. In that case, use the numbering system most familiar to your readers or the one your company prefers. Figure 3-2 shows a set of topics organized by both numbering systems.

Al Martinez decided to develop his informal outline into a formal one because he wanted a detailed plan before he began his rough draft.

Roman-Arabic Numbering System

I. Costs
  A. Equipment
    1. Purchase
    2. Maintenance
      a. Weekly
      b. Breakdown
  B. Employees
    1. Salaries
    2. Benefits
II. Locations
  A. European
    1. France

Decimal Numbering System

1.0 Costs
  1.1 Equipment
    1.1.1. Purchase
    1.1.2. Maintenance
      1.1.2.1. Weekly
      1.1.2.2. Breakdown
  1.2 Employees
    1.2.1. Salaries
    1.2.2. Benefits
2.0 Locations
  2.1 European
    2.1.1. France

Figure 3-2 Formal Outline Numbering Systems

Topic Outlines

A *topic outline* lists all the major topics and all the subtopics in a document by key words. Like the major topics, all subtopics should be in an order that the readers will find useful. Al developed his formal topic outline by adding the items he knew he had to explain under each major topic and using the Roman-Arabic numbering system. Figure 3-3 shows Al's formal topic outline.

In drafting his formal topic outline, Al knew that he wanted to begin his bulletin by explaining the process of applying for and receiving reimbursement for travel expenses because his readers needed to understand that system before they began collecting travel-expense information and preparing expense reports. He organized expense topics in the order in which they appeared on the forms his readers would have to submit and then listed specific items under each main topic. While redrafting his outline, Al revised some of his earlier organizational decisions. First, he decided that the topic "Spouse's Travel" was not really a separate item and ought to be covered under each travel expense item, such as "Hotels." He also decided that the topic "Gifts" was really a form of entertainment for clients, so he listed gifts under "Entertainment" in his formal topic outline.

Sentence Outlines

A *sentence outline* develops a topic outline by stating each point in a full sentence. When a project is long and complicated or the writer is not confident about the subject, a sentence outline helps clarify content and organization before drafting begins. Some writers prefer sentence outlines because they can use the sentences in their first drafts, thereby speeding up the initial drafting. Al Martinez also decided to expand his topic outline into a sentence outline as a way of moving closer to his first draft. Figure 3-4 shows a portion of Al's sentence outline.

Al expanded each topic into a full sentence by stating specific details. Under "Meals," Al said Tri-Tech would pay for three a day, and then he defined each according to the time period involved.

Developing Effective Paragraphs

Each paragraph in a document is a unit of sentences that focuses on one idea and acts as a visual element to break up the text into manageable chunks of information. Effective paragraphs guide readers by

- Introducing individually distinct but related topics
- Emphasizing key points

- I. Travel Authorization and Reimbursement
    - A. Approval
    - B. Forms
        - 1. Form 881
        - 2. Form 669
        - 3. Form 40-A
        - 4. Form 1389
    - C. Submission
- II. Meals
    - A. Breakfast
    - B. Lunch
    - C. Dinner
- III. Hotels
    - A. Chain Guaranteed Rates
    - B. Suites
- IV. Transportation
    - A. Automobile
        - 1. Personal Cars
            - a. Mileage
            - b. Maintenance
        - 2. Rental Cars
            - a. Approved Companies
            - b. Actual Costs Only
    - B. Airplane
        - 1. Tri-Tech Aircraft
        - 2. Other Companies' Aircraft
        - 3. Commercial Aircraft
    - C. Railroad
    - D. Taxi, Limousine, Bus
- V. Daily Incidental Expenses
    - A. Parking and Tolls
    - B. Tips
    - C. Laundry and Telephone
- VI. Entertainment
    - A. Parties
        - 1. Home
        - 2. Commercial
    - B. Meals
    - C. Gifts
        - 1. Tickets
        - 2. Objects
- VII. Loss or Damage
- VIII. Nonreimbursed Items
    - A. Personal Items
    - B. Legal, Insurance Expenses

Figure 3-3 Al's Formal Topic Outline

I. Travel Authorization and Reimbursement. Tri-Tech will pay travel expenses directly or reimburse an employee for costs of traveling on company business.
   A. All official travel for Tri-Tech must be approved in advance by supervisors.
   B. Several company forms must be completed and approved.
      1. Form 881, "Travel Authorization," is used for trip approval if travel reservations are needed.
      2. Form 669, "Petty Cash Requisition," is used if no travel reservations are needed.
      3. Form 40-A, "Domestic Travel Expenses," is used to report reimbursable expenses in the United States.
      4. Form 1389, "Foreign Expense Payment," is used to report reimbursable expenses outside the United States.
   C. Travel authorization forms should be submitted to the employee's supervisor, but posttravel forms should be sent directly to the Travel Coordinator in the Personnel Department.

II. Meals. Tri-Tech will reimburse employees for three meals a day while traveling.
   A. Breakfast expenses are reimbursed if the employee is away overnight or leaves home before 6:00 a.m.

Figure 3-4 A Portion of Al's Formal Sentence Outline

- Showing relationships between major points
- Providing visual breaks in pages to ease reading

For effective paragraphs, writers should consider unity, coherence, development, and an organizational pattern for presenting information.

Unity

Unity in a paragraph means concentration on a single topic. One sentence in the paragraph is the *summary sentence* (also called the *topic sentence*), and it establishes the main point. The other sentences explain or expand the main point. If a writer introduces several major points into one paragraph without developing them fully, the mix of ideas violates the unity of the paragraph and leaves readers confused. Here is a paragraph that lacks unity:

> Many factors influence the selection of roofing material. Only project managers and engineers have the authority to change the recommendations, and the changes should be in writing. The proper installation of roofing material is important, and contractors should be trained carefully. Roof incline,

roof deck construction, and climatic conditions must be considered in selecting roofing material.

The first sentence appears to be a topic sentence, indicating that the paragraph will focus on factors in selecting roofing. Readers therefore would expect to find individual factors enumerated and explained. However, the second and third sentences are unrelated to the opening and instead introduce two new topics: authority to select roofing and the training of contractors. The fourth sentence returns to the topic of selecting roofing material. Thus the paragraph actually introduces three topics, none of which is explained. The paragraph fails to fulfill the expectations of the reader. Here is a revision to achieve unity:

> Three factors influence the selection of roofing material. One factor is roof incline. A minimum of 1/4 in. per foot incline is recommended by all suppliers. A second factor is roof deck construction. Before installing roofing material, a contractor must cover a metal deck with rigid installation, prime a concrete deck to level depressions, or cover a wooden deck with a base sheet. Finally, climatic conditions also affect roofing selection. Roofing in very wet climates must be combined with an all-weather aluminum roof coating.

In this version, the paragraph is unified under the topic of which factors influence the selection of roofing materials. The opening sentence establishes this main topic; then the sentences that follow identify the three factors and describe how contractors should handle them. Remember, readers are better able to process information if they are presented one major point at a time.

Coherence

A paragraph is *coherent* when the sentences proceed in a sequence that supports one point at a time. Transitional, or connecting, words and phrases help coherence by showing the relationships between ideas and by creating a smooth flow of sentences. Here are the most common ways writers achieve transition:

- Repeat key words from sentence to sentence.
- Use a pronoun for a preceding key term.
- Use a synonym for a preceding key term.
- Use demonstrative adjectives (*this* report, *that* plan, *these* systems, *those* experiments).
- Use connecting words (*however, therefore, also, nevertheless, before, after, consequently, moreover, likewise, meanwhile*) or connecting phrases (*for example, as a result, in other words, in addition, in the same manner*).
- Use simple enumeration (*first, second, finally, next*).

Here is a paragraph that lacks clear transitions, and the sentences are not in the best sequence to explain the two taxes under discussion:

> Married couples should think about two taxes. When a person dies, an estate tax is levied against the value of the assets that pass on to the heirs by will or automatic transfer. Income tax on capital gains is affected by whatever form of ownership of property is chosen. Under current law, property that one spouse inherits from the other is exempt from estate tax. Married couples who remain together until death do not need to consider estate taxes in ownership of their homes. The basic rule for capital gains tax is that when property is inherited by one person from another, the financial basis of the property begins anew. For future capital gains computations, it is treated as though it were purchased at the market value at the time of inheritance. When the property is sold, tax is due on the appreciation in value since the time it was inherited. No tax is due on the increase in value from actual purchase to time of inheritance. If the property is owned jointly, one-half gets the financial new start.[1]

This paragraph attempts to describe the impact of two federal taxes on married couples who own their homes jointly. The writer begins with the topic of estate tax, but he or she quickly jumps to the capital gains tax, then returns to the estate tax, and finally concludes with the capital gains tax. Since readers expect information to be grouped according to topic, they must sort out the sentences for themselves in order to use the information efficiently. The lack of transition also forces readers to move from point to point without any clear indication of the relationships between ideas. Here is a revision of the paragraph:

> For financial planning, married couples must think about two taxes. First is the estate tax. When an individual dies, this tax is levied against the value of the assets that pass on to the heirs by will or automatic transfer. Under current law, any amount of property that one spouse inherits from the other is exempt from estate tax. Therefore, married couples who remain together until death do not need to consider estate taxes in planning ownership of their home. The second relevant tax is capital gains tax. Unlike estate tax, capital gains tax is affected by what form of ownership of property a couple chooses. The basic rule for capital gains tax is that when property is inherited by one person from another, it begins anew for tax purposes. That is, for future capital gains computation, the property is treated as though it were purchased at the market value at the time of inheritance. When the property is sold, tax is due on the appreciation in value since the time it was inherited. No tax, however, is due on the increase in value from actual purchase to time of inheritance. If the property is owned jointly, one-half begins anew for tax purposes.

In this revision, the sentences are rearranged so that the writer gives information first about the estate tax and then about the capital gains tax. Transitions come from repetition of key words, pronouns, demonstrative adjectives, and connecting words or phrases.

Coherence in paragraphs is essential if readers are to use a document effectively. When a reader stumbles through a paragraph trying to sort out the information or decide the relationships between items, the reader's understanding and patience diminish rapidly.

Development

Develop your paragraphs by including enough details so that your reader understands the main point. Generally, one- and two-sentence paragraphs are not fully developed, unless they are used for purposes of transition or emphasis or are quotations. Except in these circumstances, a paragraph should contain a summary or topic sentence and details that support the topic.

Here is a poorly developed paragraph from a brochure that explains diabetes to patients who must learn new eating habits.

> People with insulin-dependent diabetes need to plan meals for consistency. Insulin reactions can occur if meals are not balanced.

This sample paragraph is inadequately developed for a new patient who knows very little about the disease. The opening sentence suggests that the paragraph will describe meal planning for diabetics. Instead, the second sentence tells what will happen without meal planning. Either sentence could be the true topic sentence, and neither point is developed. The reader is left, therefore, with an incomplete understanding of the subject and no way to begin meal planning. Here is a revision:

> People with insulin-dependent diabetes need to plan meals for consistency. To control blood sugar levels, schedule meals for the same time every day. In addition, eat about the same amounts of carbohydrates, protein, and fat every day in the same combinations and at the same times. Your doctor will tell you the exact amounts appropriate for you. This consistency in eating is important because your insulin dose is based on a set food intake. If your meal plan is not balanced, an insulin reaction may occur.

In this revision, the writer develops the summary sentence by giving examples of what consistency means. The paragraph is developed further by an explanation of the consequences if patients do not plan balanced meals. The point about an insulin reaction is now connected to the main topic of the paragraph—meal planning.

In developing paragraphs, think first about the main idea and then determine what information the reader needs to understand that idea. Here are some ways to develop your paragraphs:

- Provide examples of the topic.
- Include facts, statistics, evidence, details, or precedents that confirm the topic.
- Quote, paraphrase, or summarize the evidence of other people on the topic.
- Describe an event that has some influence on the topic.
- Define terms connected with the topic.
- Explain how equipment operates.
- Describe the physical appearance of an object, area, or person.

In developing your paragraphs, remember that long, unbroken sections of detailed information may overwhelm readers, especially those without expert knowledge of the subject, and may interfere with the ability of readers to use the information.

Patterns for Presenting Information

The following patterns for presenting information can be effective ways to organize paragraphs or entire sections of a document. Several organizational patterns may appear in a single document. A writer preparing a manual may use one pattern to give instructions, another to explain a new concept, and a third to help readers visualize the differences between two procedures. Select the pattern that will best help your readers understand and use the information.

Ascending or Descending Order of Importance Pattern

To discuss information in order of importance to the reader, use the ascending (lowest-to-highest) or descending (highest-to-lowest) pattern. If you are describing the degrees of hazard of several procedures or the seriousness of several production problems, you may wish to use the descending pattern to alert your readers to the matter most in need of attention. In technical writing, the descending order usually is preferred by busy executives because they want to know the most important facts about a subject first and may even stop reading once they understand the main points. However, the ascending order can be effective if you are building a persuasive case on why, for example, a distribution system should be changed.

This excerpt from an advertisement for a mutual fund illustrates the descending order of importance in a list of benefits for potential investors in the fund:

Experienced and successful investors select the Davies-Meredith Equity Fund because

1. The Fund has outperformed 98% of all mutual funds in its category for the past 18 months.
2. The Fund charges no investment fees or commissions.
3. A 24-hour toll-free number is available for transfers, withdrawals, or account information.
4. Each investor receives the free monthly newsletter, "Investing for Your Future."

Although a reader may be persuaded to invest in the fund by any of these reasons, most potential investors would agree that the first item is most important and the last item least important. The writer uses the descending order of importance to attract the reader's attention.

Cause and Effect Pattern

The cause-and-effect pattern is useful when a writer wants to show readers the relationship between specific events. If you are writing a report about equipment problems, the cause-and-effect pattern can help readers see how one breakdown led to another and interrupted production. Be sure to present a clear relationship between cause and effect and give readers evidence of that relationship. If you are merely speculating about causes, make this clear:

> The probable cause of the gas leak was a blown gasket in the transformer.

Writers sometimes choose to describe events from effects back to cause. Lengthy research reports often establish the results first and then explain the causes of those results. This excerpt from an encyclopedia uses cause-and-effect organization to show how acid rain develops:

> Once produced in the atmosphere, acids may be deposited as precipitation or in dry form. Precipitation becomes acidic by incorporating acids directly or by incorporating the gaseous acid precursors which are then oxidized to acids. Nucleation is the most important process incorporating pollutants into precipitation. Here the pollutant or the particle to which it is adsorbed provides a surface, or nucleus, onto which water vapor condenses. The droplets grow by further condensation or by collisions with other droplets to form what becomes acid precipitation.[2]

Chronological Pattern

The chronological pattern is used to present material in stages or steps from first to last when readers need to understand a sequence of events or follow specific steps to perform a task. This excerpt from consumer instructions for a popcorn maker illustrates the chronological pattern:

1. Before using, wash cover, butter-measuring cup, and popping chamber in hot water. Rinse and dry. *Do not immerse housing in water.*
2. Place popcorn container under chute.
3. Preheat the unit for 3 minutes.
4. Using the butter-measuring cup, pour 1/2 cup kernels into popping chamber.

Classification Pattern

The classification pattern involves grouping items in terms of certain characteristics and showing your readers the similarities within each group. The basis for classification should be the one most useful for your purpose and your readers. You might classify foods as protein, carbohydrate, or fat for a report to dietitians. In a report on foods for a culinary society, however, you might classify them as beverages, appetizers, and desserts. Classification is useful when you have many items to discuss, but if your categories are too broad, your readers will have trouble understanding the distinctiveness of each class. Classifying all edibles under "food" for either of the two reports just mentioned would not be useful because you would have no basis for distinguishing the individual items.

This excerpt from an encyclopedia uses classification to explain motor types:

> Motors are classified in many ways. The following classifications show some of the many available variations in types of motors.
>
> 1. Size: flea, fractional, or integral horsepower.
> 2. Application: general purpose, definite purpose, special purpose, or part-winding start. May be further classified as crane, elevator, pump, and so forth.
> 3. Electrical type: alternating-current induction, synchronous, or series; direct-current series, permanent magnet, shunt, or compound.
> 4. Mechanical protection and cooling: (a) open: dripproof, splashproof, semiguarded, fully guarded.[3]

Partition Pattern

The partition pattern, in contrast to the classification pattern, involves separating a topic or system into its individual features. This division allows readers to master information about one aspect of the topic before going on to the next. Instructions are always divided into steps so that readers can perform one step at a time. The readers' purpose in needing the information should guide selection of the basis for partition. For example, a skier on the World Cup circuit would be interested in a discussion of skis based on their use in certain types of races, such as downhill or slalom. A manufacturer interested in producing skis might want the same discussion based on materials used in production. A sales representative might want the discussion based on costs of the different types of skis.

This excerpt from an encyclopedia uses the partition pattern to explain the two components of a computer's central processing unit:

> The CPU . . . contains the following two components that handle the actual processing, the control unit and the arithmetic-logic unit.
>
> *The control unit.* This component within the CPU, also known as the I/O unit or I/O control unit, is the group of circuits that controls the operation of the CPU. Instructions are fed to the control unit one at a time, where they are decoded and processed. The control unit utilizes various registers to hold data and to keep track of what the computer is doing.
>
> *The arithmetic-logic unit (ALU).* This component of the CPU carries out any mathematical or logical operations specified by a program's instructions. Examples are adding or subtracting numbers or making comparisons of values.[4]

Comparison and Contrast Pattern

The comparison and contrast pattern focuses on the similarities (comparison) or differences (contrast) between subjects. Writers often find this pattern useful because they can explain a complex topic by comparing or contrasting it with another familiar topic. For this pattern, the writer has to set up the basis for comparison according to what readers want to know. In evaluating the suitability of two locations for a new restaurant, a writer could compare the two sites according to accessibility, neighborhood competition, and costs. There are two ways to organize the comparison and contrast pattern for this situation. In one method (topical), the writer could compare and contrast location A with location B under specific topics, such as

Accessibility
1. Location A
2. Location B

Neighborhood competition
1. Location A
2. Location B

Costs
1. Location A
2. Location B

In the other method (complete subject), the writer could present an overall comparison by discussing all the features of location A and then all the features of location B, keeping the discussion in parallel order, such as

Location A
1. Accessibility
2. Neighborhood competition
3. Costs

Location B
1. Accessibility
2. Neighborhood competition
3. Costs

In choosing one method over another, consider the readers' needs. For the report on two potential restaurant locations, readers might prefer a topical comparison so as to judge the suitability of each location in terms of the three vital factors. The topical method does not force readers to move back and forth between major sections looking for one particular item as the complete subject method does. In comparing two typewriters, however, a writer may decide that readers need an overall description of the two in order to determine which seems to satisfy more office requirements. In this case, the writer would present all the features of typewriter A and then all those of typewriter B.

This excerpt showing comparison is taken from a report written by a swimming coach to a university director of athletics who needs to know about coaching techniques used on campus:

> This year the swim team has been using a nontraditional training regimen. As you may know, the traditional training method relies on swimmers developing a good aerobic base by swimming between 10,000 and 20,000 yards a day at a steady pace for two months before the meet season. The new training method we are using involves lower yardage but higher swimming pace. This new method calls for 4000 yards of swimming daily, but all swimming is to be at race pace. The two methods both develop an aerobic base, but the second emphasizes race conditions.

Definition Pattern

The purpose of definition is to explain the meaning of a term that refers to a concept, process, or object. Chapter 6 discusses writing definitions for objects or processes. Definition also can be part of any other organizational pattern when a writer decides that a particular term will not be clear to readers. An *informal definition* is a simple statement using familiar terms:

A drizzle is a light rainfall.

A *formal definition* places the term into a group and then explains the term's special features that distinguish it from the group:

term — A pronator is; group — a muscle in the forearm; special features — that turns the palm downward.

A pronator is a muscle in the forearm that turns the palm downward.

Writers use *expanded definitions* to identify terms and explain individual features when they believe readers need more than a sentence definition. Definitions can be expanded by adding examples, using an analogy, or employing one of the organization patterns, such as comparison and contrast or partition. This excerpt from a technical definition of a tornado uses partition to expand the definition:

> The tornado life cycle typically consists of five parts: dust-whirl stage (dust swirls upward from the surface, or a short pendant funnel may appear above the ground), organizing stage (the visible funnel intermittently touches the ground with a continuous damage path), mature stage (the tornado is at its largest size and most intense circulation), shrinking stage (the entire funnel decreases to a thin, ropelike column), and decaying stage (the funnel is fragmented and contorted).[5]

Spatial Pattern

In the spatial pattern, information is grouped according to the physical arrangement of the subject. A writer may describe a machine part by part from top to bottom so that readers can visualize how the parts fit together. The spatial pattern creates a path for readers to follow. Features can be described from top to bottom, side to side, inside to outside, north to south, or in any order that fits the way readers need to "see" the topic.

This excerpt is from an architect's report to a civic restoration committee about an old theater scheduled to be restored:

> The inside of the main doorway is surrounded by a Castilian castle facade and includes a red-tiled parapet at the top of the castle roof. A series of parallel brass railings just beyond the doorway creates corridors for arriving movie patrons. Along the side walls are ornamented white marble columns behind which the walls are covered with 12-ft-high mirrors. The white marble floor sweeps across the lobby to the wall opposite the entrance, where a broad split staircase curves up from both sides of the lobby to the triple doors at the mezzanine level. The top of the staircase at the mezzanine level is decorated with a life-size black marble lion on each side.

CHAPTER SUMMARY

This chapter discusses organizing information by sorting it into major topics and subtopics, constructing outlines, and developing effective paragraphs. Remember:

- Information should be grouped into major topics and subtopics based on the way the readers want to learn about the subject.
- An informal outline is a list of major topics.
- A formal outline uses a special numbering system, usually Roman and Arabic numerals or decimals, and lists all major topics and subtopics.
- A topic outline lists all the major topics and subtopics by key words.
- A sentence outline states each main topic and subtopic in a full sentence.
- Effective paragraphs need to be unified, coherent, and fully developed.
- Organizational patterns for effective paragraphs include ascending or descending order of importance, cause and effect, chronological, classification, partition, comparison and contrast, definition, and spatial.

SUPPLEMENTAL READINGS IN PART TWO

Benson, P. J. "Writing Visually: Design Considerations in Technical Publications," *Technical Communication.*

ENDNOTES

1. Adapted from Allen Bernstein, *1988 Tax Guide for College Teachers* (College Park, Md.: Academic Information Service, 1987).
2. From *Meteorology Source Book* (New York: McGraw-Hill, 1988), p. 287.
3. From *McGraw-Hill Encyclopedia of Energy,* 2d Ed. (New York: McGraw-Hill, 1981), p. 399.
4. From *The Prentice-Hall Encyclopedia of Information Technology* (Englewood Cliffs, N.J.: Prentice-Hall, 1987), pp. 98–99.
5. From *Meteorology Source Book* (New York: McGraw-Hill, 1988), p. 193.

Model 3-1 Commentary

This memo was written by the manager of an exclusive women's fashion store to all the sales clerks. The manager wants to increase sales enthusiasm for the coming holiday season. She includes some advice on making a successful sale that her readers will need to follow.

The writer has grouped her sales tips chronologically so that her readers can follow through the steps to a successful sale. The section entitled "Deferred Billing" begins with a formal sentence definition so that readers will understand the term before reading the advice connected with it.

Discussion

1. Discuss how appropriate the writer's organization is for her readers. Three of the four sections are stages of the sales process. Which section is not? Are there any changes you would make in the organization for reader convenience?

2. In groups, discuss how to develop either the "Approach" section or the "Closing" section. What other tips might the manager have included for the salespeople in this section? Draft a revised and more fully developed section. Share your draft with the other groups in class, and discuss why you developed the section the way you did.

TO: All Sales Personnel

FROM: Gabrielle Molina, Sales Manager

DATE: November 12, 1989

SUBJECT: Closing the Sale Effectively

With the coming holiday season, we all need to review the proper sales procedures and how to serve our customers efficiently in our store. I've outlined below the procedures to keep in mind when you serve our customers.

Approach--Effective customer service begins with the approach you use. Be prompt in offering assistance, offer information about the merchandise the customer is interested in, and give the impression we care about her special needs. The approach begins the process of selling. Once you initiate positive contact with the customer, you can determine her special style, focus on which of our selections will fulfill her needs, and assist her in making a final decision.

Deferred Billing--Our deferred billing option is a payment plan that allows the customer to postpone payment for 90 days after purchase. The deferred billing option should be used as a sales tool to close a difficult sale. Do not offer the deferred billing to every customer. Only sales of $2000 or more qualify. Deferred billing is a means of encouraging a customer to make a decision on the spot, particularly about the expensive selections available. If a customer has already decided to buy one of our selections, let's save that deferred billing incentive for customers who need something extra to make the final decision to buy.

Closing--The closing is as important as the approach for successful sales. When you believe the customer has made a decision to buy, ask her if she wants to put her selection on her Fashion Frontiers account. This question will do two things: (1) remind her that she has a Fashion Frontiers account, and (2) encourage a customer without an account to open one.

Thank You--Be sure to thank all customers after you have completed the sale. At Fashion Frontiers, we want each customer to feel special. Our customers keep us in business and deserve to know how much we appreciate them.

Model 3-2 Commentary

This report was written by the supervisor of a test center at a large manufacturing company to the company vice president of operations. The report was written because the vice president asked for an account of the test center responsibilities as part of an efficiency audit of all company departments. The vice president intends to use the report in making decisions about shifting responsibilities among the company's three test centers.

The writer groups information by type of test or activity. The final section includes several recommendations for improving the efficiency of the test center.

Discussion

1. Discuss the "Recommendations" section. How many suggestions is the writer making? Draft a new version of the "Recommendations" section to help the vice president distinguish the recommendations more readily.

2. Discuss how helpful a section entitled "Miscellaneous" is to a reader looking for specific topics. Can you think of a new descriptive heading for this section?

DATE: March 27, 1987

FROM: Timothy Beck

TO: Margaret Howard

SUBJECT: JOB REPORTS

The Environmental Laboratory is a part of the Corporate Test Center, which in turn is under the direction of Corporate Engineering. The Laboratory is currently staffed by 17 technicians varying in educational background from associate degrees to doctorates. The average years of college is 5–1/2. The following are the areas of responsibilities for the lab:

Component Environmental Testing

The E & E Laboratory is responsible for testing electrical and mechanical components such as gauges, lights, alternators, cranking motors, batteries, etc. that are used on agricultural and construction equipment. Test specifications come from the vendor, Acme engineering departments, and the Society of Automotive Engineers. Test conditions are intended to simulate the most severe actual field environmental conditions. Some of our simulated environments are high temperature (up to +400°F), low temperature (to −100°F), vibration (to 40 "g's"), sand and dust at zero visibility, and sunshine/rain. Generally the duration of test for each environment is 100 hours.

Component Durability Tests

The durability of components is established by rapid cycling of the component while it is performing its normal function. An example of this is cranking motor cycle tests. The cranking motor is mounted on a diesel engine and connected to an automatic cycler that operates 24 hours per day, 7 days per week. Safety shutdown features are incorporated in the cycler to shut down the test in case of a malfunction or test sample failure. The cranking motor cranks the engine for 5 seconds, the engine starts during the sixth second and continues to run for an additional 30 seconds--then it shuts down. This cycle is repeated every 30 seconds. A cranking motor is expected to survive 20,000 such cycles. Other components are tested in a similar manner.

Environmental Testing of Whole Machines

The whole vehicle is performance tested in specific environments much in the same manner that the components are tested. As an example, the minimum temperature at which our vehicles can be started is established in our Cold Room. The minimum temperature capability of the Cold Room is −70°F. Maximum safe operating ambient temperature (air to boil) is measured in a 40- by 80-ft room located at the 7-Mile Proving Grounds. This is accomplished by operating the tractor under load in the room and elevating the temperature of the room until the water in the radiator reaches 212°F. Additionally, rain tests are performed to check for cab leakage, as well as heater and defroster functional checks.

Noise and Vibration Testing

Local, state, and federal governments have in the past 10 years begun regulating the amount of noise exposure for the operator and the amount of exterior noise that the machine can emit. Competitive factors have caused operator noise exposure on agricultural tractors to be considerably below the regulated level. The laboratory provides assistance to the engineering departments during the design phase to ensure that the noise levels are competitive in compliance with the regulations. In addition, special assistance is provided to manufacturing in setting up instrumentation procedures and limits for noise and vibration during the manufacturing process.

Noise Reduction in Manufacturing Areas

Two men continually conduct noise exposure surveys at all plants owned by Acme, Inc., in North America. Compliance with OSHA regulations are the criteria used. Individuals who are overexposed are identified to management, and recommendations for ear protection are made.

Two other men use this report in their work, which is engineering the noise out at the source. These two men make recommendations involving the expenditure of approximately $500,000 per year.

Audiometric Testing for Acme and Universal

A program that offers computerized evaluation of audiometric tests (hearing threshold levels) for all employees at Universal companies is accomplished by the lab. The audiograms are evaluated using several techniques; some are government-developed and others have been developed in-house. The lab is reimbursed at a rate of $1.10 for each audiogram. A management report is issued annually for each plant identifying the individual employees who require attention. A rating system for all plants within a company is included that identifies those plants requiring most attention.

Miscellaneous Responsibilities

The laboratory reviews government regulations and specifications developed by technical societies and writes Corporate Specifications for Acme. Laboratory personnel testify at hearings on proposed legislation and assist in company defense during OSHA and liability litigation. We also prepare proposals for land acquisition and support corporate equipment displays with manpower.

Recommendations

The laboratory currently has no secretarial help, and a half-time secretary would increase efficiency substantially. Research, which is currently funded but not performed, should be redefined to allow work within the scope of an existing engineering program rather than the rigid requirement that it be separated from engineering programs.

A great deal of technology exists at the Test Center, but it is not incorporated in the product because of rigid, inappropriate definitions of project budgets. A means of passing this technology to the plants should be established.

Model 3-3 Commentary

This article is from a magazine written for managers and staff of low-power television stations (*The LPTV Report,* April 1989). The writer is offering advice to those readers (probably station managers) who want to expand the local news segments at their stations.

Discussion

1. How helpful are the headings in leading readers to specific topics? In groups, draft a new set of headings for the article—as many as you believe are necessary to help readers find specific information.

2. Discuss which points in the article a reader might want developed further if he or she intends to follow the writer's plan for surveying the station's audience.

The News in Community Broadcasting—The First Step: Researching Your Audience

by Bob Horner

Many community broadcasters have already discovered that their most powerful programming tools are local news, sports, and special events.

The reasons are simple: Ratings are good. Revenues are strong because local news provides plenty of availabilities and delivers key demographics. And anchors, reporters, and photographers are walking promotion for the station as they cover events in the community.

Most importantly, local news gives your station an identity that your competitors can't match. The others can deliver the same kind of entertainment you do, but they can't cover your local community the way you can.

Local news, then, can be the single most important way you can create an identity for your station in the minds of your viewers. If you aren't doing local news, you're missing the best promotional opportunity you have.

However, doing a newscast can seem so complex that some operators may be reluctant to start one. Others may want to improve or expand their existing news programs, but they may not be sure how to proceed. In this column, we'll be addressing these and other problems facing community broadcasters who are producing, or thinking about producing, local newscasts.

Content and Research

As a broadcast news consultant, I find that station managers spend too much time worrying about equipment and production when they start a news operation. I think it's helpful, instead, to do what your audience is going to do—concentrate on the content. And good content begins with good research.

You may recall the article entitled "The LPTV Business Plan" by John Kompas and Richard P. Wiederhold in the December 1988 issue of *The LPTV Report.* The first step in LPTV business planning, they suggest, is to answer the question, "What do the viewers want?" That question is particularly crucial to local news planning; and answering it doesn't have to be expensive.

I suggest that you begin by creating a questionnaire that will determine which issues are really important to the people who live in your coverage area. First, get some general information:

- What are your current news media habits?
- What parts of the newspaper do you read?
- When do you listen to the news on the radio?
- What newscasts do you watch on television?
- What items do you find most interesting?

The answers to these questions will tell you what the audience is doing now. (Don't ignore people who say they don't watch news now. They represent your biggest opportunity, and there will be some questions for them in a moment.)

Next, ask questions that will help you design your newscast:

- What time of day would you most likely be able to watch our local newscast?
- How much time do you have to spend with local news?
- Would you want our newscast to include items from around the country, or to focus on our community exclusively?
- What types of events and subjects would you like to see us cover? What would be most useful to you?

The last question should be asked as an open-ended question, just to see what your viewers say spontaneously. Be ready to move in right away, though, with a list of items to prompt them.

Events or Issues?

The prompt list should include both events and issues. For example, ask about school board meetings and city council proceedings. But also ask about consumer news and medical news. Do your viewers want to know about accidents and fires, or about efforts to attract new industry? Or both? You are trying to determine whether events or issues should be the focus of your newscasts.

After you've written your survey questions, try them out on your staff and their families. (It's important that everyone at the station participate if you're going to get a useful cross-section of people.) Using the feedback from these test interviews, fine-tune your questions and technique. Then, perform the survey again by interviewing people, in person, in small groups of 10 to 20.

The next step is to conduct a telephone survey, with the help, if possible, of a local college or community group. Try to bring in at least 200 responses from your coverage area. This helps you to be sure that the information you developed in the focus groups is on target.

Even if you already have a newscast on the air, it can be a good idea to conduct this kind of survey regularly to see where you need to make changes, and where your opportunities are.

In the next column, we'll examine ways to turn this research into a plan for news coverage.

Chapter 3 Exercises

1. The following memo was addressed to all department supervisors at a large manufacturing company. Because the memo covered changes in payroll dates, the supervisors passed it on to the payroll clerks in each department. The clerks need to adjust their work procedures to match these temporary changes. Revise the memo for improved style and organization that will help the clerks use the information. Before revising, read the articles by Benson and Kleim in Part II for ideas about how to guide the readers through the information.

> The Computer Center will be closed from December 24, 1989 through December 25, 1989 and December 31, 1989 through January 1, 1990 for Christmas and New Year's Holidays. Due to this scheduling and field personnel not being available for consultation, it is of the utmost importance that the following time-sheet schedule be adhered to. Any delays in this schedule will result in delays of delivery of payroll checks. Following this schedule, checks will be available for payment on Friday, December 26, 1989 and Friday, January 2, 1990.
>
> For the pay period ended December 20, time sheets that would normally be forwarded to Payroll on Friday, December 19 should be forwarded on Thursday, December 18, and those time sheets that would be normally forwarded on Monday, December 22 should be forwarded on Friday, December 19.
>
> For the pay period ended December 27, time sheets that would normally be forwarded to Payroll on Friday, December 26, should be forwarded on Tuesday, December 23, and those time sheets that would normally be forwarded on Monday, December 29 should be forwarded on Friday, December 26.
>
> In order to meet this schedule, it will be necessary to estimate some time. Any changes required to correct the estimated time for the week ending December 20 should be reflected on the time sheets for the week ended December 27. Estimated time for the week ending December 27 should be reflected on the time sheets for the week ending January 3, 1990. Payroll personnel will not make telephone corrections for estimated time.
>
> In order that employees involved with estimated time are aware of the procedure that will be followed, a copy of this memo should be posted as general information.
>
> Should you have any questions concerning this schedule, please contact this office.

2. The following section from an Administrative Manual is the official company description of the responsibilities of the safety engineer. Revise the section into paragraphs for better use by readers who need to understand the safety engineer's duties. Before revising, read the article entitled "Eliminating Gender Bias in Language" in Part Two.

The Safety Engineer is responsible for developing, coordinating, and administering the local safety program. He shall cooperate with, give counsel, and render assistance to all levels of management regarding safety. He provides the supervisor with the necessary information and assistance so that the supervisor, who is responsible for the safety of his area, employees, and equipment, can fulfill his safety responsibility. Service to the supervisor—and not the usurping of the supervisor's authority and responsibility for safety—must be recognized by the Safety Engineer as his major job. He has the responsibility of inspecting the physical facilities to insure compliance with established standards and rules and to identify unsafe conditions and practices that require additional control. Follow-up is an essential element of his inspection responsibility. Insuring that corrective measures have been taken to eliminate the existence of a hazard is his duty. The Safety Engineer must have a knowledge of and be familiar with safety legislation, both proposed and in effect at the location, in order to ensure that all company operations comply. He must maintain a liaison between the Medical and Worker's Compensation Representatives and exchange mutually beneficial information with them so that the Company's safety objectives can be accomplished. The Safety Engineer must report accidents or fatalities. He must investigate the physical circumstances surrounding the event. A monthly report of injuries and illnesses is required by Occupational Safety and Health Act regulations. Reports must be maintained accurately. These are the broad areas of the Safety Engineer's duties and responsibilities. Those practices and procedures of an administrative nature that he will specifically be held accountable for are enumerated in the Administration Manual, Section 172–11.

3. You are on a student committee that is preparing an information brochure for incoming freshmen. Your assignment is to write a section of the brochure that will explain one of the following:

Fraternity and sorority activities on campus

Library facilities

Sports activities on campus

Placement and advising services on campus

Effective study habits

Counseling and health services on campus

In writing your section, consider carefully how to organize the information to make the section appealing to the freshman readers. Read the article entitled "Six Graphic Guidelines" in Part II before organizing your section. Prepare an outline of your section for your instructor before you submit your written section.

4. Select an article from Part II and create an outline of the organization the writer used. The articles in Part II primarily offer advice to people already on the job. If you were to revise the article you selected for students not yet in full-time careers, what changes would you make in the organization? Would you add, delete, or rearrange topics? Revise the outline for a student audience, and submit it and the original outline to your instructor.

5. In groups, select a problem at your school that needs to be solved. Consider something dangerous, such as a broken railing; something irritating, such as registration procedures; or something needed, such as more parking space or sports equipment. Each member of the writing team will investigate some aspect of the problem (e.g., background, cost, administrative attitudes). As a group, pool your information and decide how to organize a technical presentation of the situation to be given at the next student government meeting. Read the article by Calabrese in Part II, and plan your presentation so that each team member presents one section of the information. Prepare a written outline of your part of the presentation for your instructor.

CHAPTER 4

Document Design

Understanding Design Features

- Purpose of Design Features
- Design Principles
  - Balance
  - Proportion
  - Sequence
  - Consistency

Creating Graphic Aids

- Purpose of Graphic Aids
- Tables
- Figures
  - Bar Graphs
  - Pictographs
  - Line Graphs
  - Pie Graphs
  - Organization Charts
  - Flowcharts
  - Line Drawings
  - Cutaway Drawings
  - Exploded Drawings
  - Maps
  - Photographs

Using Format Elements

- Written Cues
  - Headings
  - Headers and Footers
  - Jumplines
  - Logos
- White Space
  - Margins
  - Heading Areas
  - Columns
  - Indentations
- Typographic Devices
  - Typefaces
  - Boldface
  - Lists
  - Boxes

Chapter Summary

Supplemental Readings in Part Two

Models

Exercises

Understanding Design Features

Document design refers to the physical appearance of a document. Because the written text and its presentation work together to provide readers with the information they need, think about the design of your documents during the planning stage—even before you select appropriate information and organize it.

Readers do not read only the printed words on a page; they also "read" the visual presentation of the text, just as a television viewer pays attention not only to the main actor and the words he or she speaks, but also to the background action, noises, music, and other actors' movements. In an effective document, as in an effective television scene, the words and visuals support each other.

Desktop publishing has expanded the writer's role in producing technical documents. A writer with sophisticated computer equipment and design software packages can create finished documents with most of the design features discussed in this chapter. Desktop publishing makes it easy to produce newsletters, catalogs, press releases, brochures, training materials, reports, and proposals—all with graphic aids, columns, headlines, mastheads, bullets, and any design feature needed by readers.

Rapidly developing computer technology increases the capabilities of desktop publishing with each new scanner, printer, and software package. It is not an exaggeration to say that desktop publishing has launched a printing revolution as more companies produce more of their own documents. No matter whether a document is produced on centuries-old movable type or on the latest computer technology, however, the writer's goal is the same—to provide readers with the information they need in a form they can use.

Purpose of Design Features

Some documents, such as business letters, have well-known, conventional formats, but letters and other documents also benefit from additional design features—graphic aids and the format elements of written cues, white space, and typographic devices. These design features increase the usefulness of documents in several ways.

1. They guide readers through the text by directing attention to individual topics and increasing the ability of readers to remember the important, highlighted sections.

2. They increase reader interest in the document. Unbroken blocks of type have a numbing effect on most readers, but eye-catching graphic aids and attention-getting format devices keep readers focused on the information they need.
3. They create a document that reflects the image you wish readers to have. A conservative law firm may want to project a solid, traditional, no-nonsense image with its documents, whereas a video equipment company may prefer to project a trendy, dramatic image. Both images can be enhanced by specific design features.

Design Principles

The principles of design are qualities important to any visual presentation regardless of topic or audience. Experienced designers use the principles of design to create the "look" they want for a document. The general principles most designers consider in all documents are balance, proportion, sequence, and consistency.

Balance

Page balance refers to having comparable visual "weight" on both sides of a page or on opposing pages in a longer document. A page in a manual filled with text and photographs followed by a page with only a single paragraph in the center would probably jar the reader. If this unsettling effect is not your intent, avoid such imbalance in page design.

Think of page balance as similar to the scales held by the figure of Justice you have seen so often. Formal balance on a page would be the same as two evenly filled scales hanging at the same level. Informal balance, which is used more often by experienced page designers, would be represented by a heavily weighted scale on one side balanced by two smaller scales equal to the total weight of the larger scale on the other side.

One large section of a page, then, can be balanced by two smaller sections. Every time an element is added to or removed from a page, however, the balance shifts. A photograph that is dominant on one page may not be dominant on another page. Model 4-2 shows informal balance. The graphs in the left and center columns are balanced by the full column of text on the right. Remember these points about visual "weight":

- Big weighs more than small.
- Dark weighs more than light.
- Color weighs more than black and white.
- Unusual shapes weigh more than simple circles or squares.

Proportion

Proportion in page design refers to size and placement of text, graphic aids, and format elements on the page. Experienced designers rarely use an equal amount of space for text and graphics page after page. Not only would this be monotonous for readers, but it would interfere with the readers' ability to use the document. Reserving the same amount of space for one heading called "Labor" and another called "Budgetary Considerations" would result in the long heading looking cramped and the short heading looking lost in the space available. In a similar fashion, you would not want every drawing in a parts manual to be the same size regardless of the object it depicts. Each design feature should be the size that is helpful to the readers and appropriate for the subject.

Sequence

Sequence refers to the arrangement of design features so that readers see them in the best order for their use of the document. Readers usually begin reading a page at the top left corner and end at the bottom right corner. In between these two points, readers tend to scan from left to right and up to down. Readers also tend to notice the features with the most "weight" first. Effective design draws readers through the page from important point to important point.

Consistency

Consistency refers to presenting similar features in a similar style. Keep these elements consistent throughout a document:

- *Margins.* Keep uniform margins on all pages of a document.
- *Typeface.* Use the same size and style of type for similar headings and similar kinds of information.
- *Indentations.* Keep uniform indentations for such items as paragraphs, quotations, and lists.

Do not mistake consistent format for boring format. Consistency helps readers by emphasizing similar types of information and their similar importance. In Model 4-1 all headings are in boldface and each section begins with a typical "reader question," placed at the top of a page and accompanied by a small drawing. This consistency in design helps readers understand and find information quickly.

Creating Graphic Aids

Graphic aids, called *figures* and *tables,* are not merely decorative additions to documents or oral presentations. Often graphic aids are essential in helping readers understand and use the information in a document. Instructions may be easier to use when they have graphics that illustrate some steps, such as directions for gripping a tennis racquet properly or for performing artificial respiration.

Purpose of Graphic Aids

Think about which graphic aids would be appropriate for your document during the planning stage. Graphic aids are important in technical documents in these ways:

- Graphic aids provide quick access to complicated information, especially numerical data. For example, a reader can more quickly see the highs and lows of a production trend from a line graph than from a long narrative explanation.
- Graphic aids isolate the main topics in complex data and appeal particularly to general and nonexpert readers. Newspapers, for instance, usually report government statistics with graphic aids so that their readers can easily see the scope of the information.
- Graphic aids help readers see relationships among several sets of data. Two pie graphs side by side will illustrate more easily than a narrative can the differences between how the federal government spent a tax dollar in 1932 and in 1987.
- Graphic aids, such as detailed statistical tables, can offer expert readers quick access to complicated data that would take pages to explain in written text.

Readers of some documents expect graphic aids. Scientists reading a research report expect to see tables and formulas that show the experimental method and results. Do not, however, rely solely on graphic aids to explain important data. Some readers are more comfortable with written text than with graphics, and for these readers, your text should thoroughly analyze the facts, their impact on the situation or on the future, their relevance for decision making or for direct action, and their relation to other data. For consistency and reader convenience, follow these guidelines for all graphic aids:

1. Identify each graphic aid with a specific title, such as "1987–1991 Crime Rates" or "Differences in Patient Response to Analgesics."
2. Number each graphic aid. All graphics that are not tables with words or numbers in columns are called figures. Number each table or figure consecutively throughout the document, and refer to it by number in the text:

The results shown in Table 1 are from the first survey. The second survey is illustrated in Figure 2.

3. Place each graphic aid in the text as near after the first reference to it as possible. If, however, the document has many tables and charts or only some of the readers are interested in them, you may decide to put them all in an appendix to avoid breaking up the text too frequently.

The guidelines in this chapter for creating various types of graphic aids provide general advice for such illustrations. Effective graphic aids, however, can violate standard guidelines and still be useful to the readers who need them. The daily newspaper *USA Today* contains many innovative illustrations of data for its readers who are interested in a quick understanding of the latest statistical or technical information. The key to creating effective graphic aids is to analyze purpose and reader before selecting graphic formats. While an industrial psychologist may prefer a detailed statistical table of the results of a survey of construction workers, a general reader may want—and need—a simple pie graph showing the key points.

Tables

A table shows numerical or topical data in rows and columns, thus providing readers with quick access to quantitative information and allowing readers to make comparisons among items easily. Table 4–1 shows a comparison of three types of data among selected states.

Here are general guidelines for setting up tables:

1. Provide a heading for each column that identifies the items in the column.
2. Use footnotes to explain specific items in columns. Footnotes for specific numbers or columns require lowercase superscript letters (e.g., [a], [b], [c]). If the footnote is for a specific number, place the letter directly after the number (e.g., 432[a]). If the footnote applies to an entire column, place the letter directly after the column heading (e.g., Payroll[b]). List all footnotes at the left margin of the table, directly below the data.
3. Space columns sufficiently so that the data do not run together.
4. Give the source of your data below the table. If you have multiple sources and have compiled the information into a table, explain this in the text, if necessary.
5. Use decimals and round off figures to the nearest whole number.
6. Indicate in the column heading if you are using a particular measure for units, such as "millions of dollars" or "per 5000 barrels."

Table 4-1 A Typical Table Comparing Three Types of Data—Agricultural Services:[a] Number of Establishments,[b] Gross Receipts, and Payroll, by State, 1978

State	Number of Establishments	Gross Receipts (in $000)	Annual Payroll (in $000)
Great Plains:			
Texas	2,436	$ 281,493	$ 96,476
Nebraska	554	48,494	14,385
Oklahoma	562	36,447	8,821
Kansas	754	41,937	8,599
South Dakota	313	16,516	3,921
North Dakota	243	15,571	3,213
Mountain region:			
Arizona	441	104,250	41,705
Idaho	352	45,853	10,503
Colorado	355	31,981	8,477
Montana	238	13,528	2,971
New Mexico	118	8,741	2,686
Wyoming	79	4,197	1,394
Utah	98	5,905	1,120
Nevada	40	2,544	490
Pacific region:			
California	3,043	1,034,223	452,186
Washington	387	78,732	24,904
Oregon	331	31,197	8,943
17 Western States	10,344	1,801,609	690,794
United States	20,595	2,936,208	2,134,248

[a] Agricultural services consist of soil preparation services; crop services; veterinary services for cattle, hogs, sheep, goats, and poultry; animal services (except veterinary) for cattle, hogs, sheep, goats, and poultry; farm labor; and management services.
[b] Establishments having a dollar volume of business less than $2500 are omitted.
Source: U.S. Department of Commerce, Bureau of the Census, *1978 Census of Agriculture*, Vol. 3, Table 23, 1981.

Figures

All graphic aids that are not tables are considered figures. Information in tables can often be presented effectively in figures as well, and sometimes readers can use figures more readily than tables. A geologist seeking information about groundwater levels may want a table giving specific quantities in specific geologic locations. Someone reading a report in the morning newspaper on groundwater levels across the United States, however, may be better served by a set of bars of varying lengths representing groundwater levels in regions of the country, such as the Pacific Northwest. Before selecting graphic aids for your document, consider your readers and their ability to interpret quantitative information and, particularly, their need for specific or general data. The most common types of graphic aids are bar graphs, pictographs, line graphs, pie graphs, organization charts, flowcharts, line drawings, cut-away drawings, exploded drawings, maps, and photographs.

Bar Graphs

A *bar graph* uses bars of equal width in varying lengths to represent (1) a comparison of items at one particular time, (2) a comparison of items over time, (3) changes in one item over time, or (4) a comparison of portions of a single item. The horizontal and vertical axes represent the two elements being illustrated, such as time and quantity. Figure 4-1 is a typical bar graph in which the bars represent different crops and the lengths of the bars represent the millions of acres harvested.

Bars can extend from either the horizontal or the vertical axis and from below the horizontal axis or both sides of the vertical axis to show negative quantities. Figure 4-2 uses bars on both sides of the horizontal axis, indicating positive and negative quantities. Notice that the zero point in the vertical axis is about one-third above the horizontal axis and that the quantities are labeled positive above the zero point and negative below it. The bar graph also uses shading to distinguish between bars representing two data sources.

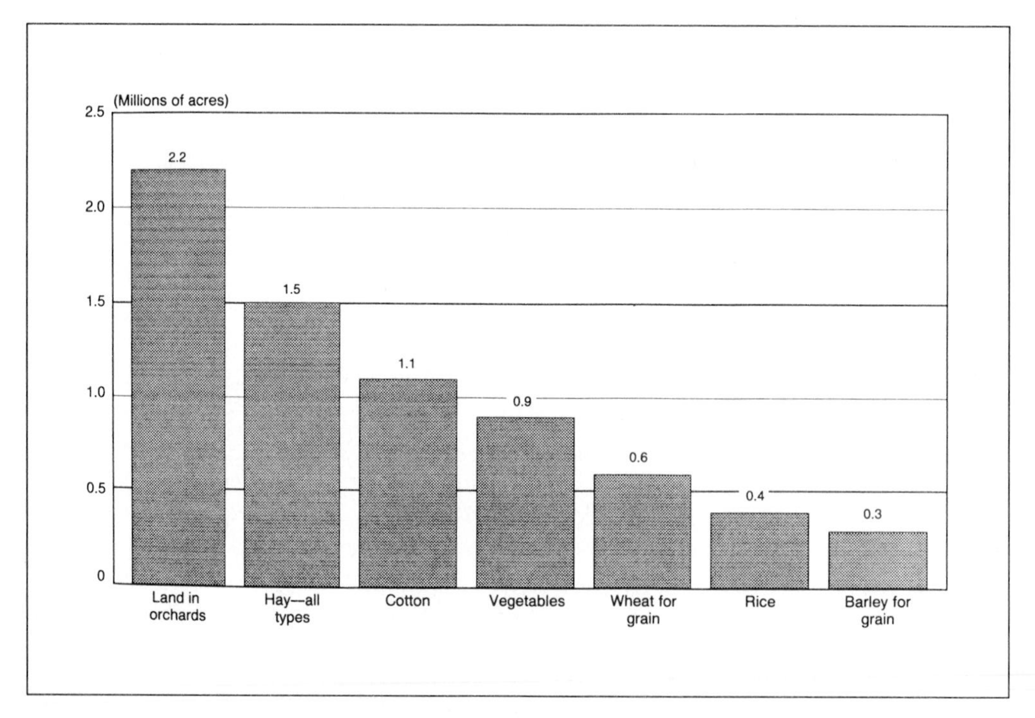

Figure 4-1 A Typical Bar Graph—Selected Crops Harvested, 1987

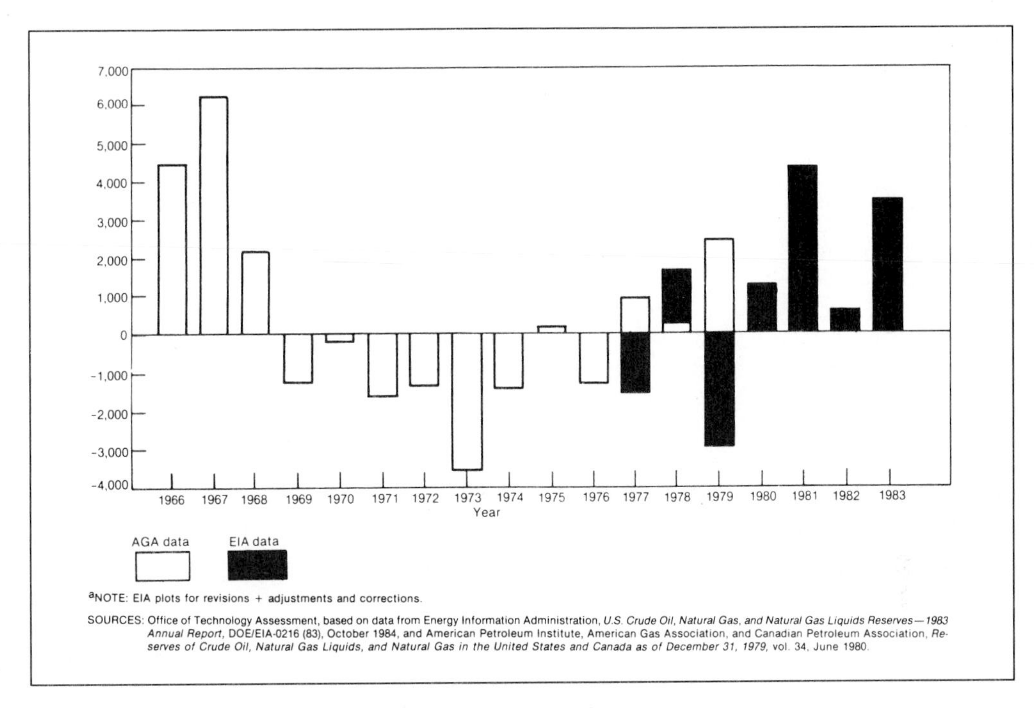

Figure 4-2 A Typical Bar Graph Presenting Positive and Negative Data—Additions to Lower 48 Natural Gas Proved Reserves: Revisions as Reported,[a] 1966–1983 (BCF)

Bar graphs cannot represent exact quantities or provide comparisons of quantities as precisely as tables can, but bar graphs are generally useful for readers who want to understand overall trends and comparisons.

Depending on the size of the graph and the shading or color distinctions, up to four bars can represent different items at any point on an axis. Label each bar or provide a key to distinguish among shadings or colors. Figure 4-3 shows two bar graphs using multiple bars. The bars in the first graph represent two research topics, whereas the bars in the second graph represent three different years.

Pictographs

A *pictograph* is a variation of a bar graph that uses symbols instead of bars to illustrate specific quantities of items. The symbols should realistically correspond to the items, such as, for example, a cow representing milk production. Pictographs provide novelty and eye-catching appeal, particularly in documents intended for consumers. Pictographs are limited, however, because symbols cannot adequately represent exact figures or fractions. When using a pictograph, (1) make all symbols the same size, (2) space the symbols equally

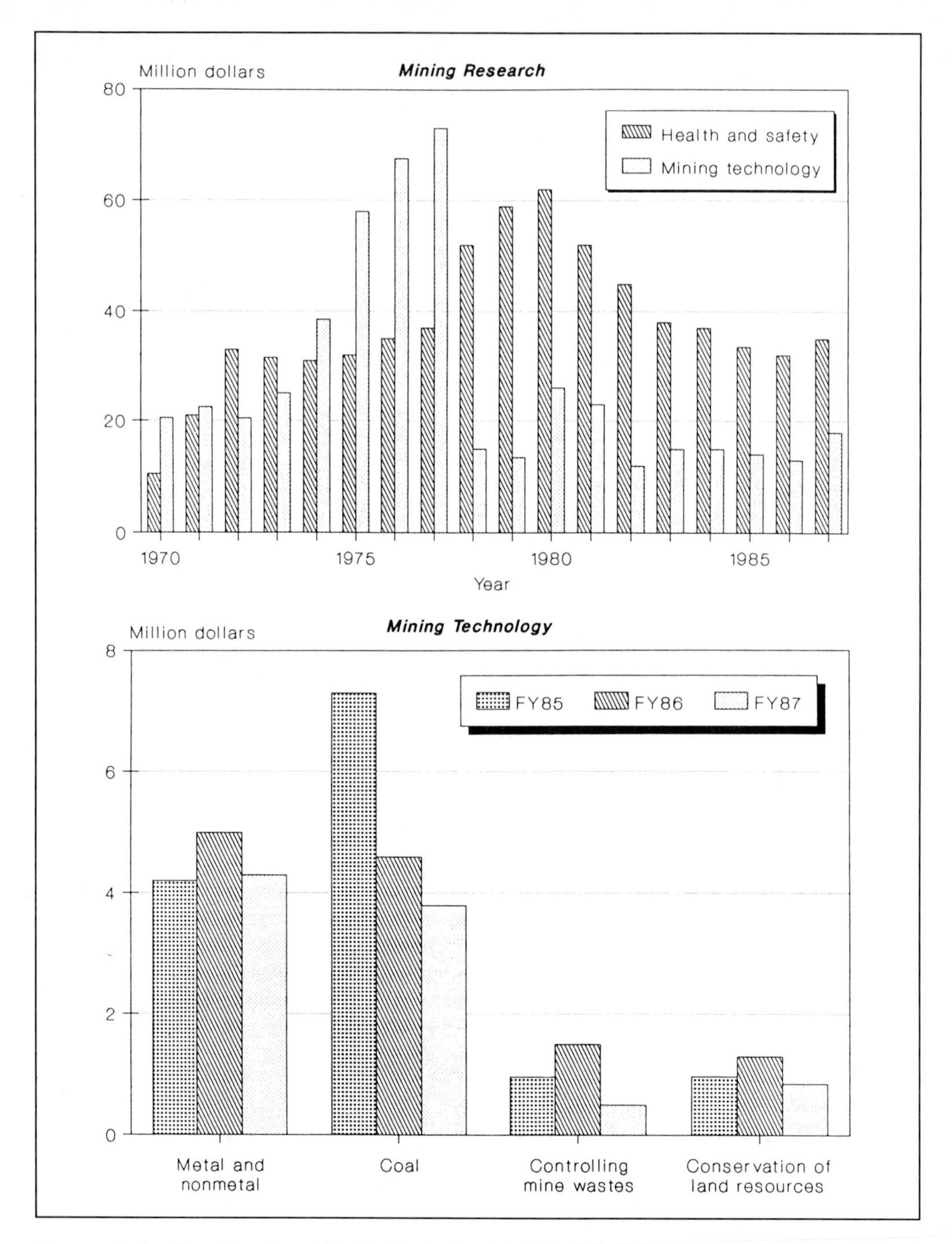

Figure 4-3 Bar Graphs with Multiple Bars—Trends in U.S. Bureau of Mines R&D

Source: U.S. Department of the Interior. Bureau of Mines. *Technical Highlights: Mining Research 1987* (Washington, DC: U.S. Government Printing Office. 1988).

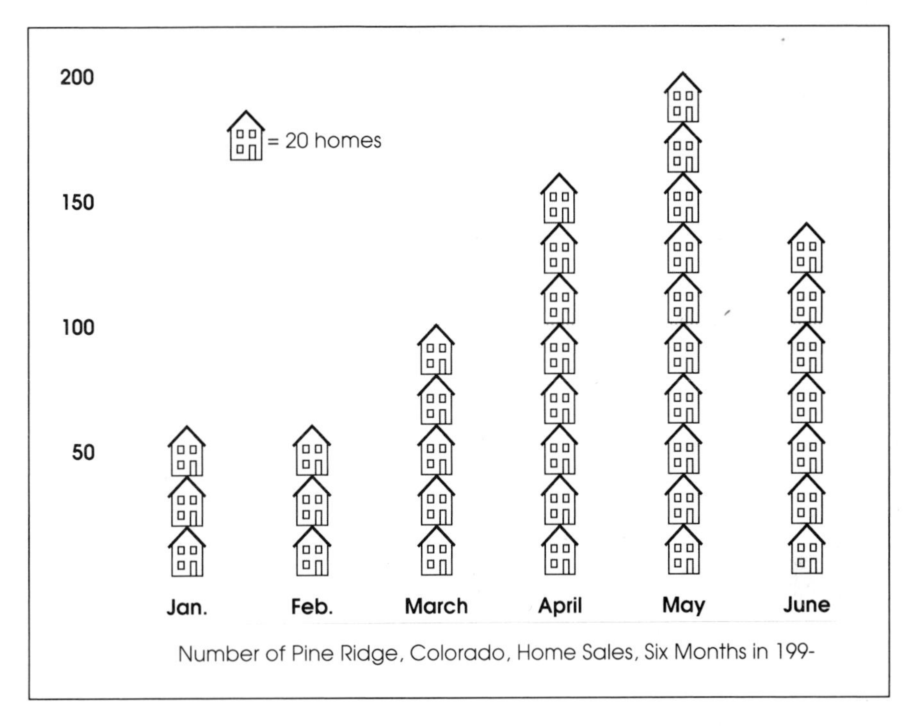

Figure 4-4 A Typical Pictograph

on the axis, (3) show increased quantity by increasing the number of symbols rather than the size of the symbol, (4) round off the quantities represented instead of using a portion of a symbol to represent a portion of a unit, (5) include a key indicating the quantity represented by a symbol. Figure 4-4 uses houses to represent the trend in housing sales during a 6-month period.

Line Graphs

A *line graph* uses a line between the horizontal and vertical axes to show changes in the relationship between the elements represented by the two axes. The line connects points on the graph that represent a quantity at a particular time or in relation to a specific topic. Line graphs usually plot changes in quantity or in position and are particularly useful for illustrating trends. Three or four lines representing different items can appear on the same graph for comparison. The lines must be distinguished by color or design, and a key must identify them. Label both vertical and horizontal axes, and be sure that the value segments on the axes are equidistant. Figure 4-5 uses three lines, each representing sales for a specific corporation. The lines plot changes in sales (vertical axis) during specific years (horizontal axis). The amounts indicated are not exact, but readers can readily see differences in the corporate sales over time.

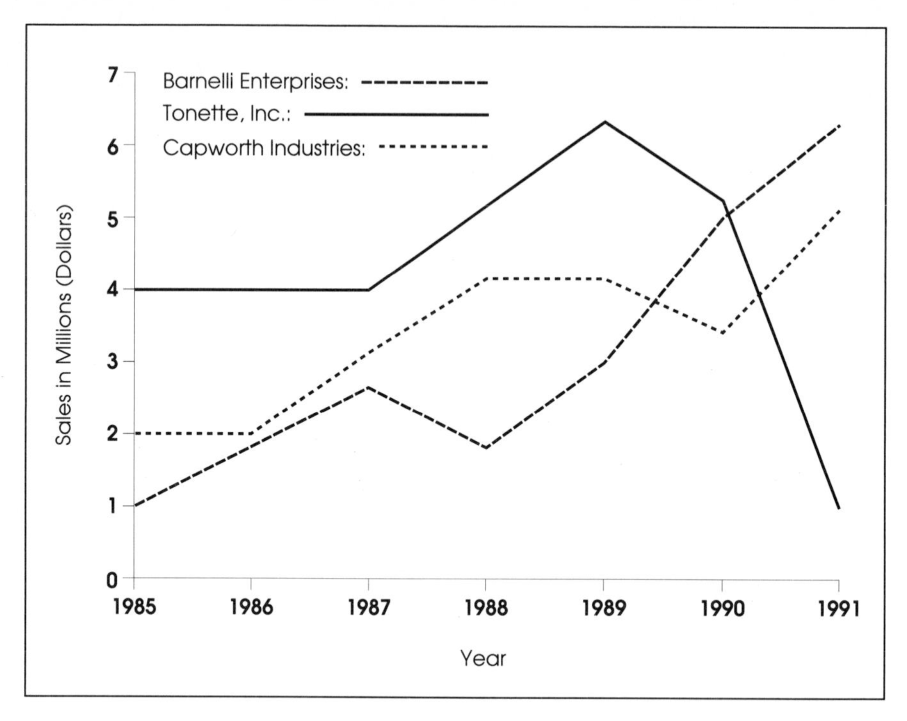

Figure 4-5 A Typical Line Graph

Pie Graphs

A *pie graph* is a circle representing a whole unit with the segments of the circle, or pie, representing portions of the whole. Pie graphs are useful if the whole unit has between three and ten segments. Use colors and shadings to highlight segments of special importance, or separate one segment from the pie for emphasis. In preparing a pie graph, start the largest segment at the 12 o'clock position and follow clockwise with the remaining segments in descending order of size. If one segment is "Miscellaneous," it should be the last. Label the segments, and be sure their values add up to 100% of the total. Figure 4-6 is a pie graph with six segments. The segments represent four individual countries—the largest individual producers of titanium pigment—and two other groups of producers. The segment representing European producers is presented before the "All others" segment because it is the more specific group.

Organization Charts

An *organization chart* illustrates the individual units in a company or any group and their relationships to each other. Organization charts are most often used to illustrate the chain of authority—the position with the most

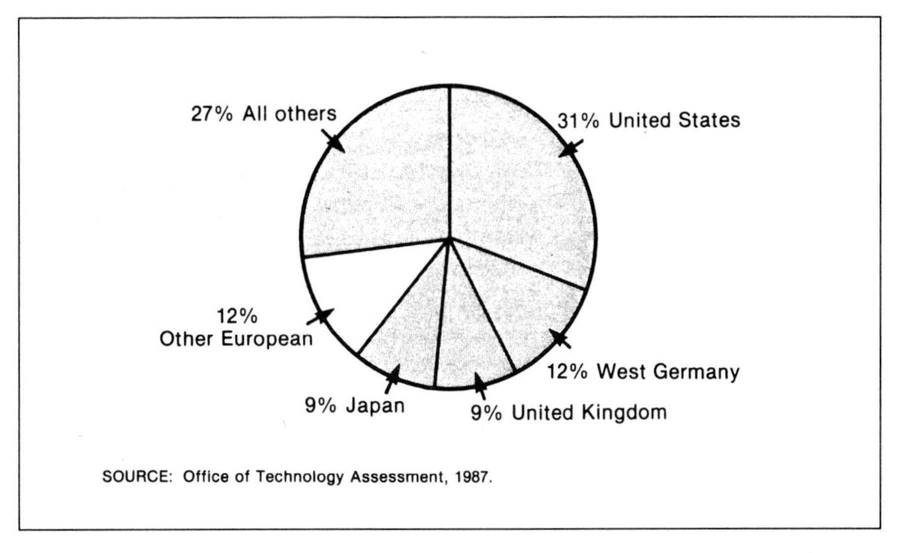

Figure 4-6 A Typical Pie Graph—World Titanium Pigment Manufacturing Capacity

authority at the top and all other positions leading to it in some way. Organization charts also indicate the lines of authority between units or positions and which positions are on the same level of authority. Rectangles or ovals usually represent the positions in an organization chart. Label each clearly. Figure 4-7 uses shading in some rectangles to illustrate which positions were most involved in a special project, identified in the key at the bottom of the chart.

Flowcharts

A *flowchart* illustrates the sequence of steps in a process. A flowchart can represent an entire process or only a specific portion of it. As a supplement to a written description of a process, a flowchart is useful in enabling readers to visualize the progression of steps. An open-system flowchart shows a process that begins at one point and ends at another. A closed-system flowchart shows a circular process that ends where it began. Use rectangles, circles, or symbols to represent the steps of the process, and label each clearly. Figure 4-8 uses simple rectangles to represent each step in the production cycle of copper. Figure 4-9 uses drawings to illustrate the stages in the process of converting offshore ore deposits to refined minerals.

Line Drawings

A *line drawing* is a simple illustration of the structure of an object or the position of a person involved in some action. The drawing may not show all the

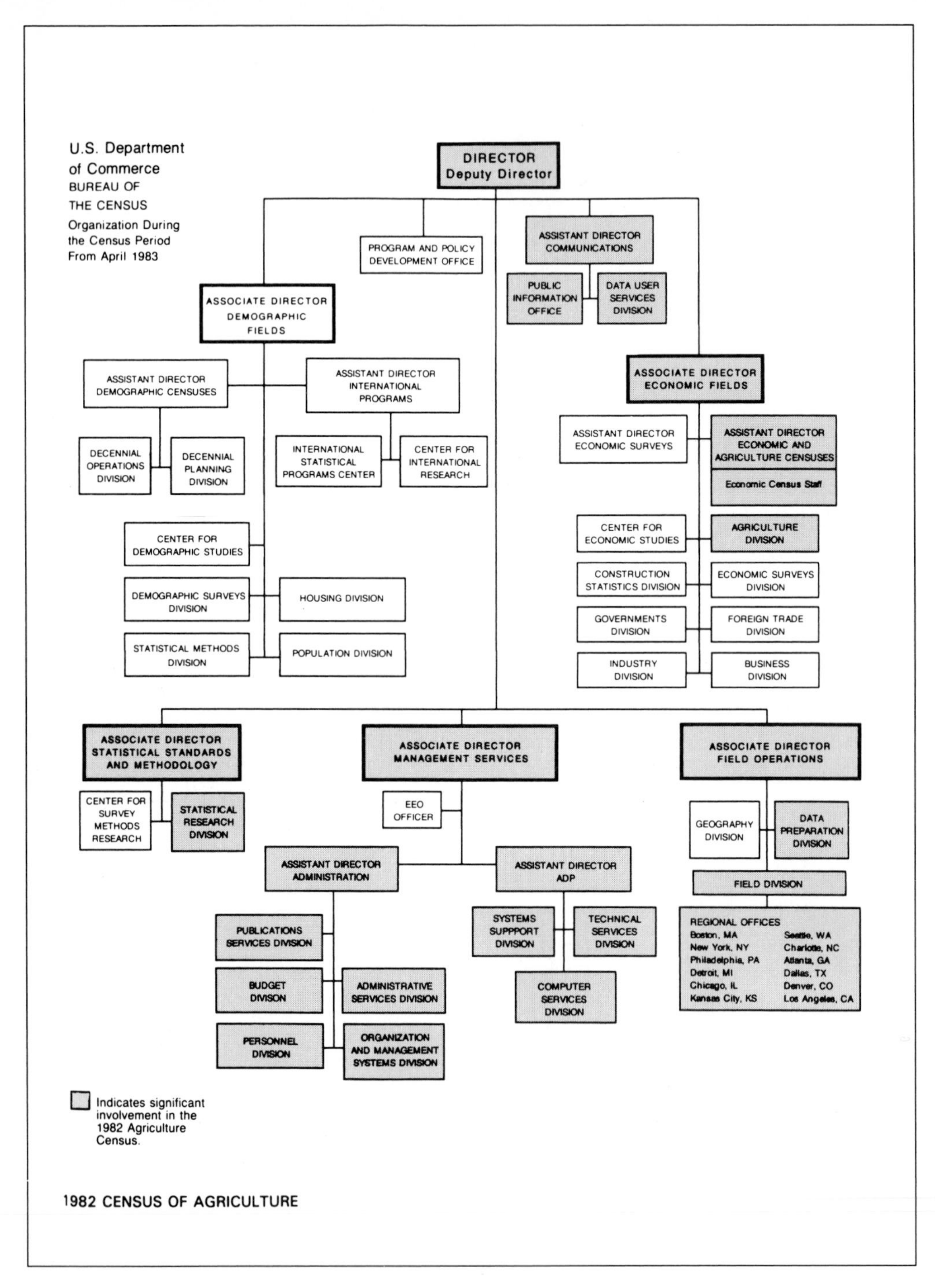

Figure 4-7 A Typical Organization Chart

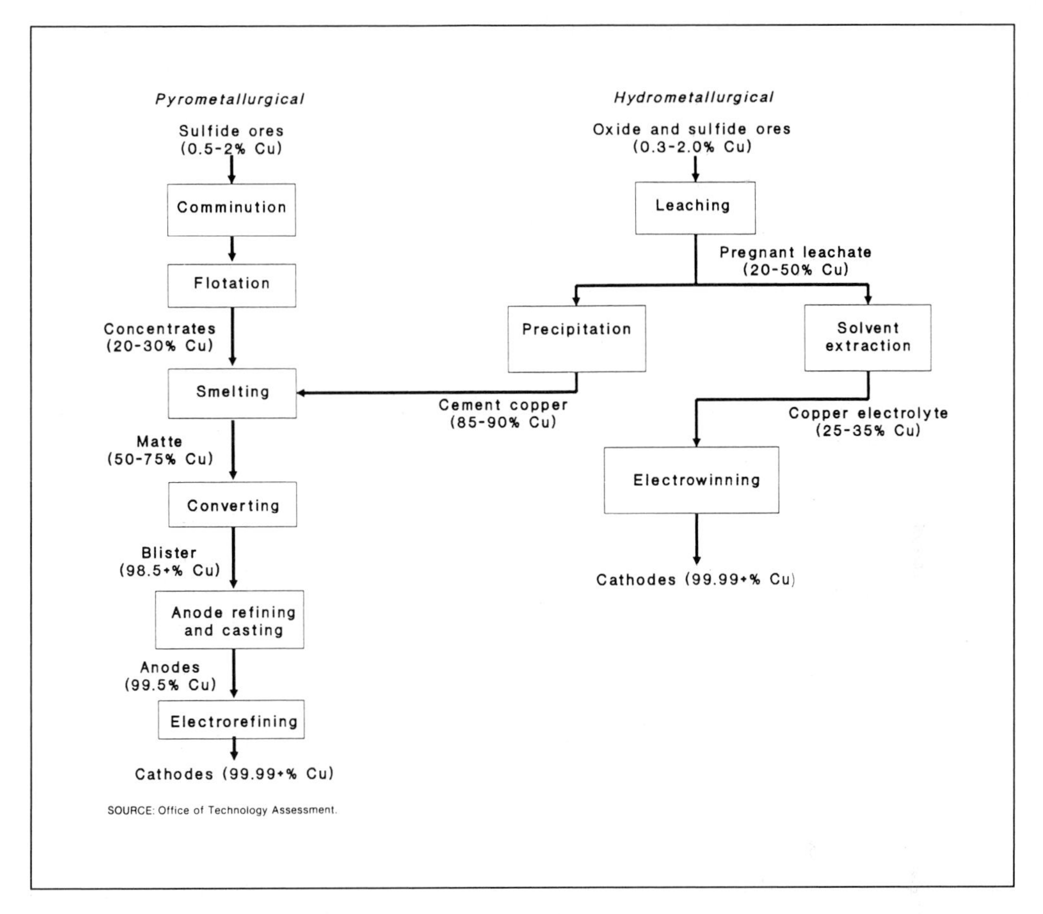

Figure 4-8 A Typical Flowchart—Flowchart for Copper Production

details but, instead, highlights certain areas or positions that are important to the discussion. If details of the interior of an object are important, a cutaway drawing is more appropriate. If the relationship of all the components is important, an exploded drawing is more appropriate. Figure 4-10 is a line drawing of a police officer conducting a driver sobriety test. The drawing shows customary positions of the driver and police officer as the driver tries to walk a straight line.

Cutaway Drawings

A *cutaway drawing* shows an interior section of an object or an object relative to a location and other objects so that readers can see a cross section below the surface. Cutaway drawings show the location, relative size, and relationships of interior components. Use both horizontal and vertical cutaway views if your readers need a complete perspective of an interior. Figure 4-11 is a

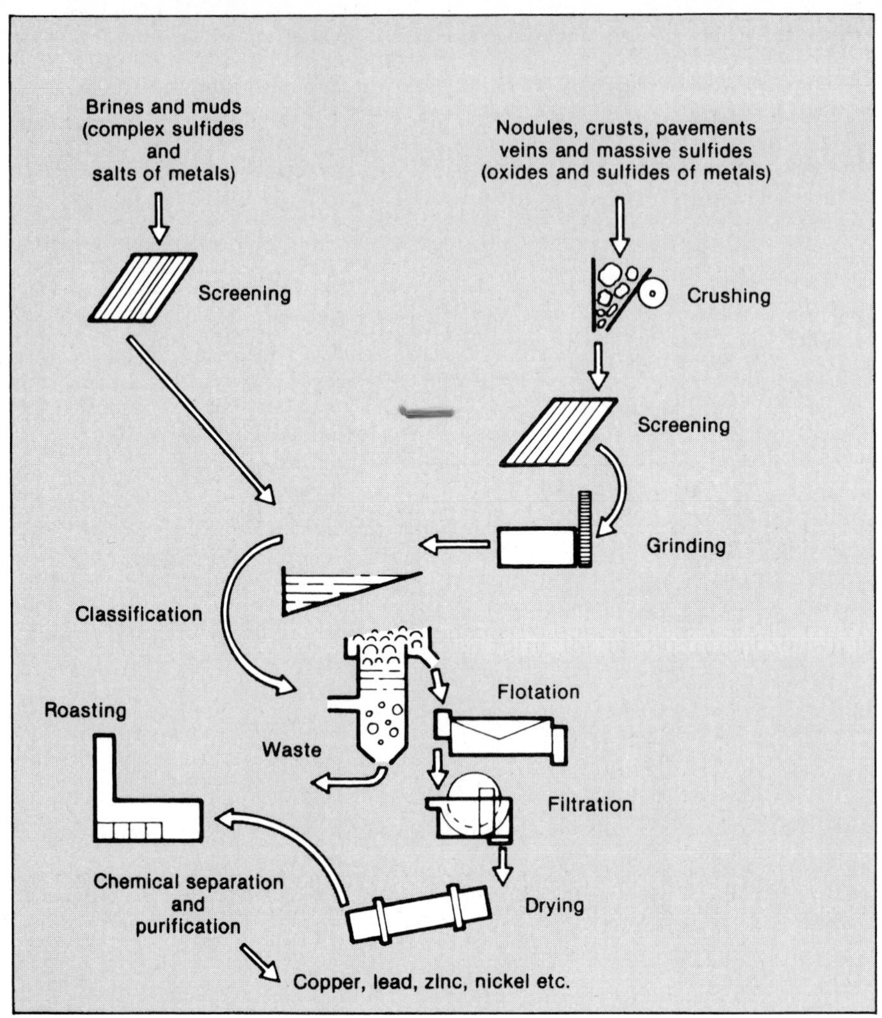

SOURCE: Office of Technology Assessment, 1987.

Figure 4-9 A Typical Flowchart Employing Drawings—To Illustrate Technologies for Processing Offshore Mineral Ores

cutaway drawing of a suction dredge revealing its parts and its position in the water.

Exploded Drawings

An *exploded drawing* shows the individual components of an object as separate but in the sequence and location they have when put together. Most often used in manuals and instruction booklets, exploded drawings help readers visualize how exterior and interior parts fit together. Figure 4-12 is an

Figure 4-10 A Line Drawing of a Police Officer Conducting a Driver Sobriety Test

exploded drawing of a cart used to illustrate assembly instructions for a consumer.

Maps

A *map* shows (1) geographic data, such as the location of rivers and highways, or (2) demographic and topical data, such as density and distribution of population or production. If demographic or topical data are featured, eliminate unnecessary geographic elements and use dots, shadings, or symbols to show distribution or density. Include (1) a key if more than one topical or demographic item is illustrated, (2) a scale of miles if distance is important, and (3) significant boundaries separating regions. Figure 4-13 uses shadings to show differences in Hispanic population density in the 50 states.

Photographs

A *photograph* provides a surface view of an object or event. To be effective, photographs must be clear and focus on the pertinent item. Eliminate distracting backgrounds or other objects from the photograph. If the size of an

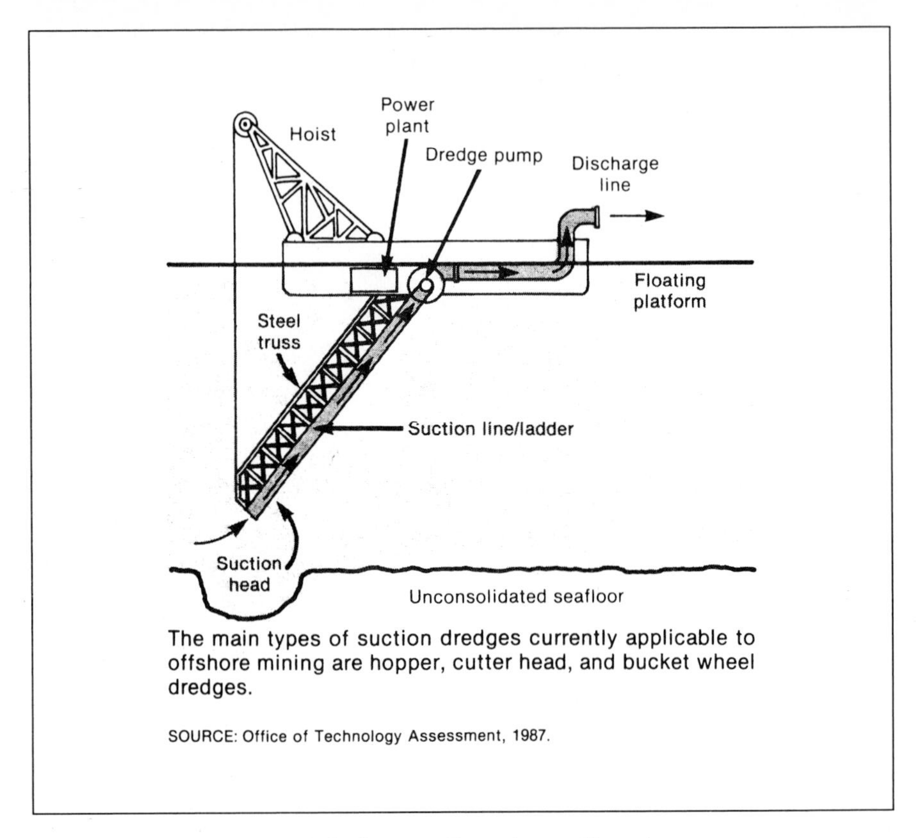

The main types of suction dredges currently applicable to offshore mining are hopper, cutter head, and bucket wheel dredges.

SOURCE: Office of Technology Assessment, 1987.

Figure 4-11 A Typical Cutaway Drawing—Showing Components of a Suction Dredge

object is important, include a person or well-known item in the photograph to illustrate the comparative size of the object. Figure 4-14 is a typical product photograph.

Using Format Elements

For many internal documents, such as reports, company procedures, and bulletins, the writer is responsible for determining the most effective format elements. Just as you use graphic aids to provide readers with easy access to complicated data, use format elements to help readers move through the document, finding and retaining important information. For printed documents, the writer often does not have the final word on which format elements to use. However, always consider these elements, and be prepared to consult with and offer suggestions to the technical editor and art director about document design. The following guidelines cover three types of format elements: (1) written cues, (2) white space, and (3) typographic devices.

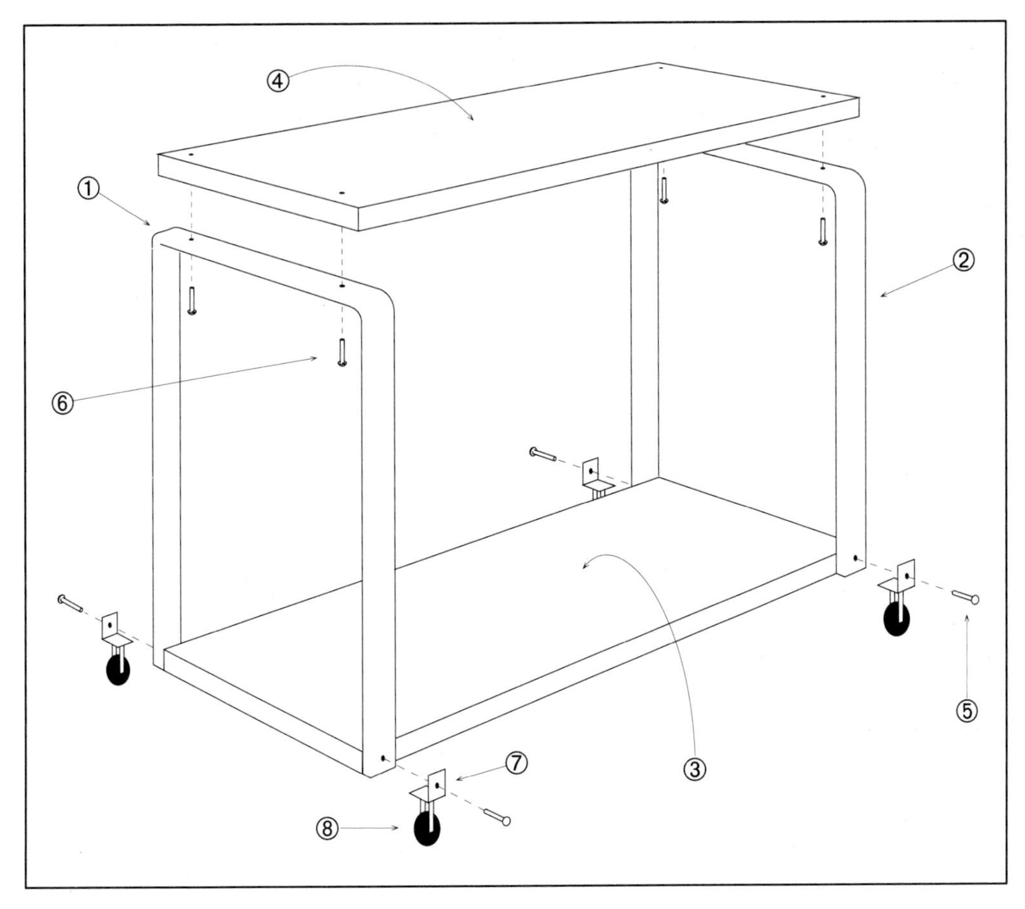

Figure 4-12 An Exploded Drawing of a Cart

Written Cues

Written cues help readers find specific information quickly. The most frequently used written cues are headings, headers and footers, jumplines, and company logos.

Headings

Headings are organizational cues that alert readers to the sequence of information in a document. Use headings

- To help readers find specific data
- To provide an outline that helps readers see the hierarchical relationship of sections

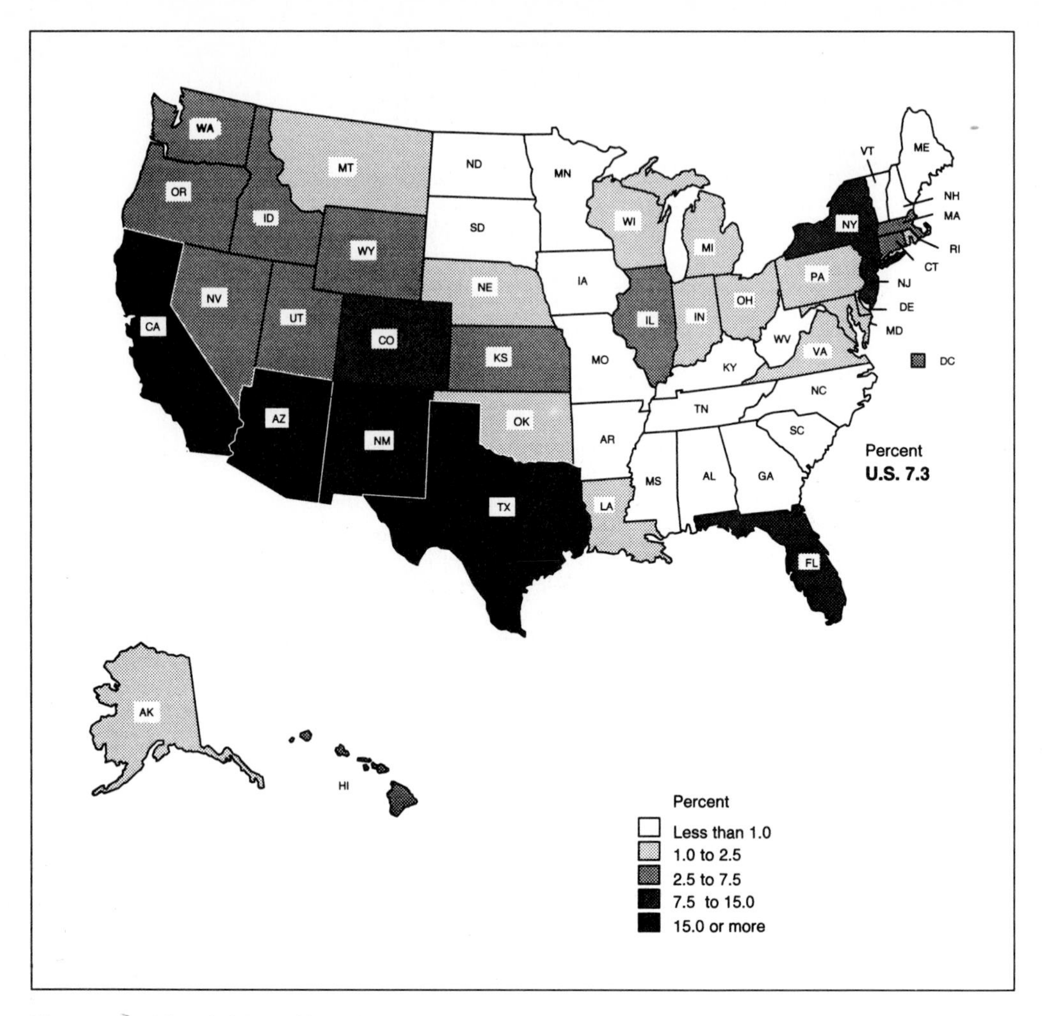

Figure 4-13 A Map Showing Hispanic Population as a Proportion of Total State Population, 1985

- To call attention to specific topics
- To show where changes in topics occur
- To break up a page so that readers are not confronted with line after line of unbroken text.

Include headings for major sections and for topics within the sections. If one heading is followed by several pages of text, divide that section into subsections with more headings. If, on the other hand, every heading is followed by only two or three sentences, regroup your information into fewer topics. Headings are distinguished by level and style according to their placement and use of format typographic devices.

Figure 4-14 A Typical Product Photograph

Levels. First-level headings represent major sections or full chapters of a document. Center such headings on the page and type them in all capital letters:

STATIC ELECTRICITY

Static electricity, a well-known byproduct of. . . .

Second-level headings represent major units of a chapter or of a long section in a document. Center such headings, type them both capital and lowercase letters, and underline them:

Flash Floods

Flash floods are the leading weather-related cause. . . .

Type third-level headings flush with the left margin in capital and lowercase letters and underline them:

Increase in Rheumatic Fever

The average number of children who contracted rheumatic. . . .

The text following any of the first three levels of headings begins two lines below the heading. Type fourth-level headings in capital and lowercase letters, flush with the left margin, underlined, and with a period following them. The text for fourth-level headings begins on the same line:

Changing Your Beneficiary. Use Form RBE/26 to change the beneficiary on your. . . .

Computers make it easy to use boldface type as another distinguishing characteristic of headings. Some writers use boldface for first-level headings; others make all headings boldface and use placement to distinguish them. Some organizations have specific guidelines for headings in documents, and professional journals often have guidelines for headings in technical articles. Check appropriate style guides.

Style. Use nouns, phrases, or sentences for headings. All headings of the same level, however, should be the same part of speech. If you wish to use a noun for a heading, all headings of that level should be nouns. If one heading is a phrase, all headings of that level should be phrases. Here are three types of headings:

Nouns:	Protein
	Fat
	Carbohydrate
Phrases:	Use of Tests
	Development of Standards
	Rise in Cost
Sentences:	Accidents Are Rising.
	Support Is Not Available.
	Reductions Are Planned.

Nouns and phrases are the most commonly used headings. Be sure also that your headings clearly indicate the topic of the section they head so that readers do not waste time reading through unneeded information. Keep sentence headings short. Headings such as "Miscellaneous" or "General Principles" provide no guide to the information covered in the section. Write headings with specific terms and informative phrases so that readers can scan the document for the sections of most interest.

Headers and Footers

A *header* or a *footer* is a notation on each page of a document that identifies it for the reader and includes one or more of the following: (1) document date, (2) title, (3) document number, (4) topic, (5) author, or (6) title of the larger publication the document is part of. Place headers at the tops of pages and footers at the bottoms flush with either the right or left margin. Here is an example:

> Interactive Systems Check, No. 24-X,
> May 1, 1990.

Jumplines

A *jumpline* is a notation indicating that a section is continued in another part of a document. Place a jumpline directly below the last line of text on the page. Here is an example:

> Continued on page 39

Logos

A *company logo* is a specific design used as a symbol of a company. It may be a written cue or a visual one or a combination. Examples include the red cross of the American Red Cross, the golden arches of McDonald's, and the line drawing of a bell for the regional telephone companies. A logo identifies an organization for readers, and it also may unify the document by appearing at the beginning of major sections or on each page.

White Space

White space is the term for areas on a page that have no text or graphics. Far from being just a wasted spot on a page, white space helps readers process the text efficiently. In documents with complicated data and lots of detail, white space rests readers' eyes and directs them to important information. Think of white space as having a definite shape on a page, and use it to create balance and proportion in your documents. White space commonly appears in margins, heading areas, columns, and indentations.

Margins

Margins of white space provide a frame for the text and graphics (called the "live area") on a page. To avoid monotony, experienced designers usually vary the size of margins in printed documents. The bottom margin of a page is generally the widest, and the top margin is slightly smaller. If a document,

such as a manual, is bound with facing pages, the inside margin is usually the narrowest, and the outside margin is slightly wider than the top margin.

Heading Areas

White space provides the background and surrounding area to set off headings. Readers need sufficient white space around headings to be able to separate them from the rest of the text.

Columns

In documents with columns of text, the white space between the columns both frames and separates the columns. Too much or too little space here will upset the balance and proportion of the page. Generally, the wider the columns, the more space that is needed between columns to help readers see them as separate sections of text.

Indentations

The white space at the beginning of an indented sentence and after the last word of a paragraph sets off a unit of text for readers, as does the white space between lines, words, and single-spaced paragraphs. Add extra space between paragraphs to increase readability. Youhaveonlytotrytoreadaline withoutwhitespacebetweenwordstoseehowimportantthistinywhitespaceis.

White space is actually the most important format element, because without it most readers would quickly give up trying to get information from a document.

Typographic Devices

Typographic devices are used to highlight specific details or specific sections of a document. Highlighting involves selecting different typefaces or using boldface type, lists, and boxes.

Typefaces

A *typeface* is a particular design for the type on a page. Typeface should be appropriate to readers and purpose. An ornate, scroll typeface suitable for a holiday greeting card would be out of place in a technical manual.

Poor: *Feasibility Tests in Mining Ventures*

Better: **Feasibility Tests in Mining Ventures**

Follow these general guidelines if you are involved in selecting the typeface for a printed document:

1. Stay in one typeface group, such as Times New Roman, throughout each document. Most typefaces have enough variety in size and weight to suit the requirements of one document.
2. Use the fewest possible sizes and weights of typeface in one document. A multitude of type sizes and weights, even from the same typeface, distracts readers.
3. Avoid type smaller than 10 points because readers over age 40 will have trouble reading it.
4. Use both capital and lowercase letters for text. This combination is more readable than all capitals or all lowercase letters.
5. Use italics sparingly. The Appendix in this book explains the conventional use of italics. Avoid using them for emphasis because they are harder to read than regular type.

Here are some samples of common typefaces:

Times New Roman	Bembo
Bodoni	Baskerville
Optima	Caledonia

Boldface

Boldface refers to type with extra weight or darkness. Use boldface type to add emphasis to (1) headings, (2) specific words, such as warnings, or (3) significant topics. This example shows how boldface type sets off a word from a sentence:

> Depending on the type of incinerator, **gas temperatures** could vary more than 2000 degrees.

Lists

Lists highlight information and guide readers to what facts they specifically need. In general, lists are distinguished by numbers, bullets, and squares.

Numbers. Traditionally, lists in which items are numbered imply either that the items are in descending order of importance or that the items are sequential stages in a procedure. Here is a sample from an operator's manual:

1. Clean accumulated dust and debris from the surface.
2. Run the engine and check for abnormal noises and vibrations.
3. Observe three operating cycles.
4. Check operation of the brake.

Because numbered lists set off information so distinctly from the text, they are often used even when the items have only minor differences in importance. If a list has no chronology, you may use bullets to set off the items.

Bullets. *Bullets* are small black circles that appear before each item in a list just as a number would. On a typewriter or word processor without a bullet symbol, use the small o and fill it in with black ink. Use bullets where there is no distinction in importance among items and where no sequential steps are involved. Here is an example from a company safety bulletin:

A periodic survey is needed to check for the following:

- Gasoline and paint vapors
- Alkaline and acid mists
- Dust
- Smoke

Squares. Squares are used with checklists if readers are supposed to respond to questions or select items on the page itself. A small square precedes each item in the same position as a number or a bullet would appear. Here is an example from a magazine subscription form:

☐ Please bill me for the total cost.
☐ Please bill me in three separate installments.
☐ Please charge to my credit card.

Boxes

A box is a frame that separates specific information from the rest of the text. In addition to the box itself, designers often use a light color or shading to further set off a box. Here is an example of boxed information:

> Boxes are used
>
> - to add supplemental information that is related to the main subject but not part of the document's specific content.
> - to call reader attention to special items such as telephone numbers, dates, prices, and return coupons.
> - to highlight important terms or facts.

Boxes are effective typographic devices, but overuse of them will result in a cluttered, unbalanced page. Use restraint in boxing information in your text.

Several articles in Part II discuss methods of developing and using document design features in both oral presentations and written materials. Check the "Thematic Table of Contents for Readings" to find an article that discusses design features for situations similar to yours.

CHAPTER SUMMARY

This chapter discusses document design features—graphic aids and format elements. Remember:

- Design features guide readers through the text, increase reader interest, and contribute to a document "image."
- Basic page design principles include balance, proportion, sequence, and consistency.
- Graphic aids (1) provide readers with quick access to information, (2) isolate main topics, (3) help readers see relationships among sets of data, and (4) offer expert readers quick access to complicated data.
- All graphic aids should be identified with a descriptive heading and a number and should be placed as near as possible to the reference in the text.
- The most frequently used graphic aids are tables, bar graphs, pictographs, line graphs, pie graphs, organization charts, flowcharts, line drawings, cutaway drawings, exploded drawings, maps, and photographs.
- Written format elements such as headings, headers and footers, jumplines, and logos help readers find special information.
- White space, the area without text or graphics, directs readers to information sections.
- Typographic devices such as typefaces, boldface type, lists, and boxes highlight specific sections or specific details.

SUPPLEMENTAL READINGS IN PART TWO

Benson, P. J. "Writing Visually: Design Considerations in Technical Publications," *Technical Communication.*

Holcombe, M. W., and Stein, J. K. "How to Deliver Dynamic Presentations: Use Visuals for Impact," *Business Marketing.*

"Is It Worth a Thousand Words?" *Simply Stated.*

Marra, J. L. "For Writers: Understanding the Art of Layout," *Technical Communication.*

"Six Graphic Guidelines," *Simply Stated.*

Model 4-1 Commentary

These pages are from an educational booklet printed by the American Heart Association for distribution to heart patients, participants in health programs, and anyone interested in learning about how exercise strengthens the heart.

Discussion

1. Identify the format elements used in these pages.

2. Discuss how easily nonexpert readers, such as students in your class, can find and understand the information in these pages. Consider how well the headings inform and guide readers to information.

Do we get enough exercise from our daily activities?

Most Americans get little vigorous exercise at work or during leisure hours. Today, only a few jobs such as lumberjacking require vigorous physical activity. People usually ride in cars or buses during their free time rather than being physically active. Many people think they are getting enough exercise from golfing, bowling, or vacuuming the house. But these activities do not produce the benefits of regular, more vigorous exercise.

There are many ways to get exercise from activities like swimming, brisk walking, running, or jumping rope. These kinds of activities are sometimes called "aerobic" — meaning the body uses oxygen to produce the energy needed for the activity. Today many people are rediscovering the benefits of regular, vigorous exercise.

What are the benefits of exercise?

These are the benefits often experienced by people who exercise regularly.

Feeling Better

- gives you more energy
- helps in coping with stress
- increases resistance to fatigue
- helps counter anxiety and depression
- helps you to relax and feel less tense
- improves the ability to fall asleep quickly and sleep well
- provides an easy way to share an activity with friends or family and an opportunity to meet new friends

Looking Better

- tones your muscles
- burns off calories to help lose extra pounds or helps you stay at your ideal weight
- helps control your appetite

You need to burn off 3,500 calories more than you take in to lose 1 pound. If you want to lose weight, regular exercise can help you in either of two ways.

First, you can eat your usual amount of calories, but exercise more. For example: A 200-pound person who keeps on eating the same amount of calories, but decides to walk briskly each day for 1½ miles will lose about 14 pounds in 1 year. Or second, you can eat fewer calories and exercise more. This is an even better way to lose weight.

A person burns up only a minimal amount of calories with daily activities such as sitting. Any physical activity in addition to what you normally do will burn up extra calories.

Below are the average calories spent per hour by a 150-pound person. (A lighter person burns fewer calories; a heavier person burns more.) Since precise calorie figures are not available for

most activities, the figures below are averaged from several sources and show the relative vigor of the activities.

Bicycling 6 mph	240 cals.
Bicycling 12 mph	410 cals.
Cross-country skiing	700 cals.
Jogging 5½ mph	740 cals.
Jogging 7 mph	920 cals.
Jumping rope	750 cals.
Running in place	650 cals.
Running 10 mph	1280 cals.
Swimming 25 yds/min.	275 cals.
Swimming 50 yds/min.	500 cals.
Tennis-singles	400 cals.
Walking 2 mph	240 cals.
Walking 3 mph	320 cals.
Walking 4½ mph	440 cals.

The calories spent in a particular activity vary in proportion to one's body weight. For example, for a 100-pound person, reduce the calories by 1/3; for a 200-pound person, multiply by 1⅓.

Exercising harder or faster for a given activity will only slightly increase the calories spent. A better way to burn up more calories is by exercising *longer* and/or covering more *distance.*

Working Better

- often contributes to more productivity at work
- increases your capacity for physical work
- builds stamina for other physical activities
- helps increase muscle strength
- helps your heart and lungs work more efficiently

Consider the benefits of a well-conditioned heart:

In 1 minute with 45 to 50 beats, the heart of a well-conditioned person pumps the same amount of blood as the average person's heart pumps in 70 to 75 beats. Compared to the well-conditioned heart, the average heart pumps up to 36,000 *more* times per day, 13.1 million *more* times per year.

Feeling, looking, and working better — all these benefits can help you enjoy your life more fully.

Can exercise reduce my chances of getting a heart attack?

Coronary artery disease is the major cause of heart disease and heart attack in America. It develops when fatty deposits build up on the inner walls of the blood vessels feeding the heart *(coronary arteries).* Eventually one or more of the major coronary arteries may become blocked — either by the buildup of deposits or by a blood clot forming in the artery's narrowed passageway. The result is a heart attack.

We know that there are several factors that can increase your risk for developing coronary artery disease — and thus the chances for a heart attack. Fortunately, many of these risk factors can be reduced or eliminated.

The most important of these risk factors.

Cigarette Smoking, High Blood Pressure, High Blood Cholesterol. The higher their levels, the greater the risk. On the average each risk factor doubles the chances of getting a heart attack. Also the more risk factors you have, the greater your risk.

Some other important associated risk factors.

Diabetes. Having diabetes doubles the risk for coronary artery disease.

Overweight. Excess weight increases the likelihood of developing high blood pressure, high blood cholesterol and diabetes.

Low Levels of HDL. Cholesterol in the blood is transported by different types of particles. One of these particles is a protein called *high density lipoprotein* or HDL. Low levels of HDL in the blood are related to an increased risk of coronary artery disease.

Physical Inactivity. Several studies have suggested that lack of exercise (a sedentary lifestyle) is associated with increased risk for heart attack. Most of the scientific research has found that compared to physically active people, inactive people have 1½ to 2 times the risk of having a heart attack.

Model 4-2 Commentary

This page is from the 1988 Annual Report of the Chesapeake Corporation. The "Financial Review 1986–1988" section explains the company's financial status over a 2-year period. The page contains three bar graphs illustrating information in the text. The bar graph entitled "Capital Structure" uses three shadings on each bar to represent the three financial categories identified in the key.

Discussion

1. Discuss the use of white space here. Is there any benefit from retaining white space in the bottom left margin?

2. Discuss the image readers will have of the company's financial status if they glance quickly at the bar graphs without reading the text.

3. Discuss what might be added to the third bar graph to make it more consistent with the first two.

Financial Review 1986-1988

Earnings Overview

In 1988 Chesapeake established new records for sales and operating, pretax and net incomes as a result of excellent demand for its major products, favorable economic conditions in the United States, favorable foreign trade and currency exchange rates and outstanding operating performances at our company facilities. Quarterly earnings records were achieved for each quarter of 1988. Net income and earnings per share for the fourth quarter of 1988 slightly exceeded levels attained in the previous record quarter — the fourth quarter of 1987. Earnings per share for 1988 were 71% greater than the previous record year of 1987. The accomplishments of 1988 were outstanding. To meet or exceed 1988 results during 1989 will be challenging because market conditions, a critical success factor that the company does not control, must remain favorable. From an operational perspective, the company is well prepared for 1989.

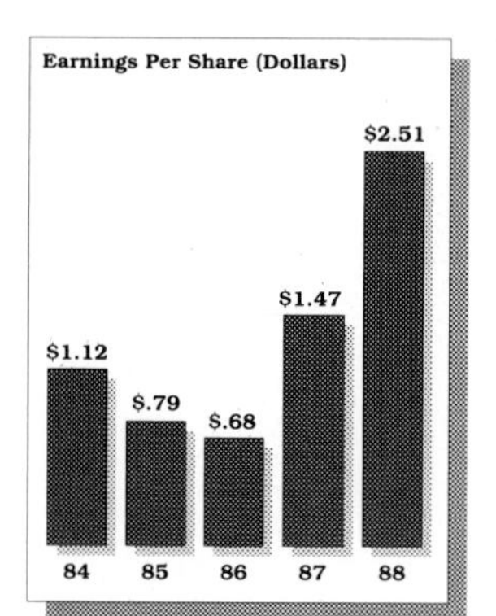

Earnings Per Share (Dollars)

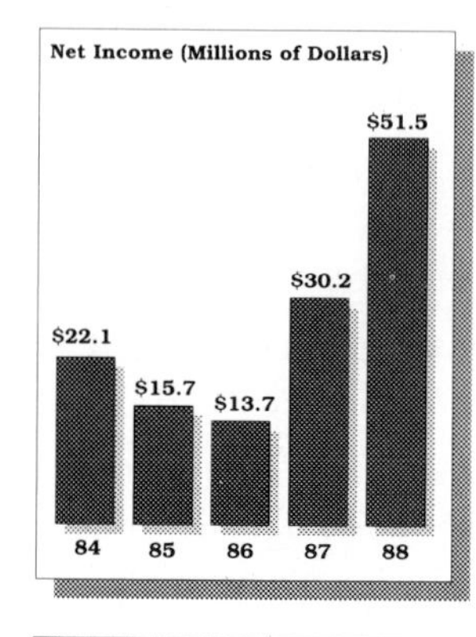

Net Income (Millions of Dollars)

Liquidity and Capital Structure

During 1988 the company continued its objective to reduce long-term debt incurred in 1985 and 1986 to acquire Wisconsin Tissue and construct the No. 3 TRI-KRAFT™ paper machine in West Point. For most of the year, long-term debt was at least $50 million below comparable 1987 levels. However, during the fourth quarter, capital expenditures and working capital requirements exceeded internal cash sources. By the end of 1988, long-term debt increased to $231.8 million, slightly ahead of the 1987 year-end level of $225.5 million.

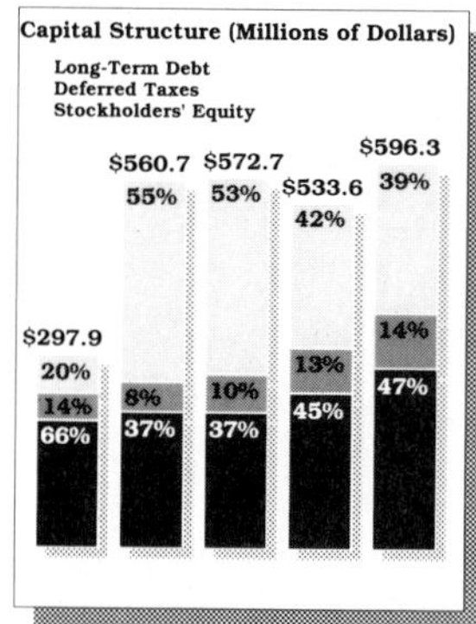

Capital Structure (Millions of Dollars)

Because of the record increase in stockholders' equity resulting from 1988's earnings, the ratio of long-term debt to total capital decreased to 38.9% in 1988 from 42.3% in 1987 and 52.8% in 1986. The ratio is now slightly below the midpoint of the long-term target range of 35-45%.

During 1988 the company ended its revolving credit agreement and, at the same time, renewed and added committed and uncommitted credit lines which will accommodate projected short-term borrowing requirements. Only $46 million of the $185 million total credit lines was utilized at the end of 1988. Interest charges for borrowings under these agreements were based on lower than prime bank rates throughout 1988.

At the end of 1988, additional working capital was required due to expanded operations and fourth quarter performance. Working capital, a record $99.6 million at the end of 1988, exceeded 1987's year-end level by over $31 million and the previous high of $76 million in 1986. The year-end 1988 ratio of current assets to current liabilities was 2.5 compared to 2.2 and 2.5 at the end of 1987 and 1986, respectively.

Net income, depreciation, cost of timber harvested, deferred taxes and amortization, formerly collectively defined as "funds provided from operations" and currently defined as "cash sources from operations", continued to generate substantial 1988 cash in-flow to the company of $111.3 million. This record level was caused primarily by the record net income and sustained levels of depreciation expense and deferred income taxes. Net cash provided by operating activities, $75 million for 1988, was less than cash sources from operations due to the significant accounts receivable and inventory increases. Cash sources from operations ($93.6 million in 1987 and $67.9 million in 1986) has grown each year during the past three years primarily due to increases in net income.

Model 4-3 Commentary

This page is from the consumer instructions for a Panasonic VCR.

Discussion

1. Identify the graphic aids and format elements used in this page. Discuss why the designer probably chose to use these. How would photographs instead of line drawings help or hinder readers trying to set up a new VCR?

2. Discuss the use of white space here and how it helps or hinders readers.

3. Bring in a page of consumer instructions for a home appliance or piece of electronic equipment that you believe needs improvement. Identify the format and graphic features used, and discuss the features that need to be improved.

4. Look at Models 2-1 and 2-2 and identify the graphic aids and format elements used in these models. Discuss how effectively these features help readers use the information. Would you make any format changes in these models or in the models in this chapter?

TO GET STARTED

Taking the steps shown on this page will prepare your VCR, so you can start a playback to watch a tape. Steps 1 and 2 are one time system setup steps.

VCR TO TV SYSTEM CONNECTION

1 Connect the supplied VHF Connecting Cable from the VCR to the TV as shown here.

Follow illustration A or B depending on the type of terminals available on the back of your TV.

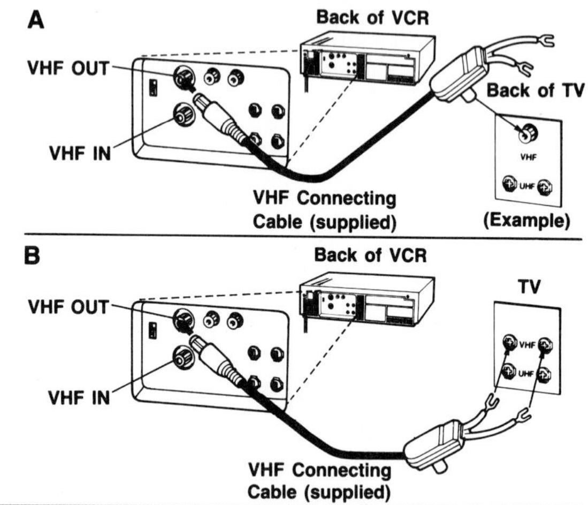

SELECT VCR CHANNEL (CHANNEL 3 OR 4)

2 Your VCR must control one TV channel permanently just as a TV broadcast station does. To decide which channel to use, select either channel 3 or 4 with the Channel 3/4 switch on the back of the VCR.
To avoid local interference, the channel 3/4 switch position you choose should match the number of the TV channel that is not normally broadcast in your area.

(VCR Monitor Channel) CHANNEL 3/4 Switch

CH3

CH4

Back of VCR

Now, tune your TV to the same channel that you selected for the channel 3 or 4 switch.

SELECT YOUR VCR OR TV FOR VIEWING

3 Set the VCR/TV Selector to "VCR".

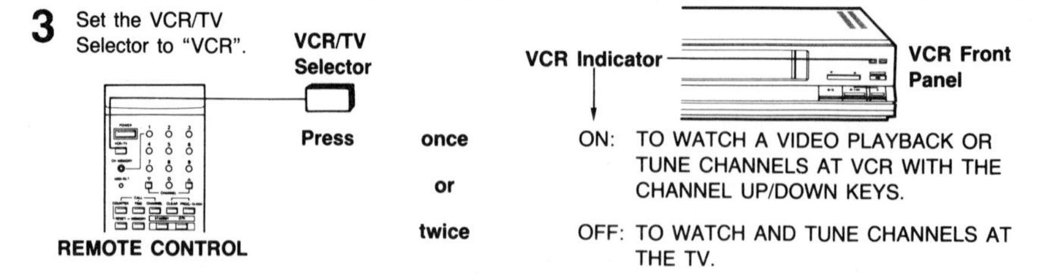

ON: TO WATCH A VIDEO PLAYBACK OR TUNE CHANNELS AT VCR WITH THE CHANNEL UP/DOWN KEYS.

OFF: TO WATCH AND TUNE CHANNELS AT THE TV.

Refer to the separate Basic Guide For VCR Connections and select the section with details on the particular hookup you want to use.

Chapter 4 Exercises

1. For practice, your instructor will assign one or more of the following exercises to be done individually or in groups. Identify a specific reader and purpose for each exercise before you begin.

a. Create a flowchart showing a technical process in your field.
b. Create a flowchart showing a process that you perform on your job.
c. Create an organization chart for a group to which you belong or for the company or department for which you work.
d. Collect information on expenditures for college expenses from five other students, and create a pie graph showing how a typical student allocates money for tuition, books, living costs, and so on.
e. Create a line drawing of a piece of equipment used in your field. Include labels for the important parts.
f. Collect information from ten other students on where they ate breakfast, lunch, and dinner yesterday. If they ate out, classify the eating places as fast-food, cafeteria, and so on. Create a table showing the data you collect.

2. Find a technical document from a company and make enough copies for the other students in class. Prepare an oral critique of the document relative to the design guidelines given in this chapter. Explain how you would revise the document for a more effective format.

3. Create an effective visual aid for one or more of the following sets of information. Identify a specific reader and purpose for each before you begin.

a. In the annual report to stockholders, Mason Barr Industries reported expenditures on charitable and social programs: in 1985, $450,000; in 1986, $470,000; in 1987, $490,000; in 1988, $420,000; in 1989, $500,000; and in 1990, $480,000.
b. Belden Enterprises, Inc., reported production volume and number of shipments for 1989 at its six oil refineries: Matchstick, Texas, 24 million barrels, 5 shipments; Sirenas Island, Georgia, 105 million barrels, 62 shipments; Center City, Missouri, 175 million barrels, 102 shipments; Pine Grove, Colorado, 78 million barrels, 10 shipments; Coles Landing, Texas, 72 million barrels, 35 shipments; and Westmore, Oklahoma, 101 million barrels, 78 shipments.
c. A survey of people in large urban areas indicated that their trash consisted of the following: paper, 42%; cloth, 4.2%; food waste, 10%; grass and leaves, 20%; glass and metal, 7.8%; and other, 16%.
d. The U.S. trade deficit varied monthly during a 3-year period. The following figures are in billions of dollars: in 1988, January $12, February $14.5, March $13, April $13, May $14, June $15.5, July $16.5, August $15.5, September $14, October $17.5, November $13.5, December $12.5; in 1989, January $14.5, February $15, March $16, April $16.5, May $13, June $13.5, July $14, August $17, September $12, October $13.5, November

$15, December $16; and in 1990, January $15, February $14.5, March $16, April $16.5, May $17.5, June $18, July $16.5, August $16.5, September $17, October $15, November $16.5, December $14.

4. The following conference schedule was sent to 75 insurance agents attending a 2-day meeting. Confusion began when 46 agents arrived at the Lakeside Hotel for the first session and 29 agents arrived at Winston Community College for the first session. Later, during the conference, agents also complained that they weren't sure about speakers, times, or places for individual meetings. Reorganize the conference schedule for easier use. Make any decisions you wish about times, locations, and speakers in your revision.

Schedule
Fall IAA Conference for Health Insurance Agents
Lakeside Hotel, 750 Market Street, Port Washington, Wisconsin
Winston Community College, 1600 Exchange Avenue,
Port Washington, Wisconsin
October 6–7, 1990
Friday, October 6
1–3 p.m. Registration (Lobby, Main Entrance)
3–4:30 p.m. Session I (McKinley Room)
Marjorie Kohner, Jennifer Ruttan, "Developing
Customer Relations Before the Sale," Daniel
Fletcher, John Abbott, Mutual of Wyoming Insurance.
Coffee Break (Lobby)
4:30–5:45 p.m. Session II (Taft Room)
"Handling Special Health Concerns," Juliet Lavery,
Anna Jones, Inter-State Health Insurance Company.
6:30–7:00 p.m. Cash Bar
7:30–9:00 p.m. Dinner (Washington Hall)
"The Insurance Agent as Financial Advisor,"
Mitchell Ryan, National Board of Insurance Agents.
Introduced by Ricardo Castillo
9 p.m.–11 p.m. Cash Bar
Saturday, October 7
8–9 a.m. IAA Executive Board Business Meeting (305 Main Hall)
9:00–9:30 a.m. Continental Breakfast
10:00–11:30 a.m. Session III (Room 301)
Lucy Arbuckle, Roscoe Taggett, Joseph Hartwell, Roosevelt Perkins,
Renata Schultz, "Following the Sale to Conclusion." Million Dollar
Winners, Ace Insurance.
11:30–12:30 Group Meetings--Rooms 306, 302, 301--
Agency Managers, First-Year Sales Representatives,
Senior Sales Representatives.
Lunch (Lower Lobby)

CHAPTER 5

Revision

Creating a Final Draft
- Global Revision
- Fine-Tuning Revision

Making Global Revisions
- Content
- Organization

Making Fine-Tuning Revisions
- Overloaded Sentences
- Clipped Sentences
- Lengthy Sentences
- Passive Voice
- Jargon
- Sexist Language
- Concrete versus Abstract Words
- Gobbledygook
- Fad Words
- Wordiness

Chapter Summary

Supplemental Readings in Part Two

Models

Exercises

Creating a Final Draft

As you plan and draft a document, you will probably change your mind many times about content, organization, and style. In addition to these changes, however, you should consider revision as a separate stage in producing your final document. During drafting, you should concentrate primarily on developing your document from your initial outline. During revision, concentrate on making changes in content, organization, and style that will best serve your readers and their purpose. These final changes should result in a polished document that your readers can use efficiently.

Thinking about revision immediately after finishing a draft is difficult. Your own writing looks "perfect" to you, and every sentence seems to be the only way to state the information. Experienced writers allow some time between drafting and revising whenever possible. The longer the document, the more important it is to let your writing rest before final revisions so that you can look at the text and graphic aids with a fresh attitude and read as if you were the intended reader and had never seen the material before. Most writers divide revision into two separate stages—global or overall document revision and fine-tuning or style and surface-accuracy revision.

Global Revision

The process of *global revision* involves evaluating your document for effective content and organization. At this level of revision, you may add or delete information; reorganize paragraphs, sections, or the total document; and redraft sections.

Fine-Tuning Revision

The process of *fine-tuning revision* involves the changes in sentence structure and language choice that writers make to ensure clarity and appropriate tone. This level also involves checking for surface accuracy, such as correct grammar, punctuation, spelling, and mechanics of technical style. Many organizations have internal style guides that include special conventions writers must check for during fine-tuning revision.

With word processing, both global and fine-tuning revisions can be fast and easy. Depending on the type of word-processing program you have, you can make revisions using these capabilities:

1. *Delete and insert*. You can omit or add words, sentences, and paragraphs anywhere in the text.

2. *Search for specific words.* You can search through the full text for specific words or phrases that you want to change. If you wish to change the word *torpid* to *inactive* throughout the document, the computer will locate every place where *torpid* appears so that you can delete it and insert *inactive.*
3. *Moving and rearranging text.* You can take your text apart as if you were cutting up typed pages and arranging the pieces in a new order. You can move whole sections of text from one place to another and reorder paragraphs within sections. You also may decide that revised sentences should be reorganized within a paragraph.
4. *Re-format.* You can easily change margins, headlines, indentations, spacing, and other format elements.
5. *Check spelling and word choice.* Many software programs have spelling checkers that scan a text and flag misspelled words that are in the checker's dictionary. Spelling checkers, however, cannot find usage errors, such as *affect* for *effect,* or spelling errors in proper names and unusual terms. A thesaurus program will suggest synonyms for specific words. You can select a synonym and substitute it wherever the original word appears in the text.

Revision on a computer is so convenient that some people cannot stop making changes in drafts. Do not change words or rearrange sentences simply for the sake of change. Revise until your document fulfills your readers' needs, and then stop.

Making Global Revisions

For global revisions, rethink your reader and purpose, and then review the content and organizational decisions you made in the planning and drafting stages. Consider changes that will help your readers use the information in the document efficiently.

Content

Evaluate the content of your document and consider whether you have enough information or too little. Review details, definitions, and emphasis.

Details. Have you included enough details for your readers to understand the general principles, theories, situations, or actions? Have you included details that do not fit your readers' purpose and they do not need? Do the details serve both primary and secondary readers? Are all the facts, such as dates, amounts, names, and places, correct? Are your graphic aids appropriate for your readers? Do you need more? Have you included examples to help your readers visualize the situation or action?

Definitions. Have you included definitions of terms that your readers may not know? If your primary readers are experts and your secondary readers are not, have you included a glossary for secondary readers?

Emphasis. Do the major sections stress information suitable for the readers' main purpose? Do the conclusions logically result from the information provided? Is the urgency of the situation clear to readers? Are deadlines explained? Do headings highlight major topics of interest to readers? Can readers easily find specific topics?

After evaluating how well the document serves your readers and their need for information, you may discover that you need to gather more information or that you must omit information you have included in the document. Reaching the revision stage and finding that you need to gather more information can be frustrating, but remember, readers can only work with the information they are given. The writer's job is to provide everything readers need to use the document efficiently for a specific purpose.

Organization

Evaluate the overall organization of your document as well as the organization of individual sections and paragraphs. Review these elements:

Overall Organization. Is the information grouped into major topics that are helpful and relevant to your readers? Are the major topics presented in a logical order that will help readers understand the subject? Are the major sections divided into subsections that will help readers understand the information and highlight important subtopics? Do some topics need to be combined or further divided to highlight information or to make the information easier to understand? Do readers of this document expect or prefer a specific organization, and is that organization in place?

Introductions and Conclusions. Consider the introductions and conclusions both for the total document and for major sections. Does the introduction to the full document orient readers to the purpose and major topics covered? Do the section introductions establish the topics covered in each section? Does the conclusion to the full document summarize major facts, results, recommendations, future actions, or the overall situation? Do the section conclusions summarize the major points and show why these are important to readers and purpose?

Paragraphs. Do paragraphs focus on one major point? Are paragraphs sufficiently developed to support that major point with details, examples, and explanations? Is each paragraph organized according to a pattern that will help readers understand the information, such as chronological or ascending order of importance?

Headings. Are the individual topics marked with major headings and subheadings? Do the headings identify the key topics? Are there several pages in a row without headings, or is there a heading for every sentence, and therefore, do headings need to be added or omitted?

Format. Are there enough highlighting devices, such as headings, lists, boxes, white space, and boldface type, to lead readers through the information and help them see relationships among topics? Can readers quickly scan a page for key information?

After evaluating the organization of the document and of individual sections, make necessary changes. You may find that you need to write more introductory or concluding material or reorder sections and paragraphs. Remember that content alone is not enough to help readers if they have to hunt through a document for relevant facts. Effective organization is a key element in helping readers use a document quickly and efficiently.

Making Fine-Tuning Revisions

When you are satisfied that your global revisions have produced a complete and well-organized draft, you should go on to the fine-tuning revision stage, checking for sentence structure, word choice, grammar, punctuation, spelling, and mechanics of technical style. This level of revision can be difficult because by this time you probably have the document or sections of it memorized. As a result, you may read the words that should be on the page rather than the words that are on the page. For fine-tuning revisions you need to slow your reading pace so that you do not overlook problems in the text. You also should stop thinking about content and organization, and think instead about individual sentences and words. Experienced writers sometimes use these techniques to help them find trouble spots:

- *Read aloud.* Reading a text aloud will help you notice awkward or overly long sentences, insufficient transitions between sentences or paragraphs, and inappropriate language. Reading aloud slows the reading pace and helps you hear problems in the text that your eye could easily skip over.
- *Focus on one point at a time.* You cannot effectively check for sentence structure, word choice, correctness, and mechanical style all at the same time, particularly in a long document. Experienced writers usually revise in steps, checking for one or two items at a time. If you check for every problem at once, you are likely to overlook items.

- *Use a ruler.* By placing a ruler under each line as you read it and moving the ruler from line to line, you will focus on the words rather than on the content of the document. Using a ruler also slows your reading pace, and you are more likely to notice problems in correctness and technical style.
- *Read backward.* Some writers read a text from the bottom of the page up in order to concentrate only on the typed lines. Reading from the bottom up will help you focus on words and find typographic errors, but it will not help you find problems in sentence structure or grammar because sentence meaning will not be clear as you move backward through the text.

You may find a combination of these methods useful. You may use a ruler, for example, and focus on one point at a time as you look for sentence structure, word choice, and grammar problems. As a final check when you proofread, you may read backward to look for typographic errors.

The following items are those most writers check for during fine-tuning revision. For a discussion of grammar, punctuation, and mechanical style, see the Appendix.

Overloaded Sentences

An overloaded sentence includes so much information that readers cannot easily understand it. This sentence is from an announcement to all residents in a city district:

> You are hereby notified, in accordance with Section 107–27 of the Corinth City Code, that a public hearing will be held by a committee of the Common Council on the date and time, and at the place listed below, to determine whether or not to designate a residential permit parking district in the area bounded by E. Edgewood Avenue on the north, E. Belleview Place on the south, the Corinth River on the west, and Lake McCormick on the east, except the area south of E. Riverside Place and west of the north/south alley between N. Oakland Avenue and N. Bartlett Avenue and the properties fronting N. Downer Avenue south of E. Park Place.

No reader could understand and remember all the information packed into this sentence. Here is a revision:

> As provided in City Code Section 101–27, a committee of the Common Council of Corinth will meet at the time and place shown below. The purpose of the meeting is to designate a residential permit parking district for one area between the Corinth River on the west and Lake McCormick on the east. The area is also bounded by E. Edgewood Avenue on the north and E. Belleview Place on the south. Not included is the area south of E. Riverside

> Place and west of the north/south alley between N. Oakland Avenue and N. Bartlett Avenue and the properties fronting N. Downer Avenue south of E. Park Place.

To revise the original overloaded sentence, the writer separated the pieces of information and emphasized the individual points in shorter sentences. The first sentence in the revision announces the meeting. The next two sentences identify the meeting purpose and the streets included in the new parking regulations. A separate sentence then identifies streets not included. Readers will find the shorter sentences easier to understand and, therefore, the information easier to remember. Because so many streets are involved, readers will also need a graphic aid—a map with shading that shows the area in question. Avoid writing sentences that contain so many pieces of information that readers must mentally separate and sort the information as they read. Even short sentences can contain too much information for some readers. Here is a much shorter sentence than the preceding one, yet if the reader were a nonexpert interested in learning about cameras, the sentence holds more technical detail than he or she could remember easily:

> The WY–30X camera is housed in a compact but rugged diecast aluminum body, including a 14X zoom lens and 1.5-in. viewfinder, and has a signal-to-noise ratio of 56 dB, resolution of more than 620 lines on all channels, and registration of 0.6% in all zones.

In drafting sentences, consider carefully how much information your readers are likely to understand and remember at one time.

Clipped Sentences

Sometimes when writers revise overloaded sentences, the resulting sentences become so short they sound like the clipped phrases used in telegrams. Do not cut these items from your sentences just to make them shorter: (1) articles (*a, an, the*), (2) pronouns (*I, you, that, which*), (3) prepositions (*of, at, on, by*), and (4) linking verbs (*is, seems, has*). Here are some clipped sentences and their revisions:

Clipped: Attached material for insertion in Administration Manual.

Revised: The attached material is for insertion in the Administration Manual.

Clipped: Questioned Mr. Hill about compensated injuries full-time employees.

Revised: I questioned Mr. Hill about the compensated injuries of the full-time employees.

Clipped: Investigator disturbed individual merits of case.

Revised: The investigator seems disturbed by the individual merits of the case.

Do not eliminate important words to shorten sentences. The reader will try to supply the missing words and may not do it correctly, resulting in ambiguity and confusion.

Lengthy Sentences

Longer sentences demand that readers process more information, whereas shorter sentences give readers time to pause and absorb facts. A series of short sentences, however, can be monotonous, and too many long sentences in a row can overwhelm readers, especially those with nonexpert knowledge of the subject. Usually, short sentences emphasize a particular fact, and long sentences show relationships among several facts. With varying sentence lengths you can avoid the tiresome reading pace that results when writers repeat sentence style over and over. This paragraph from a letter to bank customers uses both long and short sentences effectively:

> Your accounts will be automatically transferred to the Smith Road office at the close of business on December 16, 1988. You do not need to open new accounts or order new checks. If for some reason another Federal location is more convenient, please stop by 2900 W. Manchester Road and pick up a special customer courtesy card to introduce you to the office of your choice. The Smith Road office is a full-service banking facility. Ample lobby teller stations, auto teller service, 24-hour automated banking, night depository, as well as a complete line of consumer and commercial loans are all available from your new Federal office. You will continue to enjoy the same fine banking service at the Smith Road office, and we look forward to serving you. If you have questions, please call me at 555–8876.

In this paragraph the writer uses short sentences to emphasize the fact that customers do not have to take action, as well as that the new bank office is a full-service branch. Longer sentences explain the services and how to use them. The offer of help is also in a short sentence for emphasis.

Passive Voice

Active voice and passive voice indicate whether the subject of a sentence performs the main action or receives it. Here are two sentences illustrating the differences:

Active: The bridge operators noticed severe rusting at the stone abutments.

Passive: Severe rusting at the stone abutments was noticed by the bridge operators.

The active-voice sentence has a subject (*bridge operators*) that performs the main action (*noticed*). The passive-voice sentence has a subject (*rusting*) that receives the main action (*was noticed*). Passive voice is also characterized by a form of the verb *to be* and a prepositional phrase beginning with *by* that identifies the performer of the main action—usually quite late in the sentence. Readers generally prefer active voice because it creates a faster pace and is more direct. Passive voice also may create several problems for readers:

1. Passive voice requires more words than active voice because it includes the extra verb (*is, are, was, were, will be*) and the prepositional phrase that introduces the performer of the action. The preceding passive-voice sentence has two more words than the active-voice sentence. This may not seem excessive, but if every sentence in a 10-page report has two extra words, the document will be unnecessarily long.

2. Writers often omit the *by* phrase in passive voice, decreasing the number of words in the sentence but also possibly concealing valuable information from readers. Notice how these sentences offer incomplete information without the *by* phrase:

   Incomplete: At least $10,000 was invested in custom software.

   Complete: At least $10,000 was invested in custom software by the vice president for sales.

   Incomplete: The effect of welded defects must be addressed.

   Complete: The effect of welded defects must be addressed by the consulting engineer.

3. Writers often create dangling modifiers when using passive voice. In the active-voice sentence below, the subject of the sentence (*financial analyst*) is also the subject of the introductory phrase.

   Active: Checking the overseas reports, the company financial analyst estimated tanker capacity at 2% over the previous year.

   Passive: Checking the overseas reports, tanker capacity was estimated at 2% over the previous year.

   In the second sentence, the opening phrase (*Checking . . .*) cannot modify *tanker capacity* and is a dangling modifier because the subject of the phrase does not appear in the sentence.

4. Passive voice is confusing in instructions because readers cannot tell *who* is to perform the action. Always write instructions and direct orders in active voice.

   Unclear: The cement should be applied to both surfaces.

   Revised: Apply the cement to both surfaces.

Unclear: The telephone should be answered within three rings.

Revised: Please answer the telephone within three rings.

In technical writing, passive voice is sometimes necessary. Use passive voice when

1. You do not know who or what performed the action:

   The fire was reported at 6:05 A.M.

2. Your readers are not interested in who or what performed the action:

   Ethan McClosky was elected district attorney.

3. You are describing a process performed by one person and naming that person in every sentence would be monotonous. Identify the performer in the opening and then use passive voice:

   The carpenter began assembling the base by gluing each long apron to two legs. The alignment strips were then cut for the tray bottoms to finished dimensions. The tabletop was turned upside down and. . . .

4. For some reason, perhaps courtesy, you do not want to identify the person responsible for an action:

   The copy machine jammed because a large sheaf of papers was inserted upside down.

Jargon

Jargon refers to the technical language and abbreviated terms used by people in one particular field, one company, or one department or unit. People in every workplace and every occupation use some jargon. A waitress may refer to "watering the tables" (putting water glasses on the tables), a public relations executive may ask a secretary to get "the glossies" (glossy photographs), or a psychologist may remark to another psychologist that the average woman has a "24F on the PAQ" (a femininity score of 24 on the Personal Attributes Questionnaire). In some cases, jargon may be so narrow that employees in one department cannot understand the jargon used in another department. Use jargon in professional communications only when you are certain your readers understand it.

Here is a sentence from a memo sent by a purchase coordinator to his supervisor:

> Guttenhausen (Deutsch) big R price is list less 40 to Inventory versus rest to Corporate.

When the supervisor included this information in a memo he prepared for the division head about the weekly orders, he revised it as follows:

> Guttenhausen Industries (Bonn) will pay our standard price less 40% (in deutsche marks) for the Model RR–26 transformer, charged to the Inventory account. The difference between this price and our usual cost plus 15% will be charged to the Corporate account.

The supervisor cut the original jargon and clarified the message by fully identifying the company, the transformer model, and the prices and by explaining how the costs would be charged. The revision required two sentences. Jargon allows employees to exchange briefer messages, but only if the readers understand every term and the abbreviated references. When in doubt, cut out jargon and clarify your terms.

Sexist Language

Sexist language refers to words or phrases that indicate a bias against women in terms of importance or competence. Most people would never use biased language to refer to ethnic, religious, or racial groups, but sexist language often goes unnoticed because many common terms and phrases that are sexist in nature have been part of our casual language for decades. People now realize that sexist language influences our expectations about what women can accomplish and also relegates women to an inferior status. Using such language is not acceptable any longer. In revising documents, check for these slips into sexist language:

1. Demeaning or condescending terms. Avoid using casual or slang terms for women.

    Sexist: The girls in the Records Department will prepare the reports.

    Revised: The clerks in the Records Department will prepare the reports.

    Sexist: The annual division picnic will be June 12. Bring the little woman.

    Revised: The annual division picnic will be June 12. Please bring your spouse.

2. Descriptions for women based on standards different from those for men. Avoid referring to women in one way and men in another.

    Sexist: The consultants were Dr. Dennis Tonelli, Mr. Robert Lavery, and Debbie Roberts.

    Revised: The consultants were Dr. Dennis Tonelli, Mr. Robert Lavery, and Ms. Debra Roberts.

    Sexist: There were three doctors and two lady lawyers at the meeting.

    Revised: There were three doctors and two lawyers at the meeting.

    Sexist: The new mechanical engineers are Douglas Ranson, a graduate of MIT and a Rhodes scholar, and Marcia Kane, an attractive redhead with a B.S. from the University of Michigan and an M.S. from the University of Wisconsin.

Revised: The new mechanical engineers are Douglas Ranson, a graduate of MIT and a Rhodes scholar, and Marcia Kane, with a B.S. from the University of Michigan and an M.S. from the University of Wisconsin.

3. Occupational stereotypes. Do not imply that all employees in a particular job are the same sex.

Sexist: An experienced pilot is needed. He must. . . .

Revised: An experienced pilot is needed. This person must. . . .

Sexist: The new tax laws affect all businessmen.

Revised: The new tax laws affect all businesspeople.

Although some gender-specific occupational terms remain common, such as *actor/actress, waiter/waitress,* and *host/hostess,* most such terms have been changed to neutral job titles.

Sexist: The policeman should fill out a damage report.

Revised: The police officer should fill out a damage report.

Sexist: The stewardess reported the safety problem.

Revised: The flight attendant reported the safety problem.

4. Generic *he* to refer to all people. Using the pronoun *he* to refer to unnamed people or to stand for a group of people is correct grammatically in English, but it might be offensive to some people. Avoid the generic *he* whenever possible, particularly in job descriptions that could apply to either sex. Change a singular pronoun to a plural:

Sexist: Each technical writer has his own office at Tower Industries.

Revised: All technical writers have their own offices at Tower Industries.

Eliminate the pronoun completely:

Sexist: The average real estate developer plans to start his construction projects in April.

Revised: The average real estate developer plans to start construction projects in April.

Use *he or she* or *his or hers* very sparingly:

Sexist: The x-ray technician must log his hours daily.

Revised: The x-ray technician must log his or her hours daily.

Better: All x-ray technicians must log their hours daily.

Concrete versus Abstract Words

Concrete words refer to specific items, such as objects, statistics, locations, dimensions, and actions that can be observed by some means. Here is a sentence using concrete words:

> The 48M printer has a maximum continuous speed of 4000 lines per minute, with burst speeds up to 5500 lines per minute.

Abstract words refer generally to ideas, conditions, and qualities. Here is a sentence from a quarterly report to stockholders that uses abstract words:

> The sales position of Collins, Inc., was favorable as a result of our attention to products and our response to customers.

This sentence contains little information for readers who want to know about the company activities. Here is a revision using more concrete language:

> The 6% rise in domestic sales in the third quarter of 1989 resulted from our addition of a safety catch on the Collins Washer in response to over 10,000 requests by our customers.

This revision in concrete language clarifies exactly what the company did and what the result was. Using concrete or abstract words, as always, depends on what you believe your readers need. However, in most professional situations, readers want and need the precise information provided by concrete words.

The paragraphs that follow are from two reports submitted by management trainees in a large bank after a visit to the retail division of the main branch. The trainees wrote reports to their supervisors, describing their responses to the procedures they observed. The paragraph from the first report uses primarily abstract, nonspecific language:

> The first department I visited was very interesting. The way in which the men buy and investigate the paper was very interesting and informative. This area, along with the other areas, helped me where I am right now—doing loans. Again, the ways in which the men deal with the dealers was quite informative. They were friendly, but stern, and did not let the dealers control the transaction.

The paragraph from the second report uses more concrete language:

> The manual system of entering sales draft data on the charge system seems inefficient. It is clear that seasonal surges in charge activity (for example, Christmas) or an employee illness would create a serious bottleneck in pro-

cessing. My talk with Eva Lockridge confirmed my impression that this operation is outdated and needs to be redesigned.

The first writer did not identify any specific item of interest or explain exactly what was informative about the visit. The second writer identified both the system that seemed outdated and the person with whom she talked. The second writer is providing her reader with concrete information rather than generalities.

Gobbledygook

Gobbledygook is writing that is indirect, vague, pompous, or longer and more difficult to read than necessary. Writers of gobbledygook do not care if their readers can use their documents because they are interested only in demonstrating how many large words they can cram into convoluted sentences. Such writers are primarily concerned with impressing or intimidating readers rather than helping them use information efficiently. Gobbledygook is characterized by these elements:

1. Using jargon the readers cannot understand.
2. Making words longer than necessary by adding suffixes or prefixes to short, well-known words (e.g., *finalization, marginalization*). Another way to produce gobbledygook is to use a longer synonym for a short, well-known word. A good example is the word *utilize* and its variations to substitute for *use*. *Utilize* is an unnecessary word because *use* can be substituted in any sentence and mean the same thing. Notice the revisions that follow these sentences:

   Gobbledygook: The project will utilize the Acme trucks.

   Revision: The project will use the Acme trucks.

   Gobbledygook: By utilizing the gravel already on the site, we can save up to $5000.

   Revision: By using the gravel already on the site, we can save up to $5000.

   Gobbledygook: Through the utilization of this high-speed press, we can meet the deadline.

   Revision: Through the use of this high-speed press, we can meet the deadline.

   The English language has many short, precise words. Use them.
3. Writing more elaborate words than are necessary for document purpose and readers. The writer who says "deciduous perennial, genus

Ulmus" instead of "elm tree" is probably engaged in gobbledygook unless his or her readers are biologists.

This paragraph from a memo sent by a unit supervisor to a division manager is supposed to explain why the unit needs more architects:

> In the absence of available adequate applied man-hours, and/or elapsed calendar time, the quality of the architectural result (as delineated by the drawings) has been and will continue to be put ahead of the graphic quality or detailed completeness of the documents. In order to be efficient, work should be consecutive not simultaneous, and a schedule must show design and design development as scheduled separately from working drawings on each project, so that a tighter but realistic schedule can be maintained, optimizing the productive efforts of each person.

You will not be surprised to learn that the division manager routinely threw away any memos he received from this employee without reading them.

Fad Words

Fad words are words that arise from a particular field and then are used to apply to situations not in the original context of the word. The words become popular and drift into general use. The expression *bottom line*, for example, is from accounting, where it means the final figure showing profit or loss. Now people often use the expression to mean any final determination, deadline, or result:

> The bottom line is that shipments have to begin Monday, or the company will close three refineries.

In some cases, the fad word is not connected to any field, but instead is a common word that people begin using in a new way. Frequently, the word is a noun that people have converted to a verb by adding verb endings to it. Here are examples of current fad words:

1. *Access* is a noun meaning the ability to approach something or someone:

   He has access to important people in the White House.

   The computer industry uses *access* as a verb meaning to retrieve files from a database:

   The secretary accessed the medical records.

2. *Impact* is a noun meaning the force with which one object strikes another or the influence of one thing on another:

   The impact of the boulder demolished the automobile.

   Many people in business now use *impact* as a verb meaning to affect or to influence:

   October's sales drop will impact our dealers.

   Impact used as a verb probably results from the uncertainty many people have about the difference between *affect,* meaning to influence, and effect, meaning result.

3. *Reference* is a noun meaning the mention of something or someone:

   The quality-control engineer included a reference to last year's test in his report.

   Reference is a particularly confusing fad word because it is used as a substitute for a variety of words. Here are some samples:

   Fad word: The two chemists *referenced* six previous studies in their article.

   Revision: The two chemists cited six previous studies in their article.

   Fad word: *Reference* your request for. . . .

   Revision: Referring to your request for. . . .

   Fad word: The vice president *referenced* the dam project in his speech.

   Revision: The vice president mentioned the dam project in his speech.

   Fad word: My assistant is *referencing* the manual before printing.

   Revision: My assistant is editing the manual before printing.

   Fad word: Clear referencing of load sources and allowable stress is required in each calculation.

   Revision: Clear statements of load sources and allowable stress are required in each calculation.

Fad words can be as confusing as jargon if readers are not familiar with the meaning of the words in a particular context. Avoid using the latest fad words whenever a clearer, more standard word is available.

Wordiness

Wordiness refers to writing that includes superfluous words that add no information to a document. Wordiness stems from several causes:

1. *Doubling terms.* Using two or more similar terms to make one point creates wordiness. These doubled phrases can be reduced to one of the words:

   final ultimate result (result)

the month of July (July)
finished and completed (finished or completed)
unique innovation (innovation)

2. *Long phrases for short words.* Wordiness results when writers use long phrases instead of one simple word:

   at this point in time (now)
   in the near future (soon)
   due to the fact that (because)
   in the event that (if)

3. *Unneeded repetition.* Repeating words or phrases too often increases wordiness and disturbs sentence clarity:

   The Marshman employees will follow the scheduled established hours established by their supervisors, who will establish the times required to support production.

   This sentence is difficult to read because *established* appears three times, once as an adjective and twice as a verb. Here is a revision:

   The Marshman employees will follow the schedule established by their supervisors, who will set times required to support production.

4. *Empty sentence openings.* These sentences are correct grammatically, but their structure contributes to wordiness:

   It was determined that the home office sent income statements to all pension clients.
   It is necessary that the loader valves (No. 3C126) be scrapped because of damage.
   It might be possible to open two runways within an hour.
   There will be times when the videotapes are not available.
   There are three new checking account options available from Southwest National Bank.
   There is a full kitchenette with microwave and wet bar in every resort condo.

   These sentences have empty openings because the first words (*There is/There are/It is/It was/It might be*) offer the reader no information, and the reader must reach the last half of the sentence to find the subject. Here are revisions that eliminate the empty openings:

   The home office sent income statements to all pension clients.
   The loader valves (No. 3C126) must be scrapped because of damage.
   Two runways might open within an hour.
   Sometimes the videotapes will not be available.
   Southwest National Bank is offering three new checking account options.
   Every resort condo has a full kitchenette with microwave and wet bar.

Remember that wordiness does not refer to the length of a document, but to excessive words in the text. A 40-page manual containing none of the constructions discussed here is not wordy, but a half-page memo full of them is.

CHAPTER SUMMARY

This chapter discusses two stages of revision—global revision and fine-tuning revision. Remember:

- Revision consists of two distinct stages—global revision, which focuses on content and organization, and fine-tuning revision, which focuses on sentence structure, word choice, correctness, and the mechanics of technical style.
- For global revision, writers check content to see if there are enough details, definitions, and emphasis on the major points.
- For global revision, writers also check organization of the full document and of individual sections for (1) overall organization, (2) introductions and conclusions, (3) paragraphs, (4) headings, and (5) format.
- For fine-tuning revision, writers check sentence structure for overloaded sentences, clipped sentences, overly long sentences, and overuse of passive voice.
- For fine-tuning revision, writers also check word choice for unnecessary jargon, sexist language, overly abstract words, gobbledygook, fad words, and general wordiness.

SUPPLEMENTAL READINGS IN PART TWO

Bagin, C. B., and Van Doren, J. "How to Avoid Costly Proofreading Errors," *Simply Stated*.

Buehler, M. F. "Defining Terms in Technical Editing: The Levels of Edit as a Model," *Technical Communication*.

"Eliminating Gender Bias in Language," *Simply Stated*.

"The Practical Writer," *The Royal Bank Letter*.

Model 5-1 Commentary

This model is the first draft of a company bulletin outlining procedures for security guards who discover a fire while on patrol. The bulletin will be part of the "Security Procedure Manual," which contains general procedures for potential security problems such as fire, theft, physical fights, and trespassing. The bulletins are not instructions to be used on the job during a security incident. Rather, the guards are expected to read the bulletins, become familiar with the procedures, and then follow them if a security situation occurs.

Discussion

Identify the major topics the writer covers in this bulletin. Discuss the general organization and how well it suits the purpose of the readers.

Draft 1

BULLETIN 46-RC Fire Emergencies--Plant Security Patrols

Following are procedures for fire emergencies discovered during routine security patrols. All security personnel should be familiar with these procedures.

General Inspection

Perimeter fence lines, parking lots, and yard areas should be observed by security personnel at least twice per shift. Special attention should be given to outside areas during the dark hours and nonoperating periods. It is preferable that the inspection of the yard area and parking lots be made in a security patrol car equipped with a spot light and two-way radio communications. Special attention should be given to the condition of fencing and gates and to yard lighting to assure that all necessary lights are turned on during the dark hours and that the system is fully operative. Where yard lights are noted as burned out, the guard should report these problems immediately to the Plant Engineer for corrective action and maintain follow-up until the yard lighting is in full service. Occasional roof spot checks should be made by security patrols to observe for improper use of roof areas and fire hazards, particularly around ventilating equipment.

Discovery

When a guard discovers a fire during an in-plant security patrol, he should immediately turn in an alarm. This should be done before any attempt is made to fight the fire because all too frequently guards think they can put out the fire with the equipment at hand, and large losses have resulted.

Whenever possible, the guard should turn in the alarm by using the alarm box nearest the scene of the fire. If it is necessary to report the fire on the plant telephone, he should identify the location accurately so that he may give this information to the plant fire department or to the guard on duty at the Security Office, who will summon the employee fire brigade and possibly the city fire department.

Once he has turned in the alarm, the guard should decide whether he can effectively use the available fire protection equipment to fight the fire. If the fire is beyond his control, he should proceed to the main aisle of approach where he can make contact with the fire brigade or firemen and direct them to the scene of the fire. If automatic sprinklers have been engaged, the men in charge of the fire fighting crew will make the decision to turn off the system.

Model 5-2 Commentary

This model is the writer's second draft of the security procedures bulletin in Model 5-1.

Discussion

Identify the changes the writer has made in this second draft. Discuss both the global and the fine-tuning revisions and how they will help the security guards understand and use the bulletin information.

Draft 2

BULLETIN 46-RC Fire Emergencies--Plant Security Patrols

All security guards should be familiar with the following procedures for fire emergencies that occur during routine security patrols.

Inspection Areas

Inspection of the yard area and parking lots should be made in a security patrol car equipped with a spot light and two-way radio communications. At least twice per shift, the guard should inspect (1) fence lines, (2) parking lots, and (3) yard areas. These outside areas require special attention during the dark hours and nonoperating periods. The guard should inspect carefully the condition of the fencing and gate closures. The yard lighting should be checked to ensure that all necessary lights are turned on during the dark hours and that the system is fully operative. If the guard notes that yard lights are burned out, he should report these problems to the Plant Engineer for corrective action and maintain follow-up until the yard lighting is in full service.

Roof Checks

Occasional roof spot checks should be made by the guard to observe improper use of roof areas and any fire hazards, particularly around ventilating equipment.

Fires

When a guard discovers a fire during an in-plant security patrol, he should immediately turn in an alarm before he makes any attempt to fight the fire. In the past, attempts to fight the fire without sounding an alarm have resulted in costly damage and larger fires than necessary.

Whenever possible, the guard should turn in the alarm at the alarm box nearest the scene of the fire. If he must use a plant telephone, he should identify the location accurately. This information helps the plant fire department or the guard on duty at the Security Office, who must summon the employee fire brigade or the city fire department.

Once he has turned in the alarm, the guard should decide whether he can fight the fire with the available equipment. If the fire is beyond his control, he should go to the main entrance of the area so that he can direct the fire fighters to the scene of the fire. If the automatic sprinklers are on, the supervisor of the fire fighting crew will decide when to turn off the system.

Model 5-3 Commentary

This model is the writer's third draft of the security bulletin.

Discussion

1. Identify both the global revisions and the fine-tuning revisions in this third draft, and discuss how they will help the security guards understand and use the information.

2. In groups, develop a fourth draft of this bulletin, making further changes that will aid the security guards who must use the information. *OR*, if your instructor prefers, rewrite one section of this model and make further fine-tuning revisions according to the guidelines discussed in this chapter.

3. Follow the guidelines for fine-tuning revisions in this chapter and revise one section of one of the following models: Model 1-4, Chapter 1; Model 2-2, Chapter 2.

Draft 3

BULLETIN 46-RC Fires--Plant Security Patrols

All security guards should be familiar with the following procedures for fires that occur during routine security patrols.

Inspection Areas

The guard should inspect the outside areas in a security patrol car equipped with a spot light and two-way radio. The outside areas require special attention after dark and during nonoperating hours. At least twice per shift, the following should be inspected:

Fence Lines--The guard should check fences for gaps or breaks in the chain and gate closures for tight links and hinges.

Parking Lots and Yards--The guard should monitor the lighting in all parking lots and open yards to ensure that all necessary lights are on after dark and that the system is fully operative. If any lights are burned out, the guard should report the locations to the Plant Engineer and maintain follow-up until the lighting is repaired.

Occasional spot checks during a shift should be made of the following:

Roofs--The guard should check for improper use of the roof areas and any fire hazards, particularly around ventilating equipment.

Fires

Discovery--Upon discovering a fire during an in-plant security patrol, the guard must turn in an alarm before attempting to fight the fire. In the past, attempts to fight the fire before turning in an alarm have resulted in costly damage and larger fires than necessary.

Alarms--Whenever possible, the guard should turn in the alarm at the alarm box nearest the scene of the fire. If a plant telephone is more convenient or safer, the guard should be sure to identify the fire location accurately. This information is needed by the plant fire department and the guard at the Security Office, who must summon the employee fire brigade or the city fire department.

Fire Fighting--Once the alarm is in, the guard should decide whether to fight the fire with the available equipment. If the fire is beyond control, the guard should go to the main entrance to the fire area and direct the fire brigade or city fire fighters to the scene of the fire. If the automatic sprinklers are on, the supervisor of the fire brigade or of the fire department crew will decide when to turn off the system.

CHAPTER 5 Exercises

1. For practice, identify the style problems in these excerpts from on-the-job writing, and then revise for greater clarity and ease of reading. If you find jargon, assume that it is not appropriate in this instance.
   a. It has been decided that due to the fact that we have had an extreme number of requests for Leave of Absence without pay, during this time and in the near future, no leaves will be honored by the company.
   b. There seems to be a distinct trend in the department away from appropriate dress which is appropriate for office work. Since this is a County agency, the gentlemen in the department should wear appropriate ties and other appropriate dress.
   c. Referencing the new travel arrangements for domestic travel during the month of August, the Fly Right Travel Agency will no longer be utilized until detailed payment information is reported. Costs have impacted heavily on working funds, and all travel must be authorized in the event that travel is required.
   d. The transition team will be headed by Mr. Anthony Hunt and Gina Ashton. Mr. Hunt is an international entrepreneur, recognized for his shrewd grasp of world finance. He will work with Gina closely during the next four months to direct the merger of our two companies. Gina is a consultant from Winkler Management Consultants, Inc.
   e. Checked status of air handlers. Smaller units may fit. Large size a problem. Tom checking basement duct and downsize unit to see if O.K.
   f. Our new *Benefits* newsletter has been developed and designed to meet a recognized and stated need described by employees. We wish to acknowledge this need and plan to present a clear and concise understanding of the full total benefits package. Please feel free to call our office with questions at any time.
   g. A student who was admitted to the College of Business and successfully completed Business Law I and Business Law II prior to fall 1990, has fulfilled the law requirement for all accounting majors, finance, marketing, industrial accounting, or administration, but the student must have been classified as a College of Business student at the time the courses were completed unless subsequently course waiver was received.
   h. Cardboard packing between units for shipping not heavy enough. Noted scratches on the springs caused by screw tube wearing through cardboard. A 14-ply cardboard sheet on all future shipments should be used.
   i. In response to your letter of August 6, attached find stats on tires in addition to info in letter of August 23. Actual mileage will be sent in

the near future with tread-depth measurements on tires not listed on forms forwarded to you earlier.

j. Communication with other staff members and with various members of the construction department has been, due to the physical nature of our department, forced on demand, rather than attended to at our discretion, and I suggest additional physical measures in the form of doors, partitions, etc. to control the working relationships within the department.

k. The pneumatic lift steno arm chair will accommodate any of our secretaries. The girls will find that the seat height adjusts instantaneously, using a pneumatic counterbalance system with high-density, sculptured foam cushions on a contoured frame that adjusts from 17 in. to 22 in. in height, with an arm height from the floor of 26 in. to 31 in., including five easy-rolling casters on a heavy-gauge, five-prong steel base concealing a pivot rod.

l. Emphasis should be placed on wearing earmuffs by welders who are exposed to noise levels that have been determined to be excessive, and all efforts should be made to persuade welders that earmuffs are beneficial. If a welder requests earplugs, he may have those instead, as long as he has them properly fitted.

m. It was discovered that design changes impacted our bottom line negatively, and as a result at this time, it was decided that all orders would be held until further notice.

2. This memo was written by the director of a county historical society to the society staff. The director believed the staff was not working hard enough, and he wanted his memo to encourage better work habits. Revise the memo following the guidelines discussed in this chapter.

TO: All Staff
FROM: Steven Weaver
SUBJECT: Problems

There seems to have developed during the last few weeks, a general problem with attending to our work on a regular and timely basis. First, there seems to be on at least some people's part, and particularly the gals in word processing, a general casualness about beginning work on time which is evidenced by general milling about the coffee equipment and gathering the elements for breakfast. Although this problem has been noted most often at starting time, it should be recognized that there is a considerable amount of chatter among some of the staff from time to time during the course of the day. It is acknowledged that a certain amount of verbal exchange is generally useful for maintaining a relaxed atmosphere in office surroundings, but it can interfere with normal working schedules and inhibit the accomplishment of productive business in spite of good intentions.

The 15-minute breaks during the course of the morning and the afternoon should be strictly adhered to because violating them is an unprofessional act, although possibly unintentional, and their coming to an end should be duly noted and even more carefully anticipated. If, from time to time, special circumstances arise, the need should be discussed so proper variance can be established.

Finally, the utilization of the telephonic communication for personal business is something that should be regarded as a normal benefit but abuses of this utilization will necessitate strict controls. Excessive numbers of calls and excessive times engaged in such personal business on office equipment may interfere with the work of the Archives and must not be permitted to interfere with our work. A reasonable anticipation on the part of the caller of the needs of the department regarding the flow of incoming calls will go a long way toward fulfilling the intent of the regulations established by the County Commissioner in the employee regulation handbook. I believe it is true of every member of this staff that his first concern is the quality of work and the mission to serve the public diligently and faithfully.

3. Bring to class a document you have written for another class. Identify the reader and purpose of this document on the top of the first page. Exchange documents with another student, and then check the other document for the style problems discussed in this chapter. Revise one page of the document. Submit your revision and the original to your instructor.

CHAPTER 6

Definition

Understanding Definitions
Writing Informal Definitions
Writing Formal Sentence Definitions
Writing Expanded Definitions
- Cause and Effect
- Classification
- Comparison or Contrast
- Description
- Development
- Etymology
- Examples
- Method of Operation or Process
- Negation
- Partition

Placing Definitions in Documents
- Within the Text
- In an Appendix
- In Footnotes
- In a Glossary

Chapter Summary
Supplemental Readings in Part Two
Endnotes
Models
Exercises

Understanding Definitions

Definition is essential in good technical writing because of the specialized vocabulary in many documents. All definitions are meant to distinguish one object or procedure from any that are similar and to clarify them for readers by setting precise limits on each expression. In writing definitions, use vocabulary understandable to your readers. Defining one word with others that are equally specialized will frustrate your readers; no one wants to consult a dictionary in order to understand an explanation that was supposed to make referring to a dictionary unnecessary.

Certain circumstances always call for definitions:

1. When technical information originally written for expert readers is revised for nonexpert readers, the writer must include definitions for all terms that are not common knowledge.
2. A document with readers from many disciplines or varied backgrounds must include definitions enabling readers with the lowest level of knowledge to understand the document.
3. All new or rare terms should be defined even for readers who are experts in the subject. Change is so rapid in science and technology that no one can easily keep up with every new development.
4. When a term has multiple meanings, a writer must be clear about which meaning is being used in the document. The word *slate*, for example, can refer to a kind of rock, a color, a handheld chalkboard, or a list of candidates for election.

As discussed in Chapter 3, writers frequently use definitions in reports addressed to multiple readers because some of the readers may not be familiar with the technical terminology used. Definitions in reports may range from a simple phrase in a sentence to a complete appendix at the end of the report. In all cases, the writer decides just how much definition the reader needs to be able to use the report effectively. Definitions also may be complete documents in themselves; for example, entries in technical handbooks or science dictionaries are expanded definitions. Furthermore, individual sections of manuals, technical sales literature, and information pamphlets are often expanded definitions for readers who lack expert knowledge of the field but need to understand the subject.

The three types of definitions are (1) informal, (2) formal sentence, and (3) expanded. All three can appear in the same document, and your use of one or another depends on your analysis of the needs of your readers and your purpose in using a specific item of information.

Writing Informal Definitions

Informal definitions explain a term with a word or a phrase that has the same general meaning. Here are some examples of informal definitions:

> Contrast is the difference between dark and light in a photograph.
>
> Terra cotta—a hard, fired clay—is used for pottery and ornamental architectural detail.
>
> Viscous (sticky) substances are used in manufacturing rayon.
>
> Leucine, an amino acid, is essential for human nutrition.

The first definition is a complete sentence; the other definitions are words or phrases used to define a term within a sentence. Notice that you can set off the definition with dashes, parentheses, or commas. Place the definition immediately after the first reference to the term. Informal definitions are most helpful for nonexpert readers who need an introduction to an unfamiliar term. Be sure to use a well-known word or phrase to define a difficult term.

The advantage of informal definitions is that they do not significantly interrupt the flow of a sentence or the information. Readers do not have to stop thinking about the main idea in order to understand the term. The limitation of informal definitions is that they do not thoroughly explain the term and, in fact, are really identifications rather than definitions. If a reader needs more information than that provided by a simple identification, use a formal sentence definition or an expanded definition.

Writing Formal Sentence Definitions

A *formal sentence definition* is more detailed and rigidly structured than an informal definition. The formal definition has three specific parts.

Term		*Group*	*Distinguishing features*
An *ace*	is	a tennis serve	that is successful because the opponent cannot reach the ball to return it.

The first part is the specific term you want to define, followed by the verb *is* or *are*, *was* or *were*. The second part is the group of objects or actions to which the term belongs. The third part consists of the distinguishing features that set this term off from others in the same group. In this example, the group part *a tennis serve* establishes both the sport and the type of action the term refers to. The distinguishing features part explains the specific quality that separates this tennis serve from others—the opponent cannot reach it.

These formal definitions illustrate how the group and the distinguishing features become more restrictive as a term becomes more specific:

Term		*Group*	*Distinguishing features*
A *firearm*	is	a weapon	from which a bullet or shell is discharged by gunpowder.
A *rifle*	is	a firearm	with spiral grooves in the inner surface of the gun barrel to give the bullet a rotary motion and increase its accuracy.
A *Winchester*	is	a rifle	first made about 1866, with a tubular magazine under the barrel that allows the user to fire a number of bullets without reloading.

In writing a formal sentence definition, be sure to place the term in as specific a group as possible. In the preceding samples, a rifle could be placed in the group *weapon,* but that group also includes clubs, swords, and nuclear missiles. Placing the rifle in the group *firearm* eliminates all weapons that are not also firearms. The distinguishing features part then concentrates on the characteristics that are special to rifles and not to other firearms. The distinguishing features part, however, cannot include every characteristic detail of a term. You will have to decide which features most effectively separate the term from others in the same group and which will best help your readers understand and use the information.

When writing formal sentence definitions, remember these tips:

1. Do not use the same key word in the distinguishing features part that you used in one or both of the other two units.

   Poor: A blender is an electric kitchen appliance that blends liquids or solid foods with propellerlike blades.

   Better: A blender is an electric kitchen appliance that mixes, chops, grinds, or purees liquids or solid foods with a single set of propellerlike blades at the base of the single food container.

   Poor: An odometer is a measuring instrument that measures the distance traveled by a vehicle.

   Better: An odometer is a measuring instrument that records the distance traveled by a vehicle.

   In the first example, the distinguishing features part includes the verb *blend,* which is also part of the term being defined. In the second example, the distinguishing features part includes the verb *measures,* which is used in the group part. In both cases, the writer must revise to avoid repetition that will send readers in a circle. Sometimes you can assume that your readers know what a general term means. If you are certain your readers understand the term *horse* and your purpose is to clarify the term *racehorse*, you may write, "A racehorse is a horse that is bred or trained to run in competition with other horses."

2. Do not use distinguishing features that are too general to adequately specify the meaning of the term.

   Poor: Athena was a Greek goddess who was a daughter of Zeus.

   Better: Athena was a Greek goddess who, as a daughter of Zeus, leaped fully grown out of Zeus's forehead.

   Poor: A staple is a short piece of wire that is bent so as to hold papers together.

   Better: A staple is a short piece of wire that is bent so both ends can pierce several papers and fold inward, binding the papers together.

   In both examples, the distinguishing features are not restrictive enough. Zeus had many daughters, and a paper clip is also a device that binds papers together. Remember that the distinguishing features part of a definition must provide enough information to isolate the term from its group.

3. Do not use distinguishing features that are too restrictive.

   Poor: A tent is a portable shelter made of beige canvas in the shape of a pyramid, supported by poles.

   Better: A tent is a portable shelter made of animal skins or a sturdy fabric and supported by poles.

   Poor: A videotape is a recording device made of a magnetic ribbon of material 3/4 in. wide and coated plastic that registers both audio and visual signals for reproduction.

   Better: A videotape is a recording device made of a magnetic ribbon of material, usually coated plastic, that registers both audio and visual signals for reproduction.

   In the first example, the distinguishing features are too restrictive because not all tents are pyramid-shaped or made of beige canvas. The second example establishes only one size for a videotape, but videotapes come in several sizes, so size is not an appropriate distinguishing feature. Do not restrict your definition to only one brand or one model if your term is meant to cover all models and brands of that particular object.

4. Do not use *is when, is where,* or *is what* in place of the group part in a formal definition.

Poor: A tongue depressor is what medical personnel use to hold down a patient's tongue during a throat examination.

Better: A tongue depressor is a flat, thin, wooden stick used by medical personnel to hold down a patient's tongue during a throat examination.

Poor: Genetic engineering is when scientists change the hereditary code on an organism's DNA.

Better: Genetic engineering is the set of biochemical techniques used by scientists to move fragments from the genes of one organism to the chromosomes of another to change the hereditary code on the DNA of the second organism.

In both examples, the writer initially neglects to place the term in a group before adding the distinguishing features. The group part is the first level of restriction, and it helps readers by eliminating other groups that may share some of the distinguishing features. The first example, for instance, could apply to any object shoved down a patient's throat. The poor definition of genetic engineering refers only to the result and does not clarify the term as applying to specific techniques that will produce that result.

Writing Expanded Definitions

Expanded definitions can range from one paragraph in a report or manual to an entry several pages long in a technical dictionary. Writers use expanded definitions in these circumstances:

1. When a reader must fully understand a term to successfully use the document. A patient who has just been diagnosed with hypoglycemia probably will want a more detailed definition in the patient information booklet than just a formal sentence definition. Similarly, a decision maker who is reading a feasibility study about bridge reconstruction will want expanded definitions of such terms as *cathodic protection* or *distributed anode system* in order to make an informed decision about the most appropriate method for the company project.
2. When specific terms, such as *economically disadvantaged* or *physical therapy*, refer to broad concepts and readers other than experts need to understand the scope and application of the terms.
3. When the purpose of the document, such as a technical handbook or a science dictionary, is to provide expanded definitions to readers who need to understand the terms for a variety of reasons.

A careful writer usually begins an expanded definition with a formal sentence definition and then uses one or more of the following strategies to enlarge the definition with more detail and explanation.

Cause and Effect

Writers use the cause-and-effect (or effect-and-cause) strategy to illustrate relationships among several events. This strategy is effective in expanded definitions of terms that refer to a process or a system. The following paragraph is from a student report about various types of exercises used in athletic training. The definition explains the effect aerobic exercise has on the body.

> Aerobic exercise is a sustained physical activity that increases the body's ability to obtain oxygen, thereby strengthening the heart and lungs. Aerobic effect begins when an exercise is vigorous enough to produce a sustained heart rate of at least 150 beats a minute. This exercise, if continued for at least 20 minutes, produces a change in a person's body. The lungs begin processing more air with less effort, while the heart grows stronger, pumping more blood with fewer strokes. Overall, the body's total blood volume increases and the blood supply to the muscles improves.

Classification

Classification is used in expanded definitions when a writer needs to break a term into types or categories and discuss the similarities and differences among the categories. The following expanded definition is from a report written by a public television station manager to a donors group, soliciting funds for equipment purchases. The writer uses classification to define videotape machines. Note that the writer does not begin with a formal sentence definition and that the distinguishing features actually appear in the second sentence.

> The videotape recorder/player is an electronic device that is one of the most important components of television station operation. The videotape recorder creates distinct video and audio tracks during recording; then, during playback, the videotape player "reads" the images and sound information from the recorded tracks. Videotape machines fall into three categories—production, editing, and playback. A production recorder preserves images on tape for later editing. The editing recorder plays back an image on one machine and records it on a second machine. The playback machine must reproduce the best signal possible from a recorded tape so that the edit recorder can capture the best possible image and record it on another tape. The requirements for each of these VCRs vary because each is expected to do a different job.

Comparison or Contrast

Writers use comparison or contrast to show readers the similarities or differences between the term being defined and another commonly understood term. When using this strategy, be sure that the term you are using for comparison or contrast is familiar to your readers. The following definition is from a sales flyer written to consumers who may purchase a computer. The flyer compares a floppy disk to a record album.

> A floppy disk is a data storage device that is thin, flat, round, $5\frac{1}{4}$ in. in diameter, and protected by a square outer casing. The disk has many similarities to a record album. Both have grooves for storing recorded information, and both have the same thin, round shape, but the record album is larger and not as flexible as a floppy disk. An album is placed on a turntable to play its information, and a floppy disk is placed in a disk drive for the same purpose. The disk drive works much like a record player in that the floppy disk spins around and a read/write head (like the record player's needle) picks up the information from the disk's grooves.

Description

A detailed description of a term being defined will expand the definition. Reading about the physical properties of a term often helps readers to visualize the concept or object and remember it more readily. Chapter 7 provides a detailed discussion of developing physical descriptions. This excerpt from an encyclopedia illustrates the use of physical description as part of an expanded definition of *bison*:

> Head and body length is 2100 to 3500 mm, tail length is 500 to 600 mm, shoulder height is 1500 to 2000 mm, and weight is 350 to 1000 kg. Males average larger than females. The pelage of the head, neck, shoulders, and forelegs is long, shaggy, and brownish black. There is usually a beard on the chin. The remainder of the body is covered with short hairs of a lighter color. Young calves are reddish brown. The forehead is short and broad, the head is heavy, the neck is short, and the shoulders have a high hump. The horns, borne by both sexes, are short, upcurving, and sharp. Females have smaller humps, thinner necks, and thinner horns than males.[1]

Development

Explaining how the meaning of a term has changed over time may be an effective strategy for expanding a definition. You may include (1) discovery or invention of the concept or object, (2) changes in the components or design of the concept or object, or (3) changes in the use or function of the concept or object over time. This excerpt from an encyclopedia definition of smog explains the changes in the meaning of the term over the past 100 years:

> The word *smog* can be defined in several ways. It was coined near the beginning of the century by H. A. Des Voeux to refer to a combination of coal smoke and fog, the particles of the smoke often serving as nuclei on which the water vapor condensed. Later, the term was used to refer to any dirty urban atmosphere. A unique type of dirty atmosphere was recognized in the late 1940s and early 1950s and came to be known as photochemical smog. It can be defined as that type of air pollution which owes may of its properties to the products of photochemical reactions involving the vapor of various organic substances, especially hydrocarbons, oxides of nitrogen, and atmospheric oxygen.[2]

Etymology

Etymology is the study of the linguistic history of individual words. Although etymology is rarely used as the only strategy in an expanded definition, writers sometimes include it with other strategies. The following definition from a reference book on word histories is written for the general reader. Note that the writer does not begin with a formal sentence definition, probably because the general reader is assumed to be familiar with a tulip.

> Although we associate tulips with Holland windmills, the history of the word takes us on an odyssey to the Middle East, where the tulip is associated with turbans and whence the flower was brought to Europe in the sixteenth century. The word *tulip,* which earlier in English appeared in such forms as *tulipa* or *tulipant,* comes to us by way of French *tulipe* and its obsolete form *tulipan* or from Modern Latin *tulipa,* from Turkish *tülbend,* "tulip, turban," derived from Persian *dulband,* "turban." (Our word *turban,* first recorded in English in the sixteenth century, can also be traced back through Turkish to Persian *dulband.*) The word for turban in Turkish was used for the flower because a fully opened tulip was thought to resemble a turban.[3]

Examples

Another way to expand a definition is to provide examples of the term. This strategy is particularly effective in expanding definitions for nonexpert readers who need to understand the variety included in one term. This definition of a combat medal is taken from a student examination on military history:

> A combat medal is a military award given to commemorate an individual's bravery under fire. Medals are awarded by all branches of the armed forces and by civilian legislative bodies. The Congressional Medal of Honor, for in-

stance, is awarded by the President in the name of Congress to military personnel who have distinguished themselves in combat beyond the call of duty. The Purple Heart is awarded by the branches of the armed services to all military personnel who sustain wounds during combat. The Navy Cross is awarded to naval personnel for outstanding heroism against an enemy.

Method of Operation or Process

Another effective strategy for expanding a definition is to explain how the object represented by the term works, such as how a scanner reads images, stores them on disks, and prints them. Also, a writer may expand a definition of a system or natural process by describing the steps in the process, such as how a drug is produced or how a hurricane occurs. Chapter 8 covers developing process explanations.

This excerpt, from a pamphlet about first aid procedures, expands the definition of mouth-to-mouth artificial respiration by explaining the process:

> The purpose of mouth-to-mouth artificial respiration is to maintain an open air passage to the lungs. The rescuer first places the patient on his back and clears the mouth of any foreign objects. Then the patient's head is tilted back until the chin juts upward, thus opening the air passage. The rescuer next places his mouth over the patient's, holds the patient's nostrils shut, and blows a deep breath into the patient's mouth. The patient's mouth is then uncovered to allow exhalation. This cycle is repeated 12 times a minute for an adult victim. If the victim is a child, the rescuer blows shallow breaths 15–20 times a minute.

Negation

Writers occasionally expand a definition by explaining what a term *does not* include. This definition of a company warranty uses negation to explain the limits of the warranty:

> This warranty for your HelpRite freezer is our guarantee that this product is free of all mechanical and electrical defects for one full year from the date of purchase. The manufacturer will make any necessary repairs during that period without charge. This warranty does not cover damage from (a) negligence, (b) misuse, (c) improper voltage, and (d) improper disassembly, repair, or alteration by unauthorized persons.

Partition

Separating a term into its main parts and explaining each part individually also can effectively expand a definition. This excerpt from an encyclopedia uses partition to explain the important parts of COBOL, a widely used programming language:

> A COBOL program contains the following four divisions:
>
> 1. The Identification Division, which identifies the program, its author (programmer), and date and provides other background information.
> 2. The Environment Division, which specifies the computer being used and the various input and output devices that are connected to it.
> 3. The Data Division, which describes the data to be processed by the program. It also describes the files from which any data is to be taken or to which it is to be transferred after processing.
> 4. The Procedure Division, which provides the detailed programming instructions that are to be followed by the computer in executing the program. This is the division that does the actual processing and is usually the largest division within the program.[4]

Placing Definitions in Documents

If a definition is part of a longer document, you need to place it so that readers will find it useful and nondisruptive. Include definitions (1) within the text, (2) in an appendix, (3) in footnotes, and/or (4) in a glossary.

Within the Text

Informal definitions or formal sentence definitions can be incorporated easily into the text of a document. If an expanded definition is crucial to the success of a document, it can also be included in the text. However, expanded definitions may interrupt a reader's concentration on the main topic. If the expanded definitions are not crucial to the main topic, consider placing them in an appendix or glossary.

In an Appendix

In a document intended for multiple readers, lengthy expanded definitions for nonexpert readers may be necessary. Rather than interrupt the text, place such expanded definitions in an appendix at the end of the document. Readers who do not need the definitions can ignore them, but readers who do need them can easily find them. Expanded definitions longer than one paragraph usually should be in an appendix unless they are essential to helping readers understand the information in the document.

In Footnotes

Writers often put expanded definitions in footnotes at the bottoms of pages or in endnotes listed at the conclusion of a long document. Such notes do not interrupt the text and are convenient for readers because they appear on the same page or within a few pages of the first reference to the term. If a definition is longer than one paragraph, however, it is best placed in an appendix.

In a Glossary

If your document requires both formal sentence definitions and expanded definitions, a glossary or list of definitions may be the most appropriate way to present them. Glossaries are convenient for readers because the terms are listed alphabetically and all definitions appear in the same location. Chapter 11 discusses preparation of glossaries.

CHAPTER SUMMARY

This chapter discusses how to write definitions of technical terms that readers need to understand. Remember:

- Definitions distinguish an object or concept from similar objects or concepts and clarify for readers the limits of the term.
- Definitions are necessary when (1) technical information for expert readers must be rewritten for nonexpert readers, (2) not all readers will understand the technical terms used in a document, (3) rare or new technical terms are used, and (4) a term has more than one meaning.
- Informal definitions explain a term with another word or a phrase that has the same general meaning.
- Formal sentence definitions explain a term by placing it in a group and identifying the features that distinguish it from other members of the same group.
- Expanded definitions explain the meaning of a term in a full paragraph or longer.
- Writers expand definitions by using one or more of these strategies: cause and effect, classification, comparison or contrast, description, development, etymology, examples, method of operation or process, negation, and partition.
- Definitions may be placed within the text, in an appendix, in footnotes, or in a glossary, depending on the needs of readers.

SUPPLEMENTAL READINGS IN PART TWO

Benson, P. J. "Writing Visually: Design Considerations in Technical Publications," *Technical Communication*.

"Is It Worth a Thousand Words?" *Simply Stated*.

"Six Graphic Guidelines," *Simply Stated*.

ENDNOTES

1. From *Walker's Mammals of the World* (Baltimore, Md.: Johns Hopkins University Press, 1983), p. 1254.
2. From *Meteorology Source Book* (New York: McGraw-Hill, 1988), p. 282.
3. From *Word Mysteries and Histories* (Boston: Houghton Mifflin Company, 1986), p. 256.
4. From *The Prentice-Hall Encyclopedia of Information Technology* (Englewood Cliffs, N.J.: Prentice-Hall, 1987), pp. 51–52.

Model 6-1 Commentary

This expanded definition of *hacking* is from *The Prentice-Hall Encyclopedia of Information Technology* (1987). The writer begins with a formal sentence definition, but it is rather general. The writer then extends the definition to include another distinguishing feature—unauthorized attempts to breach computer security systems.

To illustrate *hacking,* the writer includes examples of two real-life instances and one fictional instance. In the final section of the definition, the writer provides a new, informal definition of *hacking* that is more closely related to the examples than the original definition was.

Discussion

1. Consider why people use encyclopedias to get information, and discuss why the writer probably chose these examples of *hacking*. How do they support the conclusion of this expanded definition?

2. Write a more technical definition of *hacking* for students in computer science. Include one example of the term that shows in detail how a hacker proceeds.

Hacking

Hacking, in its broadest definition, is a term that applies to any use of a computer or computer system that has no predefined objective. The definition of hacking has been extended to apply to unauthorized attempts to breach security techniques set up to protect computer systems. Hacking is performed by a very large number of individuals, both computer amateurs and professionals. Hacking is usually performed without mischievous or criminal intent in mind; it is generally done simply because the hacker is curious or likes technical challenges.

Our primary focus in this topic is on the smaller group of individuals whose activities are a cause for genuine concern to computer professionals, users, and business and governmental organizations.

EXAMPLES OF HACKING

The cookie monster. A few years ago, students at MIT, working at their computer terminals, were occasionally startled when the word "COOKIE" mysteriously appeared on their display. Unless a student knew just how to respond, the "cookie monster" ran amuck. It proceeded to write "COOKIE" over and over again, until the display was literally filled. The underlying data was obliterated. In a final act of desperation, the cookie monster finally displayed "gimme cookie." By typing the word "cookie" on the keyboard, the student was finally able to persuade the monster to go away, only to reappear unpredictably at a later time at a different terminal.

The 414s. The cookie monster was a harmless "hack," well worth the distraction and a good laugh it gave to pressured students. However, not so humorous was the case of a group of Milwaukee youths whose exploits captured nationwide attention during the dog days of the summer of 1983. This group of teenagers became known as the 414s after their local telephone area code. They amused themselves by regularly penetrating a large number of computer systems throughout the United States and Canada. The "victims" included several banks, the Sloan-Kettering Cancer Center, and the Los Alamos National Laboratory. While no apparent damage was caused by this electronic joy ride, the vulnerability of major computer installations became painfully apparent to many business executives, government officials, military leaders, and the general public.

War games. As if to drive home the point, at about the time when the activities of the 414s was being publicized, the movie *War Games* was released throughout the country. In this film, a young hacker penetrates a U.S. air defense computer system, nearly triggering a thermonuclear war with the Soviet Union. While the scenario portrayed was immediately dismissed as utterly improbable, the message hit a raw nerve in many people, much as the film *Towering Inferno* a few years earlier had made suspect the fire resistive features of high-rise buildings.

HACKING REDEFINED

To redefine the term, hacking is simply the unauthorized access to computer facilities and computer-based data. For nearly all of its many participants, the objective of hacking in this sense is simply to respond to a challenge or to satisfy one's curiosity. Large computer systems are usually the intended target. No malfeasance is normally intended. When one of the Milwaukee 414s was queried by federal investigators about this point, he admitted that the first time he realized he might have been morally wrong was when the FBI agents knocked on his door. Youthful hackers do, for the most part, go on and lead useful and productive lives.

It is worth noting, with perhaps some dismay, that the best response to hackers is not to attempt to outsmart them with layers of electronic barriers. Such obstacles will merely increase the challenge. In fact, it has been observed that the hacker may be more disposed to cause damage if the stakes of the game have been raised.

Hacking has become much more of a factor in recent years. This is largely due to the proliferation of personal computers, countrywide information networks, and on-line computer systems. Hackers have established their own informal networks in which they share telephone numbers and passwords that allow them to gain access to computer systems. Sometimes this information is even posted on electronic bulletin boards for anyone to read.

Model 6-2 Commentary

This expanded definition of *greenhouse effect* is from *Meteorology Source Book* (New York: McGraw-Hill, 1988). The format of the encyclopedia requires that the title and the opening clause together constitute a formal sentence definition that includes the term, group, and distinguishing features.

Discussion

1. Identify the strategies the writer uses to expand this definition of *greenhouse effect*.

2. Notice the formal sentence definition created by the title and the opening clause. What rule of effective formal sentence definitions is violated? Write a formal sentence definition for *greenhouse effect* that fulfills all the principles of effective formal sentence definitions.

3. Underline all the terms you believe would have to be defined if you rewrote this expanded definition for freshmen in an introductory high-school science class. Write a formal sentence definition of one or more of these terms, as your instructor directs.

Greenhouse Effect

Lewis D. Kaplan

The effect created by Earth's atmosphere acting as the glass walls and roof of a greenhouse in trapping heat from the sun. The atmosphere is largely transparent to solar radiation, but it strongly absorbs the longer-wavelength (infrared) radiation from the Earth's surface. Much of this long-wave radiation is reemitted down to the surface, with the paradoxical result that the Earth's surface receives more radiation than it would if the atmosphere were not between it and the Sun.

The absorption of long-wave radiation is affected by small amounts of water vapor, carbon dioxide, ozone, nitrous oxide, methane, and other minor constituents of air, and by clouds. Clouds absorb, on average, about one-fifth of the solar radiation striking them, but unless they are extremely thin, they are almost completely opaque to infrared radiation. The appearance even of cirrus clouds after a period of clear sky at night is enough to cause the surface air temperature to increase rapidly by several degrees because of long-wave radiation emitted by the cloud.

The greenhouse effect is most marked at night, and usually keeps the diurnal temperature range below 20°F (11°C). Over regions where the water-vapor content of the air is low, however, the atmosphere is more transparent to infrared radiation, and cool nights may follow hot days.

The effect, described by John Tyndall in 1861, was among the earliest discoveries resulting from the rapid development of quantitative spectroscopy in the 1850s. The greenhouse analogy was attached to the effect at a much later date. Tyndall showed that water vapor was the major contributor to the effect, since its contribution to absorption in a cell of laboratory air was an order-of-magnitude greater than that of carbon dioxide and than that of the remainder of the air.

Although the term greenhouse effect has generally been used for the role of the whole atmosphere (mainly water vapor and clouds) in keeping the surface of the Earth warm, it has been increasingly associated in the twentieth century with the contribution of carbon dioxide, since Svante Arrhenius raised the problem in 1896 of increasing temperatures at the Earth's surface due to production of carbon dioxide by industrial combustion of fossil fuel. Arrhenius calculated that a doubling of carbon dioxide would raise the average temperature by about 9°F (5°C). Four decades later G. S. Callendar estimated that the industrial production of carbon dioxide since the 1880s could account for the continual rise in surface temperatures that had been observed thereafter.

The continual rise in surface temperature did not persist after Callendar's papers, despite a continuous increase in atmospheric carbon dioxide, raising the possibility that much of the earlier temperature rise could have been due to fallout of stratospheric dust from the great Krakatoa eruption. Some other uncertainties are the rate of absorption of excess carbon dioxide by the ecosystem and oceans, the rate of production of carbon dioxide by destruction of forests, the cooling effect of increasing aerosols, both anthropogenic and from revived volcanic activity, and feedback effects of water vapor and clouds.

In recent times the energy crisis and increasing sophistication of numerical atmospheric circulation models have led to greatly increased interest and investigation of the carbon dioxide greenhouse effect. In 1975 V. Ramanathan pointed out that the greenhouse effect is enhanced by the continued release into the atmosphere of chlorofluorocarbons, even though their combined concentration is less than a part per billion by volume. This is because of the spectral location of several of their absorption bands in the 8- to 14-micrometer region, where the atmosphere is relatively transparent and blackbody emission at terrestrial temperatures is high.

In contrast, the much more abundant constituent carbon monoxide, which is increasing from automobile exhausts and other combustion processes, does not have absorption bands in spectral regions where it can make a direct significant contribution to the greenhouse effect. It does play an indirect role in enhancing the greenhouse effect, however, by serving as a sink for the hydroxyl radical, which acts as a catalyst in modulating the increase of nitrous oxide from combustion processes and use of chemical fertilizer, and the increase of methane from biogenic and industrial production.

Model 6-3 Commentary

This memo is a request for two new pieces of equipment. Because the writer needs the reader's approval for any equipment purchases, she must clearly identify what she needs and why she needs it.

The writer begins by explaining why she needs the new equipment and names the items she wants to purchase. Then she provides an expanded definition of each machine.

The first definition begins with a formal sentence definition and then uses description to explain the videocassette cleaner's capabilities. The writer also uses partition when she describes separately one part of the machine—the printer.

The second definition also begins with a formal sentence definition. The writer again uses description to expand the definition. Because the reader might think the two machines overlap in function and, therefore, duplicate each other, the writer also adds a short comparison to point out how the two machines differ.

Discussion

1. Decide which terms in the two expanded definitions would have to be defined for a reader who knows nothing about video equipment. Write a formal sentence definition for one or more terms, as your instructor directs.

2. Discuss why the writer probably chose to include expanded definitions in her memo rather than attach the sales literature. How do the definitions support the writer's request?

TO: George Youngblood
Director, Human Resources
FROM: Phyllis Thaxton
Training Specialist
DATE: June 10, 1989
SUBJECT: New Equipment Purchases

As you know, the training unit is expanding rapidly, and in fall 1989 we will begin offering twice the number of training sessions we have offered in the past for both management and technical personnel. To handle this increase in activity and provide the same high-quality programs, I believe we need to purchase two new pieces of equipment—a videocassette cleaner and a bulk tape eraser.

The videocassette cleaner is a device that automatically inspects, cleans, polishes, and rewinds videotapes at 30 times the standard rewind speed. Most models have optional erase capability as well. With this machine, we can inspect and clean or erase, or any combination of these three functions. The unit can be used on both prerecorded and blank tapes and is available in 3/4-inch U-Matic, Beta/BetaCam, and VHS/M formats. The printer, which is part of the unit, provides an automatic paper tape printout, identifying the number and location of tape defects on audio and video tracks. Because we have a large collection of training tapes that are used frequently, the videocassette cleaner is essential to keep our training materials in good condition.

The bulk tape eraser is a machine that clears all reel-to-reel magnetic tapes from 150-mil to 2-inch widths. In addition, the machine handles cassettes, 16-inch reels, and all magnetic film stock. Models are available which feature overheat protection and eraser field control to minimize residual noise caused by the turn-off transients. This bulk eraser will enable us to clear tapes that we no longer use, leaving blanks. Blank tapes are preferable to recording over old material because they produce higher-quality video and audio, and there is no distracting "bleed" into old programs at the end of the freshly recorded program. Although most of our current programs are on videotape, we expect to use more varied media in the future, and having an eraser that can handle film as well as tape will be useful. In addition, the bulk eraser, unlike the videocassette cleaner, can handle various size tapes. Therefore, the two machines serve different purposes and do not duplicate each other significantly.

Both machines are available from Audio-Video Technology in Los Angeles. The videocassette cleaner costs $2400, and the bulk tape eraser costs $1295. I have brochures discussing the features of the machines, and I would be glad to discuss these with you.

Model 6-4 Commentary

This model is part of a brochure for patients who have just been diagnosed with multiple sclerosis. The brochure answers common questions that new patients who are not experts in the field are likely to have.

The opening is a formal sentence definition. The writer then expands the definition by describing how the disease attacks a victim—the process of the disease. The writer further expands the definition with examples of the symptoms connected with multiple sclerosis. Throughout the expanded definition, the writer also uses informal definitions by stating a condition in lay terms and then providing the technical term in parentheses—a reversal of the usual method.

Discussion

1. Discuss why the writer probably chose to reverse the usual method of informal definition in this expanded definition.

2. Read this excerpt and locate other definitions. What inconsistencies do you find in the style of presenting informal definitions? Discuss why these inconsistencies may have occurred.

3. In groups, draft an expanded definition of one of these conditions that commonly affect students—writer's block, test anxiety, or oral presentation panic. Assume you are writing the definition to be included in a brochure entitled "Common Student Problems" to be distributed to incoming freshmen by the campus counseling office.

What is multiple sclerosis?

MS is a neurological disease the cause of which is as yet undetermined. It attacks the myelin sheath, the coating or insulation around the message-carrying nerve fibers in the brain and spinal cord. Where myelin has been destroyed, it is replaced by plaques of hardened tissue *(sclerosis)*; this occurs in *multiple* places within the nervous system. At first, nerve impulses are transmitted with minor interruptions; later plaques may completely obstruct impulses along certain nerves.

MS varies tremendously from patient to patient in its symptoms, severity, and course. For many, it involves a series of attacks *(exacerbations)* and partial or complete recoveries *(remissions)*.

Symptoms vary according to the area of the nervous system affected. Different people suffer different symptoms, which may include one or more of the following: weakness, tingling, numbness, impaired sensation, lack of coordination, disturbances in equilibrium, double vision, involuntary, rapid movement of the eyes (nystagmus), slurred speech, tremor, stiffness or spasticity, weakness of limbs, and in more severe cases paralysis of the extremities or impaired bladder and bowel function.

AT WHAT AGE IS THE ONSET OF MS MOST COMMON?

Approximately two-thirds of those who have MS experience their first symptoms between the ages of twenty and forty. Sometimes, however, the diagnosis is not made until a person is in his forties or even fifties; frequently in these cases a detailed medical history reveals that symptoms did appear previously but that they were not long-lasting or bothersome enough to warrant medical attention and hence a diagnosis at the time. In the remaining cases, the onset may occur before age twenty or after age forty.

DO PEOPLE DIE FROM MS?

MS is not a fatal disease. A 1971 study carried out by Dr. Alan K. Percy in Rochester, Minnesota, followed a group of MS patients for twenty-five years after diagnosis. At the end of this period, 74 percent of the group were still living, compared with an expected survival rate of 86 percent for the general population. The study demonstrated that life expectancy is reduced by no more than 15 percent.*

A person with MS may be more susceptible to other diseases such as respiratory or urinary infections, while his body's ability to fight off such infections may be reduced. If untreated, these infections may bring about premature death.

IS MS CONTAGIOUS?

No. No one can catch MS.

IS MS INHERITED?

MS is not a hereditary disease; parents do not pass it on to their children.

There are two groups of factors which influence whether or not a person develops MS: exogenous factors (factors present within the environment, for example, a virus, to which a person may be exposed in childhood) and endogenous factors (factors within a person which predispose him to get MS). Because family members share many of both kinds of factors, MS occurs with slightly greater frequency among family members than in the general population. This is true of other diseases as well. (In 4 percent of the families affected by MS there is more than one case of the disease.) Thus susceptibility to MS may run in families, but the disease is not hereditary in the strict sense of the term.

HOW DISABLED WILL I BECOME?

Unfortunately there are no guarantees in this area, and there is a natural inclination to imagine the worst. Statistics, however, are in your favor. In the study quoted above, after twenty-five years two-thirds of the MS patients were still ambulatory (able to walk with or without the assistance of walking aids).

---

*Publisher's Note: As a result of improvements in symptoms management, by 1986 life expectancy had increased to better than 93% of the expected life span.

Chapter 6 Exercises

1. The following formal sentence definitions are poorly written. Identify the problem, and rewrite each into an effective formal sentence definition.

a. Manslaughter is when one person kills another without planning it ahead of time.
b. Garlic is a plant used in cooking.
c. A gas tank is used for storing gas.
d. The Distinguished Service Cross is awarded for heroism against an armed enemy.
e. An obstacle course is a military training area that has obstacles which must be crossed in succession.
f. Anorexia is a disease that causes a person to lose weight.
g. Oscillation is the movement by an oscillating body.
h. A tricycle is a vehicle for a child.
i. A contact lens is one of a pair of plastic lenses worn to improve sight.
j. Grout is what is used to fill cracks in joints or between tiles.

2. In groups, write one formal sentence definition for each of the following terms. Identify a specific audience before you begin.

lap	mole	recess
bridge	knot	scab
soil	note	horn

Compare your definitions with those written by the other groups in class. Discuss how many definitions are possible for each term.

3. In groups, write an expanded definition for one of the following terms. Identify your specific audience and purpose before you write. Use one of the strategies discussed in this chapter. After completing a draft, share your results with the other groups in class.

parking ticket	ballpoint pen
roller skates	hair brush
peanut butter	stapler
rubber band	weight lifting

4. Read one of these articles in Part II, as selected by your instructor: (1) Kliem, "Writing Technical Procedures," or (2) Buehler, "Defining Terms in Technical Editing: The Levels of Edit as a Model." Identify every definition you find in the article as informal, formal sentence, or expanded. Also identify the definition strategies the author uses in the article.

5. Prepare one of these exercises:

a. Read the article by Bagin and Van Doren in Part II, and write a one-paragraph expanded definition of proofreading for a student who is not taking this class and does not know what his or her instructor meant by "Proofread more." Begin your paragraph with a formal sentence definition.

b. Read the article by Perryman in Part II. Assume that you must adapt the article for a reader who knows nothing at all about computers. Identify the terms you should define in such an adaptation. Then write sentence definitions for each *or* write an expanded definition of one term, if your instructor prefers.

6. Write an expanded definition of one of the terms listed below. Before you begin writing, identify a specific audience and purpose for your definition. Use at least three of the strategies for expanding definitions discussed in this chapter, and list them at the bottom of the last page of your definition.

rock and roll music
any sport
fast-food restaurants
job interviews
a piece of gardening equipment
a piece of sports equipment
carbohydrate
floppy disk
auction
any home appliance
a university
ice cream

7. Write an expanded definition of a piece of equipment you use in your major or of an important concept in your major. Write the definition for a student just beginning in your major who needs to understand the equipment or concept. Use at least two of the strategies for expanding definitions discussed in this chapter. At the bottom of the last page of your definition, list the two strategies you used.

8. Write a memo to your boss at your present job to request a new piece of equipment or new supplies. You know that your boss is reluctant to spend money on equipment right now, but you think this equipment is necessary. Include a formal sentence definition of the item you are requesting and an expanded definition using two of the expansion strategies discussed in this chapter. *Note:* Memo format is explained in Chapter 9.

9. Write a memo to the appropriate administrator on your campus to request a piece of equipment for use in one of your classes or for a campus activity. Assume that you are the chair of a student committee that has the task of making such requests on behalf of the student body. Include a formal sentence definition of the object you are requesting and an expanded definition using two or three of the expansion strategies discussed in this chapter. *Note:* Memo format is explained in Chapter 9.

CHAPTER 7

Description

Understanding Description
Planning Descriptions
- Subjects
- Kinds of Information
- Detail and Language
- Graphic Aids

Writing Descriptions
- Organization
- Sections of a Technical Description
  - Introduction
  - Description of Parts
  - Cycle of Operation
- Sample Technical Descriptions

Chapter Summary
Supplemental Readings in Part Two
Models
Exercises

Understanding Description

Technical description provides readers with precise details about the physical features, appearance, or composition of a subject. A technical description may be a complete document in itself, such as an entry in an encyclopedia or a technical handbook. Manufacturing companies also need complete technical descriptions of each product model as an official record. Frequently, too, technical descriptions are separate sections in longer documents, such as these:

1. *Proposals and other reports.* Readers usually need descriptions of equipment and locations in a report before they can make decisions. A report discussing the environmental impact of a solid-waste landfill at a particular site, for example, would probably include a description of the location to help readers visualize the site and the potential changes.
2. *Sales literature.* Both dealers and consumers need descriptions of products—dealers so that they can advise customers and answer questions and consumers so that they can make purchase decisions. A consumer examining a new automobile in a showroom is likely to overlook certain features. A brochure containing a description of the automobile features ensures that the consumer will have enough information for decision making.
3. *Manuals.* Descriptions of equipment help operators understand the principles behind running a piece of machinery. Technicians need a record of every part of a machine and its function in order to effectively assess problems and make repairs. Consumer instruction manuals often include descriptions to help readers locate important parts of the product.
4. *Magazine articles and brochures for general readers.* Articles and brochures about science and technology often include descriptions of mechanisms, geologic sites, or natural phenomena to help readers understand the subject. An article for general readers about a new wing design on military aircraft may include a description of the wings and their position relative to the body of the aircraft to help readers visualize the design. A brochure written for visitors at a science institute describing the development of a solar heat collector may include several descriptions of the mechanism in development so that readers can understand how the design changed over time.

Plan and draft a technical description with the same attention to readers and their need for the information that you apply to other writing tasks.

Planning Descriptions

In deciding whether to include descriptions in your document, consider your readers' knowledge of the subject and why they need your information. A student in an introductory botany class has little expert knowledge about the basic parts of a flower. Such a student needs a description of these parts in order to understand the variations in structure of different flowers and how seeds are formed. A research report to botanists about a new hybrid flower, however, would probably not include a description of these basic parts because botanists already know them well. The botanists, on the other hand, do need a description of the hybrid flower because they are not familiar with this new variety and they need the description to understand the research results in the report. Both types of readers, then, need descriptions, but they differ in (1) the subjects they need described, (2) the kinds of information they need, (3) the appropriate language and detail, and (4) the appropriate graphic aids.

Subjects

As you analyze your readers' level of knowledge about a subject and their need for information, consider a variety of subjects that, if described in detail, might help your readers understand and use the information in your document. Here are some general subjects to consider as you decide what to describe for readers:

1. *Mechanism*—any machine or device with moving parts, such as a steam turbine, camera, automobile, or bicycle.
2. *Location*—any site or specific geologic area, such as the Braidwood Nuclear Power Plant, the Lake Michigan shoreline at Milwaukee, or the Moon surface.
3. *Organism*—any form or part of plant or animal life, such as an oak tree, a camel, bacteria, a kidney, or a hyacinth bulb.
4. *Substance*—any physical matter or material, such as cocaine, lard, gold, or milk.
5. *Object*—any implement without moving parts, such as a paper cup, a shoe, a photograph, or a floppy disk.
6. *Condition*—the physical state of a mechanism, location, organism, substance, or object at a specific time, such as a plane after an accident, a tumor before radiation, or a forest area after a fire.

Whenever your readers are not likely to know exactly what these subjects look like and need to know this in order to understand the information in your document, provide descriptions.

Kinds of Information

To be useful, a technical description must provide readers with a clear image of the subject. Include precise details about the following whenever such information will help your readers understand your subject better:

- Purpose or function
- Weight, shape, measurements, materials
- Major and minor parts, their locations, and how they are connected
- Texture, sound, odor, color
- Model numbers and names
- Operating cycle
- Special conditions for appropriate use, such as time, temperature, frequency

This excerpt from an entry in a sports encyclopedia illustrates how writers use specific details in describing objects and locations:

> Court and Equipment. The squash-racquets ball is made of hollow black rubber, has a 1¾-in. diameter, and weighs about 1 oz. The racquet is approximately the size of a badminton racket, but is more solidly made. Maximum length is 27 in.; the strung surface of the round wooden head has a diameter measuring about 6¾ in., and the weight is 8 to 10 oz.
>
> The American singles court is 43 ft long and 18 ft, 6 in. wide. A line from front wall to back wall splits the court in half, and there is a "short line," or "service-court line," drawn across the court 18 ft from the front wall. The ceiling is 18 ft or more above the floor, but is out of play. A play line on the walls is 16 ft above the floor on the front wall and along each side wall to the service-court line; 12 ft above the floor on the side walls behind the service-court line; and 7 ft above the floor at the back wall. The ball is out of play if it strikes a wall above this line.
>
> A "telltale" of sheet metal, 17 in. high, stretches across the base of the front wall. A service line, or cut line, is drawn across the front wall at a height of 6 ft, 6 in. At each side of the court a service box is marked; it is a quarter-circle, with a radius of 4 ft, 6 in., behind the service-court line and drawn from that line to the side wall.[1]

To determine the amount and kinds of detail appropriate for a technical description, assess your readers' purpose, their need for details, and their level of knowledge about the subject. Expert readers rarely use encyclopedias, even specialized ones, so readers for this technical description are not experts, although they may have some knowledge of the sport. Readers of this material probably are interested in learning enough about the equipment and court to distinguish them from those of other racquet sports. Notice that the writer compares the racquet to one used in badminton, a sport readers are probably more familiar with. This kind of brief comparison is useful for non-expert readers. If a description of this equipment were written for readers who want to play the game, however, more details about racquet construction would be included. How tightly strung is the racquet? Of what material? What does "solidly made" refer to? What are the exact weights and measurements?

The description of the court is organized spatially—the floor first, then the ceiling, then the side walls, and finally the service areas. This organization helps readers visualize an area they have never seen. The sports jargon, such as "telltale," is in quotation marks to indicate to readers that these terms are being used in a special way. If the description were for expert readers, such jargon would not require quotation marks.

Remember that descriptions may cover either (1) a particular mechanism, location, organism, substance, object, or condition or (2) the general type. In this excerpt, the writer describes in general all balls, racquets, and courts for this sport, so this description covers the general type. A description of the equipment and court at the Landview Country Club would be specific to those alone. In the same manner, a product description of Model XX must be specific to that model alone and, therefore, different from a description of Model YY.

Detail and Language

All readers of technical descriptions need accurate detail. However, some readers need more detail and can understand more highly technical language than others. Here is a brief description from a book about the solar system written for general readers interested in science:

> The Galaxy is flattened by its rotating motion into the shape of a disk, whose thickness is roughly one-fiftieth of its diameter. Most of the stars in the Galaxy are in this small disk, although some are located outside it. A relatively small, spherical cluster of stars, called the nucleus of the Galaxy, bulges out of the disk at the center. The entire structure resembles a double sombrero with the galactic nucleus as the crown and the disk as the brim. The Sun is located in the brim of the sombrero about three-fifths of the way out from the center to the edge. When we look into the sky in the direction of the disk we see so many stars that they are not visible as separate points of light, but blend together into a luminous band stretching across the sky. This band is called the Milky Way.[2]

In this description, the writer compares the galaxy to a sombrero, something most general readers are familiar with, and then locates the various parts of the galaxy on areas of the sombrero, thus creating an image that general readers will find easy to use. The writer also uses an informal tone, drawing readers into the description by saying, "When we look into the sky. . . ." This informal tone is appropriate for general readers but not for expert readers who need less assistance in visualizing the subject.

Here is a description of television station equipment that appeared in a magazine written for owners and technical staff of low-power television stations. The product description is used by technical staff in making decisions about future equipment purchases.

> Microdyne Corporation manufactures a complete line of fixed and motorized satellite receiving antennas, ranging in size from 1.2 meters to 7 meters and suitable for both C-band or Ku-band applications. The 3.66-meter (12-foot) antenna is specially designed for broadcast quality reception. It is a 10-piece parabolic antenna made from exceptionally strong space age fiber and polyester materials. It features a prime focus feed, superb side lobe characteristics, and high gains. The dish can view 100° of the geostationary arc and satellites within 69° west to 139° west. This feature means easy alignment with any line of sight satellite in the geostationary orbit. An optional motorized actuator is available.[3]

Clearly, the expected readers for this technical description have expert knowledge because the specialized terms, such as *geostationary arc,* are not defined, and the description stresses features of this antenna and how it will function at a television station, details that expert readers need to aid their decision making. Because this description is intended to encourage readers to purchase the equipment, the features are described in persuasive language—"specially designed," "exceptionally strong," "superb side lobe," "easy alignment." A nonexpert reader interested in learning about antennas would not find this description particularly useful. Such a reader would want an overall description of antennas in general and fewer details specific to one model. The reader would want to know, for example, what is the usual appearance of a satellite-receiving antenna? Where is it mounted? What are the major parts and their functions? What is the difference between a fixed and motorized antenna? Your readers' purpose must guide your decisions about how many details and which details to include.

Follow these guidelines for using appropriate language in technical descriptions:

1. Use specific rather than general terms. Notice these examples:

General terms	*Specific terms*
short	1 in. tall
curved	S-shaped

General terms	*Specific terms*
thin	1/8 in. thick
light	cream-colored
nearby	1/4 in. from the base
fast	400 rpm
large	7 ft high
light	1.5 oz
heavy	16 tons
noise	high-pitched shriek

2. Indicate a range in size if the description is of a general type that varies, or give an example.

   Poor: Bowling balls for adults usually weigh 12 lb.

   Better: Bowling balls for adults vary in weight from 10 to 16 lb.

3. Use precise language but not language too technical for your readers. Do not write "a combination of ferric hydroxide and ferric oxide" if "rust" will do.

4. If you must use highly technical terms or jargon and some of your readers are nonexperts, define those terms:

   The patient's lipoma (fatty tumor) had not grown since the previous examination.

5. Compare the subject to simple, well-known items and situations to help general readers visualize it:

   The difference in size between the Sun and the Earth is similar to the difference between a baseball and a grape seed.

Be specific and accurate, but do not overwhelm your readers with details they cannot use. Consider carefully the language your readers need and the number of details they can use appropriately.

Graphic Aids

Graphic aids are an essential element in descriptions because readers must develop a mental picture of the subject and words alone may not be adequate to paint that picture successfully. Chapter 4 discusses graphic aids in detail along with format devices to highlight certain facts. These graphic aids are usually effective ways to illustrate details in descriptions:

1. *Photographs.* Photographs supply a realistic view of the subject and its size, color, and structure, but they may not properly display all features or show locations clearly. Photographs are most useful for general readers who want to know the overall appearance of a subject. A guide to house plants for general readers should include photographs,

for instance, so readers can differentiate among plants. A research article for botanists, however, would probably include line drawings showing stems, leaves, whorls, and other distinctive features of the plants being discussed.

2. *Line drawings.* Line drawings provide an exterior view showing key features of mechanisms, organisms, objects, or locations and special conditions, as well as how these features are connected. Line drawings are useful when readers need to be familiar with each part or area.
3. *Cutaway and exploded drawings.* Cutaway drawings illustrate interior views, including layers of materials that cannot be seen from photographs or line drawings of the exterior. Exploded drawings emphasize the separate parts of a subject, especially those which might be concealed if the object is in one piece. Often readers need both line drawings of the exterior and interior and exploded drawings to fully understand a subject.
4. *Maps, floor plans, and architects' renderings.* Maps show geographic locations, and floor plans and architects' renderings show placement of objects within a specified area. Features that may be obscured in a photograph from any angle can be marked clearly on these illustrations, giving readers a better sense of the composition of the area than any photograph could.

Use as many graphic aids as you believe your readers need to get a clear picture of your subject. Also, be sure to direct your readers' attention to the graphic aids by references in the text, and always use the same terms in the text and in your graphic aids for specific parts or areas.

Writing Descriptions

After identifying your expected readers and the kinds of information they need, consider how to organize your descriptions.

Organization

In some cases, company policy or institution requirements dictate the organization of a description.

Model 7–1, for example, is a technical report published by NASA in its monthly series *Tech Briefs*. All reports appearing in this publication follow the same organization—purpose of the research project, physical description of the mechanism or system that is the focus of the research, and a line drawing illustrating the mechanism or a flowchart showing the system.

Another example of a predetermined description format appears in Model 7-2, a portion of a police report. Although police report formats may vary from city to city, most departments require a narrative section describing the incident that led to police intervention. The writer must follow the department's established format because readers—police, attorneys, judges, and social-service workers—handle hundreds of similar reports and rely on seeing all reports in the same format with similar information in the same place.

If you are not locked into a specific description format for the document you are writing, select an organizational pattern that will best serve your readers, usually one of these:

Spatial. The spatial pattern is most often used for descriptions of mechanisms, objects, and locations because it is a logical way to explain how a subject looks. Remember that you must select a specific direction and follow it throughout the description, such as from base to shade for a lamp, from outside to inside for a television set, and from one end to the other for a football field. Writers often also explain the function of a subject and its main parts. This additional information is particularly important if readers are not likely to know the purpose of the subject or it will not be apparent from the description of parts or composition. Thus a description of a stomach pump may include an explanation of the function of each valve and tube so that readers can understand how the parts work together.

Chronological. The chronological pattern describes features of a subject in the order they were produced or put together. This information may be central to a description of, for example, a Gothic church that took two centuries to build or an anthropological dig where layers of sediments containing primitive tools have been deposited over two million years.

In all cases, consider your readers' knowledge of the subject and the organization that will most help them use the information effectively. A general reader of a description of a Gothic church may want a chronological description that focuses on the historical changes in exterior and interior design. A structural engineer, however, may want a chronological description that focuses on changes in materials and construction techniques.

Sections of a Technical Description

Figure 7-1 presents a typical outline for a technical description of an object or mechanism. As this outline illustrates, a technical description has three main sections: an introduction, a description of the parts, and a description of the cycle of operations.

I. Introduction
   A. Definition of the object
   B. Purpose of the object
   C. Overall description of the object—size, color, weight, etc.
   D. List of main parts
II. Description of Main Parts
   A. Main part 1
      1. Definition of main part 1
      2. Purpose of main part 1
      3. Details of main part 1—color, shape, measurements, etc.
      4. Connection to main part 2 *or* list of minor parts of main part 1
      5. Minor part 1 (if relevant)
         (a) Definition of minor part 1
         (b) Purpose of minor part 1
         (c) Details of minor part 1—color, shape, measurements, etc.
      6. Minor part 2 (if relevant)
         ⋮
   B. Main part 2
      ⋮
III. Cycle of Operation
   A. How parts work together
   B. How object operates
   C. Limitations

Figure 7-1 Model Outline of Technical Description

Introduction

The introduction to a technical description orients readers to the subject and gives them general information that will help them understand the details in the body of the description. Follow these guidelines:

1. Write a formal sentence definition of the complete mechanism or object to ensure that your readers understand the subject. If the item is a very common one, however, you may omit the formal definition unless company policy calls for it.
2. Clarify the purpose if it is not obvious from the definition. In some cases, how the object or mechanism functions may be most significant to your readers, so explain this in the introduction.
3. Describe the overall appearance so that readers have a sense of size, shape, and color. For general readers, include extra details that may be helpful, such as a comparison with an ordinary object. Also include, if possible, a graphic illustration of the object or mechanism.
4. List the main parts in the order you plan to describe them.

Description of Parts

The body of a technical description includes a formal definition and the physical details of each major and minor part presented in a specific organizational pattern, such as spatial or chronological. Include as many details as your readers need to understand what the parts look like and how they are connected. If your readers plan to construct the object or mechanism, they will need to know every bolt and clamp. General readers, on the other hand, are usually interested only in major parts and important minor parts. Graphic illustrations of the individual parts or exploded drawings are helpful to readers in this section.

Cycle of Operation

Technical descriptions for in-house use or for expert readers seldom have conclusions. General readers, however, may need conclusions that explain how the parts work together and what a typical cycle of operation is like. If the description will be included in sales literature, the conclusion often stresses the special features or advantages of the product.

Sample Technical Descriptions

Figure 7-2, taken from an encyclopedia, shows a portion of a technical description of a camera. Because there are many camera styles, the description concentrates on features that cameras have in common. The graphics that accompany the description illustrate several camera styles and the parts of a simple box camera so that readers can see where these parts are located and the way they work together generally. Later in the 19-page entry, the writer describes the general types of still cameras, such as 35 mm and twin-lens reflex, and cameras for special purposes, such as aerial photography and motion pictures. In all sections of the description, however, the focus is on cameras of a general type rather than on a specific model. The organization is spatial, and the writer also describes the purpose of each part. Notice that this description of a camera contains (1) an explanation of how the basic camera operates, (2) headings to draw readers' attention to specific topics, and (3) an explanation of the individual parts of the camera and their function.

Figure 7-3 is a technical description of a specific model of a toaster written by a student as a section of a consumer manual that includes a description, operating instructions, and maintenance guidelines. Because this description is of a particular model, the writer describes the parts and their locations in detail. The organization is spatial, moving from the lowest part, the base, to the cord, which enters the housing above the base, and then to the housing. Because the housing is so large, the writer partitions it, describing the outer casing first, the heating wells second, and then the control levers. The writer ends with a brief description of the operating cycle. Notice

CAMERA. The basic function of a camera is to record a permanent image on a piece of film. When light enters a camera, it passes through a lens and converges on the film. It forms a latent image on the film by chemically altering the silver halides contained in the film emulsion. When the film is developed, the image becomes visible in the form of a negative. From the negative a positive image, or print, can be made.

Since the time of the invention of the first camera, all cameras have operated on the same fundamental principles. As photographic technology developed, however, various camera functions underwent improvement. Thus, while the basic concept of the camera remains the same today, a wide range of accessories have been created to cope with special situations. In addition, special-purpose cameras have been developed that meet a variety of needs.

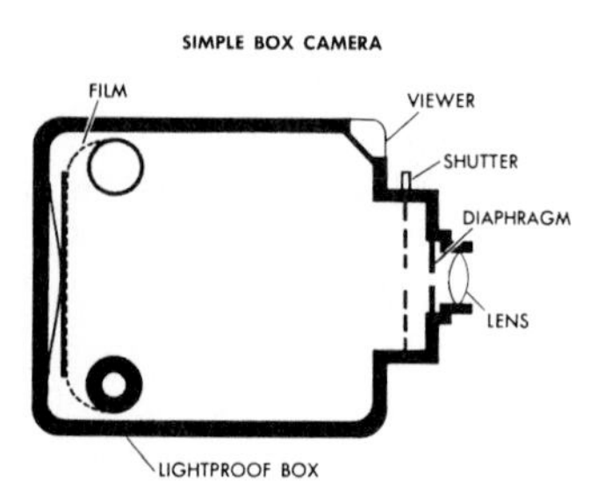

A simple box camera has one lens opening and one shutter speed; the distance between lens and film is fixed.

1. Still Cameras

All still cameras are designed to do one thing: to capture one single instant in time and space on film. Although there are many different kinds, they all have the same basic design.

PARTS OF A CAMERA

The basic parts of a camera are a lighttight body, or box, and a lens. In addition to the fundamental lens and body, a camera has a shutter, a film-holding and -transport system, focusing and viewfinding systems, and sometimes a system for determining length of exposure.

Camera Body. The simplest camera body is the one designed for the snapshot camera. It holds the lens in a fixed position at one end and the film, under lightproof conditions, at the other. In simple cameras the distance between lens and film remains fixed, but in more advanced cameras it is possible to vary the distance between the lens and film plane for precise focusing. The camera body can range

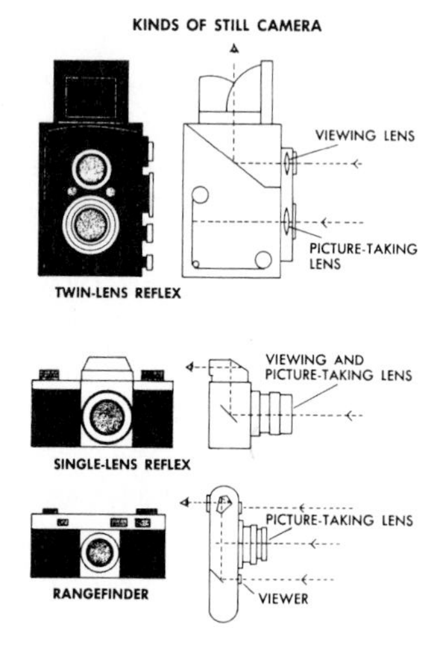

Figure 7-2 Technical Description of a Camera.

in size from the minute subminiature, the inside diameter of which may measure less than 1 inch (2.5 cm), to extremely large special-purpose machines used for cartography and engraving.

Lens. The function of the lens is to gather light and focus it on the film. A lens provides an angle of view that depends on its focal length (the distance between lens and film plane when the camera is focused on a distant object) and on the film size. On a 35mm camera, for example, a 50mm lens will provide an angle of view of 45° on the horizontal. A 100mm lens has an angle of view of 22°, while a 500mm lens covers a scant 5°. Conversely a 35mm lens covers an area of about 62° and a 28mm lens encompasses 74°. One of the most fascinating is the 8mm fisheye with a coverage of 180°.

Normal Lenses. The normal lens for a 35mm camera—a camera that uses 35mm film and produces a 24 × 36mm (1 × 1½ inch) negative—has a 50mm focal length. The normal lens for a camera using 4 × 5 inch (100 × 125mm) film is 135mm. Since the angle of view of the two lenses is approximately the same, they provide the same area coverage of the image on their respective film sizes. The image on the 4 × 5 inch film will, of course, be larger than on the 35mm. But when a 135mm lens is used in both 35mm and 4 × 5 inch cameras, and the subject is the same distance away in each case, the image of the subject will be the same size on the films in both cameras. The 135mm lens will, however, provide a much wider angle of view on the 4 × 5 camera than on the 35mm camera.

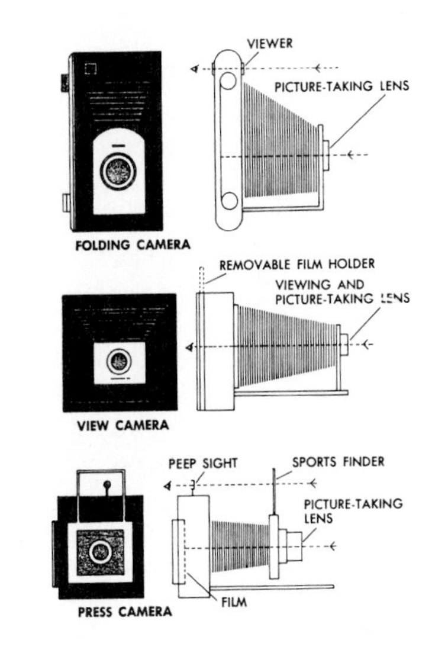

Figure 7-2 cont.

I. Introduction

The Kitchen King toaster, Model 49D, is an electric appliance that browns bread between electric coils inside a metal body. The toaster also heats and browns frozen and packaged foods designed especially to fit in the vertical heating wells of a toaster. The Kitchen King Model 49D is 6¾ in. high, 4½ in. wide, 8¾ in. long, and weighs 2 lb. There are three main parts: the base, the cord, and the housing.

II. Main Parts

The Base

The base is a flat stainless steel plate, 4½ in. wide and 8¾ in. long. It supports the housing of the toaster and is nailed to it at all four corners. The base rests on four round, black, plastic feet, each ½ in. high and ¾ in. in diameter. The feet are attached to the base ½ in. inside the four corners. The front and back of the base (4½ in. wide) are edged in black plastic strips that protrude ¾ in. past the base. A removable stainless steel plate, 3½ in. wide and 5¼ in. long, called the crumb tray, is centered in the base. One of the shorter sides of the crumb tray is hinged. The opposite side has a catch that allows the hinged plate to swing open for removal of trapped crumbs.

The Cord

The cord is an insulated cable that conducts electric current to the heating mechanism inside the body of the toaster. It is 32 in. long and ⅛ in. in diameter. Attached to the base of the toaster at the back, it enters the housing through an opening just large enough for the cable. At the other end of the cord is a two-terminal plug, measuring 1 in. long, ¾ in. wide, and ½ in. thick. The two prongs are ½ in. apart and are each ⅝ in. long, ¼ in. wide, and 1/16 in. thick. The plug must be attached to an electrical outlet before the toaster is operable.

Housing

The housing is a stainless steel boxlike cover that contains the electric heating mechanism and in which the toasting process takes place. The housing is 8¾ in. long, 6¼ in. high, and 4½ in. wide. The front and back are edged at the top with black plastic strips that protrude past the housing ¾ in. The housing also contains two heating wells and two control levers.

Heating Wells. The heating wells are vertical compartments, 5½ in. long, 1¼ in. wide, 5¾ in. deep. The wells are lined with electric coils that brown the bread or toaster food when the mechanism is activated. Inside each heating well is a platform on which the toast rests and which moves up and down the vertical length of the well.

Control Levers. The control levers are square pieces of black plastic attached to springs that regulate the heating and the degree of browning. The levers are 6½ in. square and are on the front of the toaster.

Figure 7-3 Technical Description: Kitchen King Toaster, Model 49D

Heat Control Lever. The heat control lever is centered 1 in. from the top of the housing and is attached to the platforms inside both heating wells. When the lever is pushed down, it moves along a vertical slot to the base of the toaster, lowering the platforms inside the heating wells. This movement also triggers the heating coils.

Browning Lever. The browning lever is located in the lower right corner on the front of the toaster. The browning lever moves horizontally along a 2-in. slot. The slot is marked with three positions: dark, medium, and light. This lever controls the length of time the heating mechanism is in operation and, therefore, the degree of brownness that will result.

III. Conclusion

The user must insert the cord into a 120-volt ac electrical outlet to begin the operating cycle. Next, the user inserts one or two pieces of bread or toaster food into the heating wells, moves the browning lever to the desired setting, and lowers the heat control lever until it catches and activates the heating mechanism. The toaster will turn off automatically when the heating cycle is complete, and the platforms in the heating wells will automatically raise the bread or toasted food so the user can remove it easily.

Figure 7-3 cont.

that this description contains (1) formal sentence definitions of the toaster and each main part, (2) headings to direct readers to specific topics, and (3) specific measurements for each part.

CHAPTER SUMMARY

This chapter discusses writing technical descriptions. Remember:

- Descriptions provide readers with precise details about the physical features, appearance, or composition of a subject.
- In planning a description, writers consider what subjects readers need described, the kinds of information readers need, appropriate detail and language, and helpful graphic aids.
- Subjects for descriptions may include mechanisms, locations, organisms, substances, objects, and conditions.
- Descriptions should include precise details about purpose, measurements, parts, textures, model numbers, operation cycles, and special conditions when appropriate for readers.

- The amount of detail and degree of technical language in a description depends on the readers' purpose and their technical knowledge.
- Graphic aids are essential in descriptions to help readers develop a mental picture of the subject.
- Most descriptions are organized spatially or chronologically.
- Descriptions generally have three main sections: an introduction, a description of each major and minor part, and a description of the cycle of operation.

SUPPLEMENTAL READINGS IN PART TWO

Benson, P. J. "Writing Visually: Design Considerations in Technical Publications," *Technical Communication.*

"Is It Worth a Thousand Words?" *Simply Stated.*

"Six Graphic Guidelines," *Simply Stated.*

ENDNOTES

1. From R. Hickok, "Squash Racquets," *New Encyclopedia of Sports* (New York: McGraw-Hill, 1977), p. 416.
2. From R. Jastrow, *Red Giants and White Dwarfs* (New York: W. W. Norton, 1979), p. 27.
3. From *The LPTV Report* 4 (October 1989), p. 42.

Model 7-1 Commentary

This brief technical report appeared in *NASA Tech Briefs* (April 1987), a monthly publication that announces new technology from NASA research projects. The readers are engineers and technicians who may use the technology in their own work. This model describes the design of a new meter to measure a patient's hand strength during rehabilitation therapy. The description of the meter is accompanied by a line drawing that shows major parts and measurements.

Discussion

1. Discuss the organization of this short description and how closely it matches the general guidelines for description presented in this chapter.

2. Discuss any changes, additions, or deletions you would have to make in this description if you intended to print it in a newspaper for nonexpert readers.

Hand-Strength Meter

A grip meter measures hand strength accurately and reproducibly.

Langley Research Center, Hampton, Virginia

In rehabilitation efforts, it is necessary to evaluate the therapeutic methods by monitoring the patient's progress in response to these methods. In hand rehabilitation, hand strength is an important parameter to monitor, as are dexterity and flexibility. A special grip-strength meter has been designed for accurate, reproducible measurement of hand rehabilitation.

The hand-strength meter, shown in the figure, is machined from a one-piece aluminum block. The upper and lower parts of the block are contoured for the proper grip. These two parts are connected by two parallel measuring beams to which four strain gauges are cemented. The four strain gauges are wired to form a Wheatstone bridge. Two power leads to the bridge and two signal leads from the bridge are tunneled through a small access hole in the back of the meter to an external display unit that consists of a dc power supply, a three-digit light-emitting diode (LED), and signal-conditioning circuitry. Both the meter and the display unit are compact and lightweight [0.5 lb (0.2 kg) and 3 lb (1.4 kg), respectively].

When the meter is gripped, the compressive force exerted by the hand is transmitted to the measuring beams. The beams are therefore deflected or strained, and this mechanical strain is sensed by the strain gauges and converted into an electrical signal. After amplification and conditioning, the signal is displayed on the LED as a measure of the gripping strength of the hand.

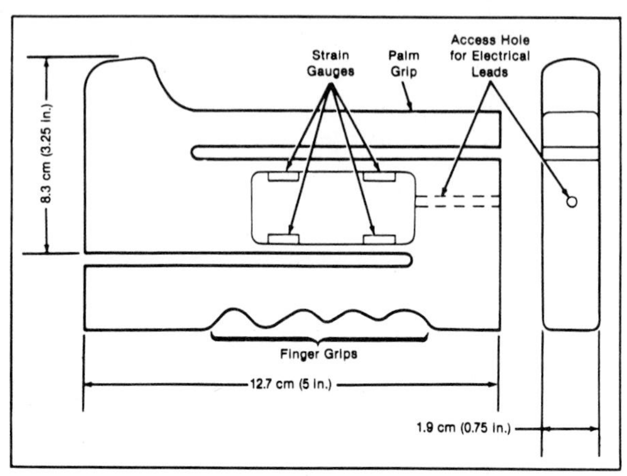

The **Hand-Strength Meter** includes four strain gauges connected in a Wheatstone bridge to measure the deflection caused by a gripping hand.

Laboratory calibration indicates that the meter is extremely linear for a gripping force ranging from 0 to 1,001 N (225 lbf), with a precision of ± 0.67 N (± 0.15 lbf). The display unit can be adjusted easily to show direct digital readings in pounds of force.

This work was done by Ping Tcheng of **Langley Research Center** *and Joe Elliot of the U.S. Army Aerostructures Directorate.*

Model 7-2 Commentary

This portion of a police accident report is the required narrative after routine identifications, license numbers, date, and times have been recorded. The readers of police reports include district attorneys, defense lawyers, police, social workers, judges, insurance investigators, and anyone who may have a legal involvement in the case. For consistency, the sections in all such reports are always in the same order.

Each section provides specific details about locations, conditions, and physical damage. The officer includes estimates of distances and notes the paths of the vehicles so that skid marks can be checked later for vehicle speed and direction. In addition, the officer identifies the ambulance service, hospital, and paramedics and lists other police reports that relate to the same case. This information will help readers who need to see all the documents in the case or who may need to interview medical personnel about injuries.

Discussion

1. Discuss the kinds of information included in each section of this report. How would this information help the district attorney prosecuting the case?

2. Assume that an accident has occurred at an intersection you know well. Write a description of the intersection that would be appropriate for the "Description of Scene" section of this type of police report.

SUPPLEMENTAL COLLISION NARRATIVE	Date of Incident: 1/7/90
Location: Newbury Blvd. and Downer Ave. Shorewood, WI	Citation: MM36-T40
Subject: Coronal Primary Collision Report 90-07-42	Officer ID: 455
	Time: 19:25

Description of Scene: Newbury Boulevard is an east-west roadway of asphalt construction with two lanes of traffic in either direction and divided by a grassy median 10 ft wide. All lanes are 14 ft wide. A curb approximately 6 in. high is on the periphery of the East 2 and West 2 lanes. This is a residential street with no business district. Downer Avenue is a north-south roadway of asphalt construction with two lanes of traffic in either direction separated by a broken yellow line. All lanes are 12 ft wide. A curb approximately 6 in. high is on the periphery of the North 2 and South 2 lanes. At the intersection with Newbury Boulevard, Downer Avenue is a residential street. A business district exists four blocks south. The speed on both streets is posted at 30 mph. The roadways are unobstructed at the intersection. Both roadways are straight at this point, and the intersection is perpendicular. At the time of the accident, weather was clear, temperature about 12°F, and the pavements were dry, without ice. Ice was present along the curb lanes extending about 4 in. from the curbs into the East 2, West 2 lanes of Newbury Boulevard.

Description of Vehicle: The Chevrolet Nova was found at rest in a westerly direction on Newbury Boulevard. The midrear of the vehicle was jammed into the street light pole at the curb of lane West 2 of Newbury Boulevard. Pole number is S–456–03. The vehicle was upright, all four wheels inflated, brakes functional, speedometer on zero, and bucket front seats pressed forward against the dashboard. The front of the vehicle showed heavy damage from rollover. The grillwork was pushed back into the radiator, which, in turn, was pushed into the fan and block. The driver's door was open, but the passenger door was wedged closed. The vehicle rear wheels were on the curb, the front wheels in lane West 2 of Newbury. The windshield was shattered, but still within the frame. Oil and radiator coolant were evident on the roadway from the curb to approximately 6 ft into lane West 2 of Newbury, directly in front of the vehicle.

Driver: Driver Coronal was standing on the curb upon my arrival. After detecting the distant odor of alcohol on his breath and clothes and noting his red, watery eyes and general unsteadiness, I administered a field sobriety test, which Coronal failed. (See Arrest Report 90–07–16.)

Passenger: Passenger Throckmorton was in the passenger seat of the vehicle, pinned between the bucket seat and the dashboard. He was in obvious pain and complained of numbness in his right leg. Passenger was removed from the vehicle and stabilized at the scene by paramedics from Station 104 and transported to Columbia Hospital by Lakeside Ambulance Service.

Physical Evidence: A skidmark approximately 432 ft long began at the center line of Downer Avenue and traveled westerly in an arc into lane West 2 of Newbury Boulevard, ending at the vehicle position on the curb. Gouge marks began in the grassy devil's strip between the curb and the sidewalk. The gouge marks ended at the light pole under the rear wheels of the vehicle. A radio antenna was located 29 ft from the curb, lying on the grassy median of Newbury Boulevard. The antenna fits the antenna base stub on the Chevrolet Nova.

Other reports: Arrest 90–07–16
Blood Alcohol Test 90–07–28
Coefficient of Friction Test 90–07–16
Hospital Report on Throckmorton

Model 7-3 Commentary

These pages are from a 56-page product-description manual written by a professional technical writer at Diebold, Inc. The manual describes a series of automatic teller machines (ATMs) with various functions. The descriptions are accompanied by line drawings of the machines and of specific areas on the machines. This manual took nearly nine months to write, during which time the writer drafted and redrafted in response to reviews by the designers, technicians, and managers involved in the development and sale of the teller machines.

Discussion

1. Identify the format elements used in these pages. Discuss how each one helps readers.

2. Discuss the kinds of information in each description of a machine and the similar organization of each section. How does this similarity in content and organization help readers?

1.3 MDS MODELS

Whether an institution desires a lobby-based ATM, a through-the-wall ATM, or even a free-standing island ATM, Diebold offers an MDS 1000 Series ATM for every application and business environment. Six MDS models are available:

- Diebold 1060 Everywhere Teller Machine (ETM)
- Diebold 1061 Consumer Transaction Terminal (CTT)
- Diebold 1062 Secure Interior ATM (Front and Rear load)
- Diebold 1072 TTW Walk-up ATM
- Diebold 1073 TTW Drive-up ATM
- Diebold 1074 Island ATM

This product description describes each MDS 1000 terminal except the 1060 ETM and the 1061 CTT. These terminals are described separately in the *Diebold 1060 Everywhere Teller Machine (ETM) Product Description,* File #79–9121, and the *Diebold Consumer Transaction Terminal (CTT) Product Description,* File #79–9249.

Diebold 1062 Secure Interior ATM (Front- or Rear-load)

The Diebold 1062 Secure Interior ATM is designed for operation in retail stores, bank lobbies, and other secure locations. Supporting two to four Dispense Cassettes, the terminal is rated for U.L. 291 (Business Hour Service).

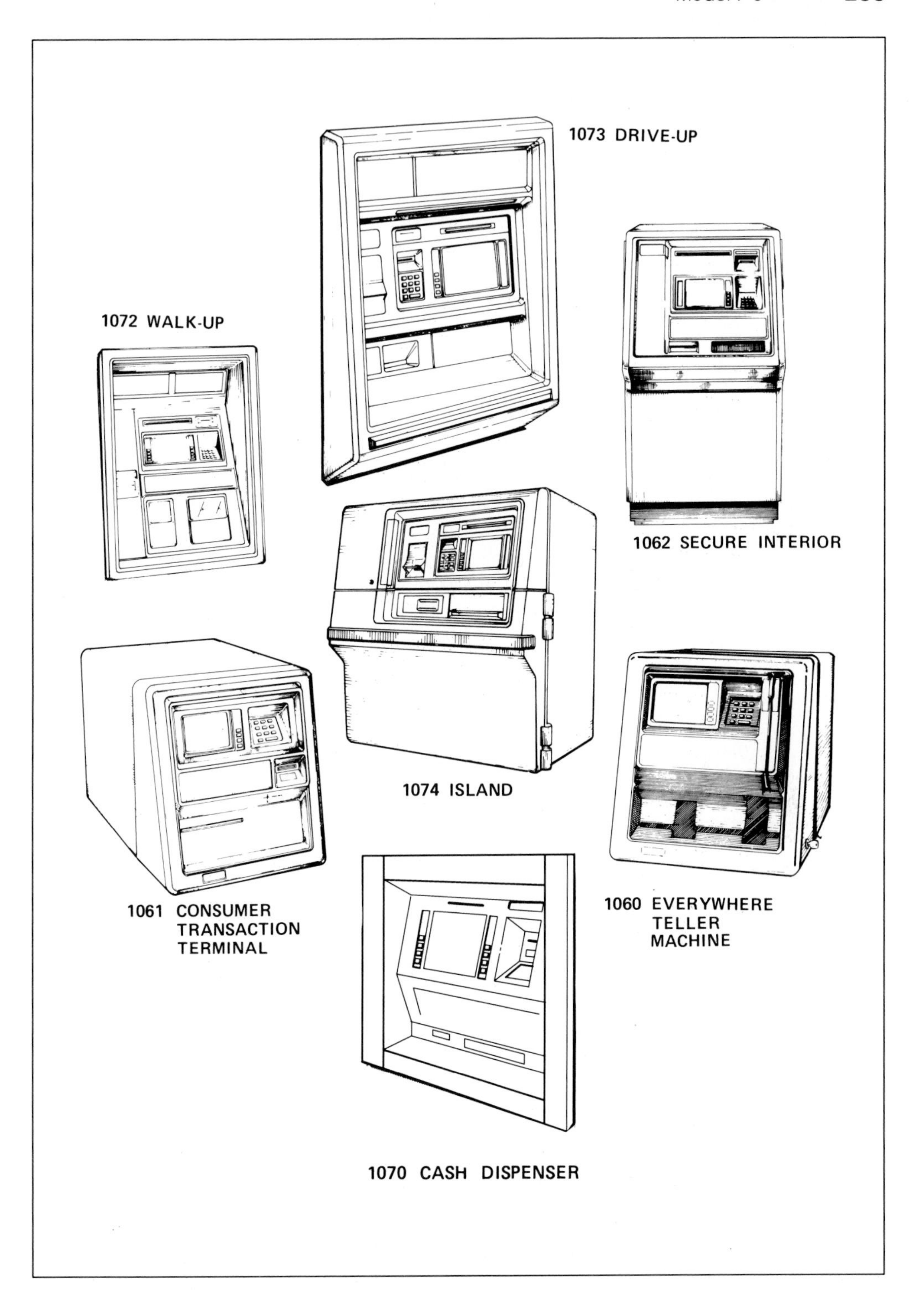
1073 DRIVE-UP
1072 WALK-UP
1062 SECURE INTERIOR
1074 ISLAND
1061 CONSUMER TRANSACTION TERMINAL
1060 EVERYWHERE TELLER MACHINE
1070 CASH DISPENSER

Depending upon the need for servicing access, both front- and rear-load models are available. The decorative cover concealing the chest door, as well as the top and side panels, can be painted in the institution's colors.

Diebold 1072 Through-the-wall Walk-up ATM

The Diebold 1072 Walk-up ATM can be installed in any wall opening that accommodates a TABS 9000 Series TTW Walk-up ATM. The chest meets the provisions of U.L. 291 (Business Hour Service), and the Dispenser supports two to four Dispense Cassettes.

The 1072 can be equipped with several fascias:

- Standard Fascia (camera or noncamera)
- Special Access Fascia
- Stainless Steel Camera Fascia
- Quick Cash Fascia

All fascia options are designed for right-handed operation. The standard fascia (both camera and noncamera) can be painted in the institution's colors.

Diebold 1073 Through-the-wall Drive-up ATM

The Diebold 1073 Drive-up ATM can be installed in any wall opening that accommodates a TABS 9000 Series TTW Drive-up ATM. The chest has a U.L. 291 (Business Hour Service), and the Dispenser supports two to four Dispense Cassettes.

The 1073 fascia is designed for left-handed operation by a consumer seated in an automobile. This arrangement is necessary so that the consumer can still read the Consumer Display during card insertion and PIN entry. Like the 1072, the 1073 terminal can optionally be equipped with a stainless steel camera fascia.

Diebold 1074 Island ATM

The Diebold 1074 Island ATM is designed for free-standing installation in drive-up locations on any standard 3.5-foot (3.10-m) island (such as those used for VAT or pneumatic delivery systems). The terminal is designed for left-handed operation by a consumer seated in an automobile.

The Island ATM's steel-reinforced, concrete chest (two inches, minimum thickness) meets the provisions of U.L. 291 for security containers in 24-hour service (Level 2). The chest, which has a U.L. TR/TL-15 × 6 rating, resists the effects of weather and most corrosive agents. However, a protective canopy is recommended for the convenience of consumers and service personnel. The Dispenser supports two or three Dispense Cassettes.

Mounted atop the 1074, the optional signage unit advertises the institution's logo for easy ATM recognition, even from a distance and at night. Further, the signage unit illuminates the terminal site for surveillance photography and added consumer safety.

An MDS-DX version of the 1074 terminal is not currently available. . . .

2.1 Dispenser

The Dispenser picks bills from Dispense Cassettes and transports them to a withdrawal area where they are retrieved by the consumer.

Each Dispense Cassette contains a single denomination. MDS 911 Native mode supports two denominations and four cassettes. MDS 912 Native mode supports up to four denominations and four cassettes.

The Dispenser reads a magnetic code from each cassette and uses this code in message exchanges with the network. Cassettes can be easily re-encoded by a Diebold technician or qualified service person for different denominations.

With Diebold software that interrogates the MDS terminal to determine what cassettes are present, institutions can change the denominations in terminals simply by re-encoding cassettes or by changing the software.

The Dispenser performs multiple-bill detection just before bill delivery. If two or more bills are stuck together, they are transported to the Divert Cassette and the Dispenser recycles by picking fresh bills.

The Dispenser reports low and out status to the network.

Currency Specifications

Bills must meet the following size ranges:

- Maximum bill size: 3.60 × 7.10 inches (91 × 180 mm)
- Minimum bill size: 2.20 × 4.30 inches (56 × 109 mm)

Cassettes

Each dispense cassette holds up to 13.4 inches (340 mm) of currency, a capacity equivalent to 3100 new U.S. bills. Cassettes are individually adjustable, so different size bills, coupons, travelers checks, etc. having the same thickness can be dispensed.

Diebold offers two types of cassettes. The Keylocking Cassette indicates tampering and provides maximum security for currency. The Convenience Cassette is available for sites where less security is required. Convenience cassettes do not have a keylock and do not indicate tampering.

Depending on the terminal model, the Dispenser supports a variable number if Dispense Cassettes:

Terminal Model	***Comes with . . .***	***Can Be Upgraded to . . .***
1060 ETM	1	1
1062 Secure Interior	2	4
1072 TTW Walk-up	2	4
1073 TTW Drive-up	2	4
1074 Island	2	3

2.2 CONSUMER INTERFACE PANEL

The Consumer Interface Panel guides the consumer through transactions and allows the consumer to request transactions and enter data. Also, the Consumer Interface Panel can perform important consumer education and advertising functions.

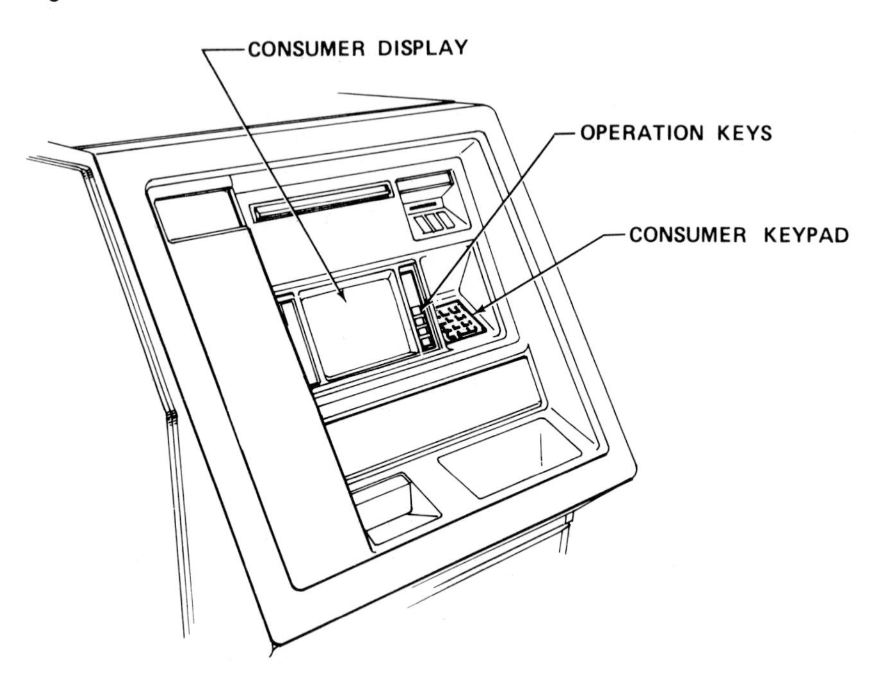

Consumer Interface Panel Components (1062 ATM)

Chapter 7 Exercises

1. Write a description of one of the following. Identify a specific reader and purpose before you begin to write. Include one graphic aid to help the reader understand your description.

a. A piece of sports equipment, e.g., a tennis racquet or an ice skate
b. A household appliance, e.g., a coffee maker or a can opener
c. A sports field or arena, e.g., a campus basketball arena or a baseball field
d. A piece of office equipment, e.g., a copy machine or a typewriter
e. The lobby area of your student center
f. One room or special area at your job
g. A piece of equipment you use in your major
h. A piece of gardening equipment, e.g., a rake or a lawn mower
i. A piece of home medical equipment, e.g., a thermometer or a blood pressure testing machine
j. An object you use daily, e.g., a razor or a blow dryer

2. Assume that you have just taken out a special insurance policy on your room against destruction of the contents and the interior. The insurance company has asked you to prepare a full description of the room and its contents to keep on file as a record of what was insured in case of loss and claim for damages. Write such a description.

3. In groups, write a description of a movable classroom desk chair with attached writing surface. Identify your specific audience and purpose before you write. Begin the description with a formal sentence definition. Compare your group's draft with the drafts of other groups. Discuss how the descriptions differ and how each draft is appropriate for the specific reader and purpose.

4. The following description of a stapler was written by a student in response to Exercise 1. The student identified the reader as a visitor to a museum of mechanical devices who is reading an information brochure distributed there. Individually, or in groups if your instructor prefers, evaluate this description both for appropriateness for the nonexpert reader and for how well it follows the guidelines presented in this chapter for effective description. Then rewrite the description (using a stapler as a model) more appropriately for the same audience and purpose.

Stapler

A stapler is a mechanical device usually found in an office setting that is used to fasten papers together. The stapler is made of metal painted black. It is 8 in. long, 2 in. high in the front, and slopes downward 60 degrees to the back of the base. Its main parts are the base, the carriage, and the arm.

The base is a piece of metal that supports the arm and the carriage. The base is 8 in. long, 1½ in. wide at the front, tapering to 1 in. at the back, and ½ in. high at its highest point. The base curves downward at all four sides. Underneath the base are two rubber-padded feet, one at the back, measuring 1 in. wide by ¾ in. long. Connected to the front rubber foot is a strip of metal ¼ in. wide and 4 in. long that supports the metal U-shaped depressor. When the depressor is pushed downward, the arm and the carriage are released to allow the user to tack papers to an upright flat surface. At the top of the base is a 1½-in. by 1-in. metal plate used to control the staple direction. The plate is located ½ in. from the front. On top of the plate are two sets of grooves. The back grooves together measure ¼ in. long and slope downward to a maximum of 1/16 in. They are parallel to the front of the stapler. This allows a staple to curve inward (the normal staple direction). The front grooves are ¼ in. long each and ¼ in. apart, parallel to the back grooves. When the plate is pushed back ⅛ in., the front grooves are positioned so that the staple can be turned outward. At the back of the base are two pieces of metal, 1 in. apart, that hold the carriage and the arm together with one metal bolt 1/16 in. in diameter.

5. Assume that you are the consulting architect for the local historical society. The director, Felicia Southworth, has asked you to prepare a memo report recommending a building in the community as the society's next restoration project. You will need to include a description of the exterior of this building in your report. (The description of interiors will be handled by Gregory Chang, interior design specialist, if the society decides to pursue your recommendation.) For this assignment, select a building with some historical value —a building on campus, an old theater, an old Victorian-style or frontier-style home, a railroad station, or a civic building—and then write a description of its exterior that could be included in your report to the historical society. *Or,* if your instructor prefers and you have already covered short report style (explained in Chapter 10), write the complete memo report to Southworth. For a memo report, include the historical background of the building itself or of the era it represents so that Southworth can readily understand why this building should be preserved.

6. Prepare a description of your campus that could be included in a brochure about your college that will be mailed to prospective freshmen as a recruiting tool. Include in your description important buildings, their locations, intersecting streets, and other important features. Also include one or more graphic aids that will help your readers visualize the campus.

7. Locate a description of a specialized topic in your field or of one of the following topics in an encyclopedia or handbook directed to readers with some knowledge of the topic and the terminology used in that field. Rewrite the description for a nonexpert reader who is interested in learning about the subject but who does not know much about its function or use.

a. Specific structural form, e.g., escalator
b. Specific flower or plant, e.g., Tetratheca
c. Human body part, e.g., skin
d. Specific wild animal, e.g., eland
e. Specific geologic site, e.g., Olduvai Gorge, Tanzania

Attach a copy of the original description to your revision, and submit both to your instructor.

CHAPTER 8

Instructions, Procedures, and Process Explanations

Understanding Instructions, Procedures, and Process Explanations
- Instructions
- Procedures
- Process Explanations
- Graphic Aids

Writing Instructions
- Readers
- Organization
  - Title
  - Introduction
  - Sequential Steps
  - Conclusion
- Details and Language
- Troubleshooting Instructions

Writing Procedures
- Procedures for One Employee
- Procedures for Several Employees
- Procedures for Handling Equipment and Systems

Writing Process Explanations
- Readers
- Organization
  - Introduction
  - Stages of the Process
  - Conclusion
- Details and Language

Chapter Summary

Supplemental Readings in Part Two

Models

Exercises

Understanding Instructions, Procedures, and Process Explanations

This chapter discusses three related but distinct strategies for explaining the stages or steps in a specific course of action. Instructions, procedures, and process explanations all inform readers about the correct sequence of steps in an action or how to handle certain materials, but readers and purpose differ.

Instructions

Instructions provide a set of steps that readers can follow to perform a specific action, such as operate a fork lift or build a sundeck. Readers of instructions are concerned with performing each step themselves in order to complete the action successfully.

Procedures

Procedures in business and industry provide guidelines for three possible situations: (1) steps in a system to be followed by one employee, (2) steps in a system to be followed by several employees, and (3) standards and rules for handling specific equipment or work systems. Readers of procedures include (1) those who must perform the actions, (2) those who supervise employees performing the actions, and (3) people, such as upper management, legal staff, and government inspectors, who need to understand the procedures in order to make decisions and perform their own jobs.

Process Explanations

Process explanations describe the stages of an action or system either in general (how photosynthesis occurs) or in a specific situation (how an experiment was conducted). Readers of a process explanation do not intend to perform the action themselves, but they need to understand it for a variety of reasons.

In some situations, these three methods of presenting information about systems may overlap, but in order to adequately serve your readers, consider each strategy separately when writing a document. Because so many people rely on instructions and procedures and companies may be liable legally for injuries or damage that results from unclear documents, directions of any sort are among the most important a company produces.

Graphic Aids

Instructions, procedures, and process explanations all benefit greatly from graphic aids illustrating significant aspects of the subject or those aspects which readers find difficult to visualize. Readers need such graphic aids as (1) line drawings showing where parts are located, (2) exploded drawings showing how parts fit together, (3) flowcharts illustrating the stages of a process, and (4) drawings or photographs showing how actions should be performed or how a finished product should look.

Do not, however, rely on graphic aids alone to guide readers. Companies have been liable for damages because instructions or procedures did not contain enough written text to adequately guide readers through the steps. Visual perceptions vary, and all readers may not view a drawing or diagram the same way. Written text, therefore, must cover every step and every necessary detail.

Writing Instructions

To write effective instructions, consider (1) your readers and their knowledge of the subject, (2) an organizational pattern that helps readers perform the action, and (3) appropriate details and language. In addition to the initial instructions, readers may need troubleshooting instructions that tell them what to do if, after performing the action, the mechanism fails to work properly or the expected results do not appear.

Readers

As you do before writing anything, consider who your readers will be. People read instructions in one of three ways. Some, but only a few, read instructions all the way through before beginning to follow any of the steps. Others read and perform each step without looking ahead to the next. And still others begin a task without reading any instructions and turn to them only when difficulties arise. Since you have no way to control this third group, assume that you are writing for readers who will read and perform each step without looking ahead. Therefore, all instructions must be in strict chronological order.

Everyone is a reader of instructions at some point, whether on the job or in private. Generally, readers fall into one of two categories:

1. *People on the job who use instructions to perform a work-related task.* Employees have various specialties and levels of technical knowledge, but all employees use instructions at some time to guide them in doing a job. Design your instructions for on-the-job tasks for the specific groups that will use them. Research chemists in the laboratory, crane operators, maintenance workers, and clerks all need instructions that match their needs and capabilities.

2. *Consumers.* Consumers frequently use instructions when they (a) install a new product in their homes, such as a VCR or a light fixture, (b) put an object together, such as a toy or a piece of furniture, and (c) perform an activity, such as cook a meal or refinish a table.

Consumers, unlike employees assigned to a particular task, represent a diverse audience in capabilities and situations.

Varied Capabilities. A consumer audience usually includes people of different ages, education, knowledge, and skills. Readers of a pamphlet about how to operate a new gasoline lawn mower may be 12 years old or 80, may have elementary education only or postgraduate degrees, may be skilled amateur gardeners, or may have no experience with a lawn mower. When writing for such a large audience, aim your instructions at the level of those with the least education and experience, because they need your help most. Include what may seem like overly obvious directions, such as, "Make sure you plug in the television set," or obvious warnings, such as, "Do not put your fingers in the turning blades."

Varied Situations. Consider the circumstances under which consumers are likely to use your instructions. A person building a bookcase in the basement is probably relaxed, has time to study the steps, and may appreciate additional information about options and variations. In contrast, someone in a coffee shop trying to follow the instructions on the wall for the Heimlich maneuver to rescue a choking victim is probably nervous and has time only for a quick glance at the instructions.

In all cases, assume that your readers do not know how to perform the action in question and that they need guidance for every step in the sequence.

Organization

Readers have come to expect consistency in all instructions they use—strict chronological order and numbered steps. Figure 8-1 presents a typical outline for a set of instructions for either consumers or employees. As the outline illustrates, instructions generally have a descriptive title and three main sections.

Title

Use a specific title that accurately names the action covered in the instructions.

Poor: Snow Removal

Better: Using Your Acme Snow Blower

I. Introduction
  A. Purpose of instructions
  B. Audience
  C. List of parts
  D. List of materials/tools/conditions needed
  E. Overview of the chronological steps
  F. Description of the mechanism or object, if needed
  G. Definitions of terms, if needed
  H. Warnings, cautions, notes, if needed
II. Sequential Steps
  A. Step 1
    1. Purpose
    2. Warning or caution, if needed
    3. Instruction in imperative mood
    4. Note on condition or result
  B. Step 2
    ⋮
III. Conclusion
  A. Expected result
  B. When or how to use

Figure 8-1 Model Outline for Instructions

Poor: Good Practices
Better: Welding on Pipelines

Introduction

Depending on how much information your readers need to get started, your introduction may be as brief as a sentence or as long as a page or more. If appropriate for your readers, include these elements:

Purpose and Audience. Explain the purpose of the instructions unless it is obvious. The purpose of a coffee maker, for instance, is evident from the name of the product. In other cases, a statement of purpose and audience might be helpful to readers:

> This manual provides instructions for mechanics who install Birkins fine-wire spark plugs.
> These instructions are for nurses who must inject dye into a vein through a balloon-tipped catheter.
> This safe practices booklet is for employees who operate cranes, riggers, and hookers.

Lists of Parts, Materials, Tools, and Conditions. If readers need to gather items in order to follow the instructions, put lists of those items in the introduction. *Parts* are the components needed, for example, to assemble and finish a table, such as legs, screws, and frame. *Materials* are the items needed to perform the task, such as sandpaper, shellac, and glue. *Tools* are the implements needed, such as a screwdriver, pliers, and a small paintbrush. Do not combine these items into one list unless the items are simple and the list is very short. If the lists are long, you may place them after the introduction for emphasis. *Conditions* are special circumstances that are important for completing the task successfully, such as using a dry, well-ventilated room. Be specific in listing items, such as

- One 3/4-in. videotape
- One Phillips screwdriver
- Three pieces of fine sandpaper
- Five 9-volt batteries
- Maintain a 70°F to 80°F temperature

Overview of Steps. If the instructions are complicated and involve many steps, summarize them for the reader:

> The following sections cover installation, operation, start-up and adjustment, maintenance, and overhaul of the Trendometer DR-33 motors.

Description of the Mechanism. A complete operating manual often includes a technical description of the equipment so that readers can locate specific parts and understand their function. Chapter 7 explains how to prepare technical descriptions.

Definitions. If you use any terms your readers may not understand, define them. Chapter 6 explains how to prepare formal and informal definitions.

Warnings, Cautions, and Notes. Readers must be alerted to potential danger or damage before they begin following instructions. If the potential danger or damage pertains to the whole set of instructions, place these in the introduction. When only one particular step is involved, place the warning or caution *before* the step. *Warnings* refer to possible injury:

> WARNING: To prevent electrical shock, do not use this unit in water.

Cautions refer to possible damage to equipment:

> CAUTION: Failure to latch servicing tray completely may damage the printer.

Notes give readers extra information about choices or conditions:

> *Note:* A 12-in. cord is included with your automatic slicer. An extension cord may be used if it is 120 volt, 10 amp.

Sequential Steps

Explain the steps in the exact sequence readers must perform them, and number each step. Explain only one step per number. If two steps must be performed simultaneously, explain the proper sequence:

> While pushing the button, release the lever slowly.

Include the reason for a particular action if you believe readers need it for more efficient use of the instructions:

> Tighten the belt. *Note:* A taut belt will prevent a shift in balance.

Use headings to separate the steps into categories. Identify the primary stages, if appropriate:

- Separating
- Mixing
- Applying

Identify the primary areas, if appropriate:

- Top drawer
- Side panel
- Third level

Refer readers to other steps when necessary:

> If the valve sticks, go to Step 12.

Then tell readers whether to go on from Step 12 or go back to a previous step:

> When the valve is clear, go back to Step 6.

For regular maintenance instructions, indicate the suggested frequency of performing certain steps:

Once a month:
- Check tire pressure
- Check lights
- Check hoses

Once a year:
- Inspect brakes
- Change fuel filter
- Clean choke

Conclusion

If you include a conclusion, tell readers what to expect after following the instructions and suggest other uses and options if appropriate:

> Your food will be hot, but not as brown as if heated in the oven rather than in the microwave. A few minutes of standing time will complete the cooking cycle and distribute the heat uniformly.

Figure 8-2 shows a typical set of consumer instructions for changing a typewriter ribbon. These instructions have a descriptive title that names the action readers want to do. The step-by-step instructions are numbered and include references to line drawings that show readers how to insert the new ribbon. Notice that Step 1 has an extra note telling readers what will happen after performing this step. Without this note, readers might think that the typewriter had broken and be afraid to continue with the instructions. Steps 1 and 6 might seem rather elementary, but for widely used products, you must

Changing the Ribbon Cassette

1. Raise the typewriter cover.
   - When this cover is raised, the typewriter is deactivated.
2. Remove the ribbon cassette by lifting it upward (Figure A).
3. Before inserting the new cassette, tighten the ribbon by turning the drive gear in the direction of the arrow (Figure B).
   Important: Be careful not to crease the ribbon while rolling it into the cassette.
4. While inserting the cassette, hold it tilted toward the keyboard, inserting the two guides on the ribbon cassette into the mounting slots (Figure C).
5. Press the cassette down (Figure D).
6. Close the cover.

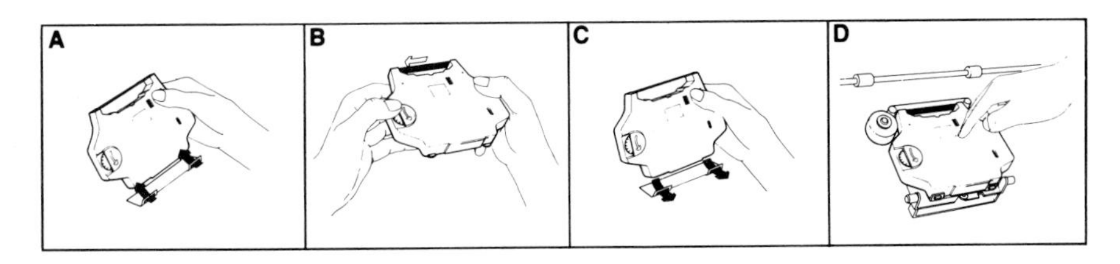

Figure 8-2 Instructions for Changing a Typewriter Ribbon

always assume that consumers do not know anything about the proper action. Therefore, include every necessary step.

Details and Language

Whether you are writing for consumers or for highly trained technicians, follow these guidelines:

1. Keep parallel structure in lists. Be sure that each item in a list is in the same grammatical form.

   Nonparallel: This warranty does not cover:
   a. Brakes
   b. Battery
   c. Using too much oil

   This list is not parallel because the first two items are names of things (nouns) and the last item is an action (verbal phrase). Rewrite so that all items fit the same pattern.

   Parallel: This warranty does not cover:
   a. Brakes
   b. Battery
   c. Excess oil consumption

   In this revision, all items are things—brakes, battery, and consumption. Lists also can confuse readers if the listed items do not all fit the same category. Here is an example:

   Poor: Parts Included for Assembly:
   End frames
   Caster inserts
   Casters
   Screwdriver
   Sleeve screws
   Detachable side handles

   The screwdriver is not a part included in the assembly package, but the consumer might think it is and waste time looking for it. List the screwdriver in a different category, such as "Tools Needed."

2. Maintain the same terminology for each part throughout the instructions, and be sure the part is labeled similarly in illustrations. Do not refer to "Control Button," "Program Button," and "On/Off Button" when you mean the same control.

3. Use headings to help readers find specific information. Identify sections of the instructions with descriptive headings:

   Assembly Kit Contents

Materials Needed
Preparation
Removing the Cover
Checking the Fuel Filter

4. Use the imperative mood for sequential steps. Readers understand instruction steps better when each is a command:

   Adjust the lever.
   Turn the handle.
   Clean the seal.

Never use passive voice in instructions. The passive voice does not make clear who is to perform the action in the step.

Poor: The shellac should be applied.

Better: Apply the shellac.

5. State specific details. Use precise language in all sections.

   Poor: Turn the lever to the right.

   Better: Turn the lever clockwise one full revolution.

   Poor: Keep the mixture relatively cool.

   Better: Keep the mixture at 50°F or less.

   Poor: Place the pan near the tube.

   Better: Place the pan 1 in. from the bottom of the tube.

   Poor: Attach the wire to the terminal.

   Better: Attach the green wire to the AUDIO OUT terminal.

6. Write complete sentences. Long sentences in instructions can be confusing, but do not write fragments or clip out articles and write "telegrams."

   Poor: Repairs excluded.

   Better: The warranty does not cover repairs to the electronic engine controls.

   Poor: Adjust time filter PS.

   Better: Adjust the time filter to the PS position.

7. Do not use *should* and *would* to mean *must* and *is.*

   Poor: The gauge should be on empty.

   Better: The gauge must be on empty.

   Poor: The seal would be unbroken.

   Better: The seal is unbroken.

   Both *should* and *would* are less direct than *must* and *is* and may be interpreted as representing possible conditions rather than absolute situations. A reader may interpret the first sentence as saying, "The gauge should be on empty, although it might not be." Always state facts definitely.

Troubleshooting Instructions

Troubleshooting instructions tell readers what to do if the mechanism fails to work properly or results do not match expectations. Figure 8-3 shows a typical three-part table for troubleshooting instructions. These troubleshooting instructions (1) describe the problem, (2) suggest a cause, and (3) offer a solution. Readers find the problem in the left column and then read across to the suggested solution. Troubleshooting instructions are most helpful when simple adjustments can solve problems.

Problem	***Possible Cause***	***Solution***
Snow in picture	Antenna	Check antenna connection
	Interference	Turn off dishwasher, microwave, nearby appliances
Fuzzy picture	Focus pilot	Adjust Sharpness Control clockwise
Multiple images	Antenna	Check antenna connection
	Lead-in wire	Check wire condition
Too much color	Color saturation	Adjust Color Control counterclockwise
Too little color	Color saturation	Adjust Color Control clockwise
Too much one color	Tint level	Adjust Tint Control clockwise

Figure 8-3 Troubleshooting Guide for a Television Set

Writing Procedures

Procedures often appear similar to instructions, and people sometimes assume that the terms mean the same thing. However, the term *procedures* most often and most accurately refers to official company guidelines covering three situations: (1) a system with sequential steps that must be completed by one employee, (2) a system with sequential steps that involves several employees interacting and supporting each other, and (3) the standards and methods for handling equipment or events with or without sequential steps.

Because these documents contain the rules and appropriate methods for the proper completion of tasks, they have multiple readers, including (1) employees who perform the procedures, (2) supervisors who must understand the system and oversee employees working within it, (3) company lawyers concerned about liability protection through the use of appropriate warnings, cautions, and guidelines, (4) government inspectors from such agencies as the Occupational Safety and Health Administration who check procedures to see if they comply with government standards, and (5) company management who must be aware of and understand all company policies and guidelines. Although the primary readers of procedures are those who follow them, remember that other readers inside and outside the company also use them as part of their work.

Procedures for One Employee

Procedures for a task that one employee will complete are similar to basic instructions. These procedures are organized, like instructions, in numbered steps following the general outline in Figure 8-1. Include the same items in the introduction: purpose, lists of parts and materials, warnings and cautions, and any information that will help readers perform the task more efficiently. Most company procedures include a description of the principles behind the system, such as a company policy or government regulation, research results, or technologic advances. When procedures involve equipment, they often include a technical description in the introduction.

Procedures for one employee also share some basic style elements with instructions:

- Descriptive title
- Precise details
- Complete sentences
- Parallel structure in lists
- Consistent terms for parts
- Headings to guide readers

The steps in the procedure are always in chronological order. Steps may be written as commands, in passive voice, or in indicative mood:

> To prevent movement of highway trucks and trailers while loading or unloading, set brakes and block wheels. (Command)
>
> The brakes should be set and the wheels should be blocked to prevent movement of highway trucks and trailers during loading and unloading. (Passive voice)
>
> To prevent movement of highway trucks and trailers while loading or unloading, the driver sets brakes and blocks wheels. (Indicative mood)

The indicative mood can become monotonous quickly if each step begins with the same words, such as "The driver places, . . ." or "The driver then sets. . . ." Commands and passive voice are less repetitious. Do not use the past tense for procedures because readers may interpret this to mean that the procedures are no longer in effect.

Procedures for Several Employees

When procedures involve several employees performing separate but related steps in a sequential system, the *playscript organization* is most effective. This organization is similar to a television script that shows the dialogue in the order the actors speak it. Playscript procedures show the steps in the system and indicate which employee performs each step. The usual pattern is to place the job title on the left and the action on the same line on the right:

Operator:	1. Open flow switch to full setting.
	2. Set remote switch to temperature sensor.
Inspector:	3. Check for water flow to both chiller and boiler.
Operator:	4. Change flow switch to half setting.
	5. Tighten mounting bolt.
Inspector:	6. Check for water flow to both boiler and chiller.
Operator:	7. Shut off chiller flow.
Inspector:	8. Test boiler flow for volume per second.
	9. Complete test records.
Operator:	10. Return flow switch to appropriate setting.

Generally, the steps are numbered sequentially no matter how many employees are involved. One employee may perform several steps in a row before another employee participates in the sequence. The playscript organization ensures that all employees understand how they fit into the system and how they support other employees also participating in the system. Use job titles to identify employee roles because individuals may come and go, but the tasks remain linked to specific jobs.

Include the same information in the introduction to playscript procedures as you do in step-by-step procedures for a single employee, such as warnings, lists of materials, overview of the procedure, and the principles behind the system. The steps in playscript organization usually are commands, although the indicative mood and passive voice are sometimes used.

Procedures for Handling Equipment and Systems

Procedures for handling equipment and systems include guidelines for (1) repairs, (2) installations, (3) maintenance, (4) assembly, (5) safety practices, and

(6) systems for conducting business, such as processing a bank loan or fingerprinting suspected criminals. These procedures include lists of tools, definitions, warnings and cautions, statements of purpose, and the basic principles. They differ from other procedures in an important way: The steps are not always sequential. Often the individual steps may be performed in any order, and they are usually grouped by topic rather than by sequence.

Figure 8-4 shows a typical organization for company procedures about safe practices on a construction site. The introduction establishes the purpose of the procedures and warns readers about possibly injury if these and other company guidelines are not followed. The individual items are grouped under topic headings. Notice that the items are not numbered because they are not meant to be sequential. In this sample, the items are written as com-

PROCEDURES FOR SAFE EXCAVATIONS

This safety bulletin specifies correct procedures for all excavation operations. In addition, all Clinton Engineering standards for materials and methods should be strictly observed. Failure to do so could result in injury to workers and to passersby.

Personal protection:

- Wear safety hard hat at all times.
- Keep away from overhead digging equipment.
- Keep ladders in trenches at all times.

Excavation protection:

- Protect all sites by substantial board railing or fence at least 48 in. high or standard horse-style barriers.
- Place warning lights at night.

Excavated materials:

- Do not place excavated materials within 2 ft of a trench or excavation.
- Use toe boards if excavated materials could fall back into a trench.

Digging equipment:

- Use mats or heavy planking to support digging equipment on soft ground.
- If a shovel or crane is placed on the excavation bank, install shoring and bracing to prevent a cave-in.

Figure 8-4 A Typical Company Procedures Bulletin Outlining Safe Practices on a Construction Site

mands, but like other procedures, they may be in the passive voice or indicative mood as well.

Writing Process Explanations

A *process explanation* is a description of how a series of actions leads to a specific result. Process explanations differ from instructions and procedures in that they are not intended to guide readers in performing actions but only in understanding them. As a result, a process explanation is written as a narrative, without listed steps or commands, describing four possible types of actions:

- Actions that occur in nature, such as how diamonds form, how the liver functions, or how a typhoon develops.
- Actions that produce a product, such as how steel, light bulbs, or baseballs are made.
- Actions that make up a particular task, such as how gold is mined, how blood is tested for cholesterol levels, or how a highway is paved.
- Actions in the past, such as how the Romans built their aqueducts or how the Grand Canyon was formed.

Readers

A process explanation may be a separate document, such as a science pamphlet, or a section in a manual or report. Although readers of process explanations do not intend to perform the steps themselves, they do need the information for specific purposes, and one process explanation will not serve all readers.

A general reader reading an explanation in a newspaper of how police officers test drivers for intoxication wants to know the main stages of the process and how a police officer determines if a driver is indeed intoxicated. A student in a criminology class needs more details in order to understand each major and minor stage of the process and to pass a test about those details. An official from the National Highway Traffic Safety Administration may read the narrative to see how closely it matches the agency's official guidelines, while a judge may want to be sure that a driver's rights are not violated by the process. Process explanations for these readers must serve their specific needs as well as describe the sequential stages of the process. Figure 8-5 shows how process explanations of sobriety testing may differ for general readers and for students. Notice that the version for students includes more details about what constitutes imbalance and more explanation about the conditions under which the test should or should not be given.

For a general reader:

The Walk-and-Turn Test

The police officer begins the test by asking the suspect to place the left foot on a straight line and the right foot in front of it. The suspect then takes nine heel-to-toe steps down the line, turns around, and takes nine heel-to-toe steps back. The suspect is given one point each for eight possible behaviors showing imbalance, such as stepping off the line and losing balance while turning. A score of two or more indicates the suspect is probably legally intoxicated.

For a student reader:

The Walk-and-Turn Test

The test is administered on a level, dry surface. People who are over 60 years old, more than 50 pounds overweight, or have physical impairments that interfere with walking are not given this test.

The police officer begins the test by asking the suspect to place the left foot on a straight line and the right foot in front of it. The suspect must maintain balance while listening to the officer's directions for the test and must not begin until the officer so indicates. The suspect then takes nine heel-to-toe steps down the line, keeping hands at the sides, eyes on the feet, and counting aloud. After nine steps, the suspect turns and takes nine heel-to-toe steps back in the same manner. The officer scores one point for each of the following behaviors: failing to keep balance while listening to directions, starting before told to, stopping to regain balance, not touching heel to toe, stepping off the line, using arms to balance, losing balance while turning, and taking more or less than nine steps each way. If the suspect falls or cannot perform the test at all, the officer scores nine points. A suspect who receives two or more points is probably legally intoxicated.

Figure 8-5 Two Process Explanations of Sobriety Testing

For all process explanations, analyze the intended reader's technical knowledge of the subject and why the reader needs to know about the process. Then decide the number of details, which details, and the appropriate technical terms to include in your narrative.

Organization

Figure 8-6 presents a model outline for a process explanation. As the outline illustrates, process explanations have three main sections: an introduction, the stages in the process, and a conclusion.

I. Introduction
   A. Definition
   B. Theory behind process
   C. Purpose
   D. Historical background, if appropriate
   E. Equipment, materials, special natural conditions
   F. Major stages
II. Stages in the Process
   A. Major stage 1
      1. Definition
      2. Purpose
      3. Special materials or conditions
      4. Description of major stage 1
      5. Minor stage 1
         a. Definition
         b. Purpose
         c. Special materials or conditions
         d. Description of minor stage 1
      6. Minor stage 2
         ⋮
   B. Major stage 2
      ⋮
III. Conclusion
   A. Summary of major stages and results
   B. Significance of process

Figure 8-6 Model Outline for a Process Explanation

Introduction

The introduction should include enough details about the process so that readers understand the principles underlying it and the conditions under which it takes place. Depending on your readers' technical expertise and how they intend to use the process explanation, include these elements:

- *Definition.* If the subject is highly technical or readers are not likely to recognize it, provide a formal definition.
- *Theory behind the process.* Explain the scientific principles behind the process, particularly if you are describing a research process.
- *Purpose.* Explain the purpose unless it is obvious from the title or the readers already know it.

- *Historical background.* Readers may need to know the history of a process and how it has changed in order to understand its current form.
- *Equipment, materials, and tools.* Explain what types of equipment, materials, and tools are essential for proper completion of the process.
- *Major stages.* Name the major stages of the process so that readers know what to expect.

Stages of the Process

Explain the stages of the process in the exact sequence in which they normally occur. In some cases, one sentence may adequately explain a stage. In other cases, each stage may actually be a separate process that contributes to the whole. In describing each stage, (1) define it, (2) explain how it fits the overall process, (3) note any dangers or special conditions, (4) describe exactly how it occurs, including who or what does the action, and (5) state the results at the end of that particular stage. If a major stage is made up of several minor actions, explain each of these and how they contribute to the major stage. Include the reasons for the actions in each stage:

> The valve is closed before takeoff because. . . .
>
> The technician uses stainless steel implements because. . . .

Conclusion

The conclusion of a process explanation should explain the expected results, what they mean, and how this process influences or interacts with others if it does.

Figure 8-7 explains one stage—smelting—in the process of refining copper. Notice that the writer explains the purpose of smelting and how it fits into the copper refining process, while the accompanying flowchart illustrates the smelting position in the sequence of stages.

Details and Language

Process explanations, like instructions, can include these elements:

- Descriptive title
- Precise details
- Complete sentences
- Consistent terms for actions or parts
- Headings in long narratives to guide readers
- Simple comparisons for general readers

Smelting. Smelting is concerned with the substantial removal of unwanted sulfur and metallic impurities from the concentrated ore. The removal of the sulfur is accomplished by heating the ore in the presence of air and thereby oxidizing the sulfur, which is released as sulfur dioxide. This particular phase of smelting is often referred to as *roasting*. It yields a product quite suitable for additional smelting, which can then be carried out in an economical manner. The smelting operation also makes use of a silicate flux, which removes some of the iron (present as an oxide) from the ore, by combining with it to form a slag. The slag is less dense than the concentrate, and is periodically removed from the smelting operation.

The material that remains in the furnace—copper, some iron, and smaller amounts of sulfur—is referred to as a *matte*. The matte is fed into another special furnace called a converter, and there it is reduced to relatively pure copper in a two-step process. The first step consists of eliminating any remaining sulfur and iron, again through an oxidation process, after which the remaining copper sulfide is reduced by reaction of the sulfide with air. After this phase of the operation is completed, large poles of green wood are used to stir the melt. The wooden poles reduce any remaining oxides by the reaction of the carbon in the poles with oxygen, leaving a relatively pure copper product.

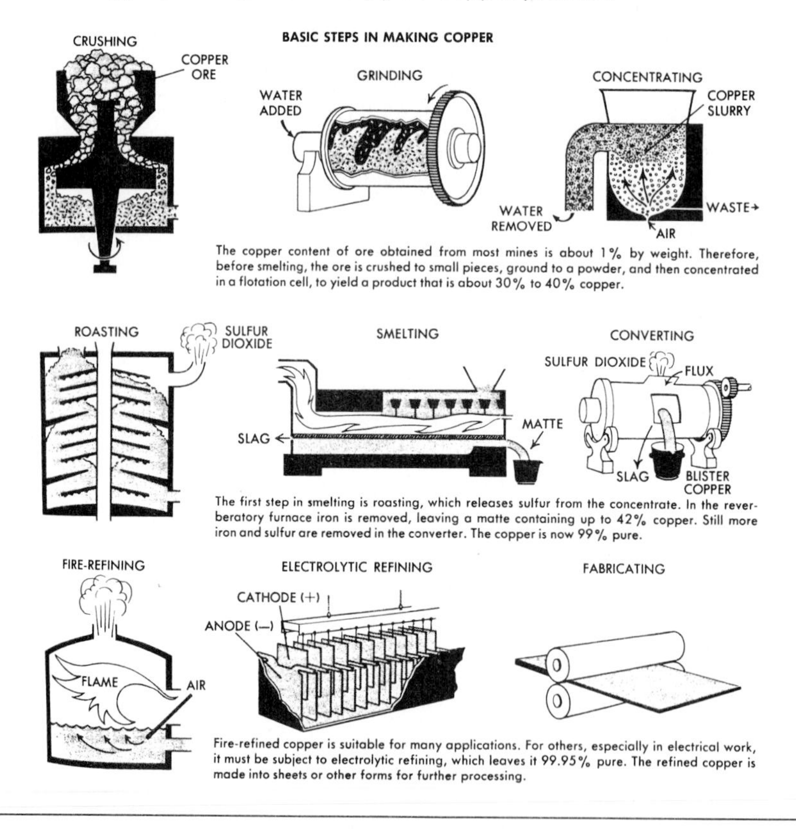

Figure 8-7 A Process Explanation of Smelting

In addition, remember these style guidelines:

- *Do not use commands.* Because readers do not intend to perform the actions, commands are not appropriate.
- *Use either passive voice or the indicative mood.* Passive voice is usually preferred for process explanations that involve the same person performing each action in order to eliminate the monotony of repeating "The technician" over and over. However, natural processes or processes involving several people are easier to read if they are in the indicative mood (see Figure 8-5).
- If you are writing a narrative about a process you were involved in, such as an incident report or a research report, use of the first person is appropriate:

  I arrested the suspect. . . .
  I called for medical attention. . . .
  I spliced both genes. . . .

  Second person (*you*) is never appropriate for a process explanation.
- Use transition words and phrases to indicate shifts in time, location, or situation in individual stages of the process:

  Shifts in time: *then, next, first, second, before*
  Shifts in location: *above, below, adjacent, top*
  Shifts in situation: *however, because, in spite of, as a result, therefore*

CHAPTER SUMMARY

This chapter discusses writing effective instructions, procedures, and process explanations. Remember:

- Instructions, procedures, and process explanations are related but distinct strategies for explaining the steps in a specific action.
- Instructions provide the steps in an action for readers who intend to perform it, either for a work-related task or for a private-interest task.
- Instructions typically have an introduction that includes purpose; lists of parts, materials, tools, and conditions; overview of the steps; description of a mechanism (if appropriate); and warnings, cautions, and notes.
- The steps in instructions should be in the exact sequence in which a reader will perform them.
- The conclusion in instructions, if included, explains the expected results.

- Instructions should be in the imperative mood and include precise details.
- Troubleshooting instructions explain what to do if problems arise after readers have performed all the steps in an action.
- Procedures are company guidelines for (1) a system with sequential steps that must be completed by one employee, (2) a system with sequential steps that involves several employees, and (3) handling equipment or events with or without sequential steps.
- Procedures for a task that one employee will perform are similar to instructions with numbered steps.
- Procedures for several employees are best written in playscript organization.
- Procedures for handling equipment and systems are often organized topically rather than sequentially because the steps do not have to be performed in a specific order.
- Process explanations describe how a series of actions lead to a specific result.
- The stages of the process are explained in sequence in process explanations, but because readers do not intend to perform the process, the steps are not written as commands.

SUPPLEMENTAL READINGS IN PART TWO

Berry, E. "How to Get Users to Follow Procedures," *IEEE Transactions on Professional Communication.*

Kleim, R. L. "Writing Technical Procedures," *Journal of Systems Management.*

Marra, J. L. "For Writers: Understanding the Art of Layout," *Technical Communication.*

Model 8-1 Commentary

These pages are from a 66-page Operator's Guide for the Diebold 1060 Everywhere Teller Machine (Model II). The instructions have descriptive headings and numbered steps. The instructions also include line drawings showing major parts and, for replacing the ribbon, the hand position appropriate for the action.

Warnings and cautions appear throughout the operator's guide and appear in boldface type to make them stand out.

Discussion

1. Compare the instructions for replacing a ribbon with those in Figure 8-1. Discuss the differences and similarities in the two.

2. Write a process explanation for "Clearing a Paper Jam" based on the instructions in this model. Assume that you are including this process explanation in a report to upper management and that your readers need to be aware of the process because the secretaries have been complaining about time wasted in clearing paper jams. The entire process takes eight minutes.

5.2
REPLACING THE 40-COLUMN CONSUMER PRINTER RIBBON

To replace the Consumer Printer ribbon cartridge, refer to the following instructions.

1. Place the ETM in the Maintenance mode (Section 3.2).
2. Select SUPPLY on the Main menu.
3. Locate the Consumer Printer (Figure 5-1).
4. Grasp the ribbon cartridge on both sides (Figure 5-4).
5. Pull the ribbon cartridge toward you (Figure 5-4).

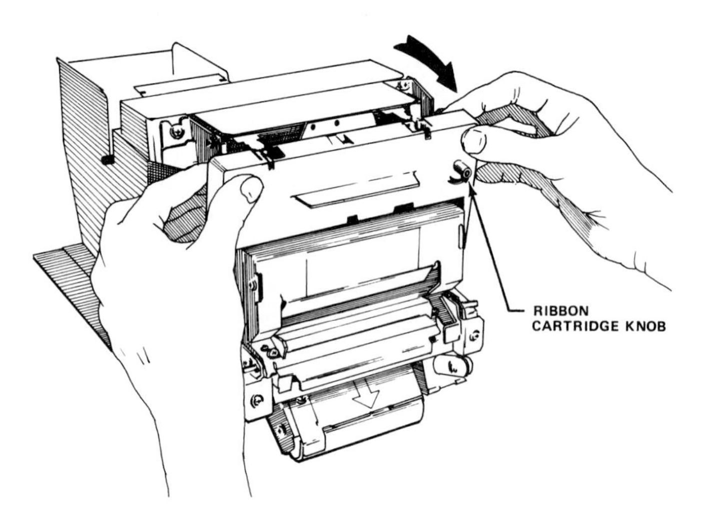

Figure 5-4 Removing the Ribbon Cartridge

6. Discard the used ribbon cartridge.
7. Remove ribbon slack on the replacement cartridge by turning the ribbon cartridge knob counterclockwise.
8. Position the new ribbon cartridge so that the ribbon rests in front of the print head.
9. Align the ribbon cartridge with the pins on the Consumer Printer.
10. Push the cartridge into place. The cartridge snaps in place when interlocked.

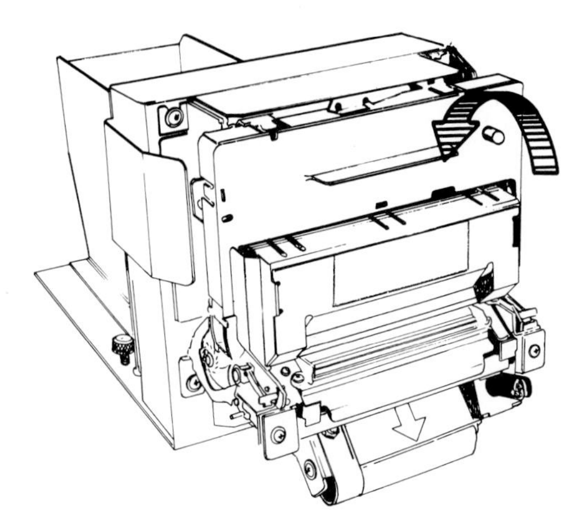

Figure 5-5 Removing Ribbon Slack

11. Turn the knob on the front of the ribbon cartridge counterclockwise until the ribbon rests under the print head (Figure 5-5).

12. Press the paper feed button. Check that the ribbon cartridge knob turns and the ribbon is not twisted under the print head.

13. Return the ETM to the Operating mode (Section 3.4).

5.3 CLEARING A PAPER JAM

If your printer has a paper jam, try to clear the printer mechanism by performing the procedure in this section.

1. Power down the ETM and remove the forms from the paper path.

   - *Turn off* the ETM by pressing the ON/OFF switch (Figure 5-1).

Make certain that the ETM is turned off. The electronics in the terminal contain enough voltage to cause serious injury.

   - Locate the Consumer Printer (Figure 5-1).
   - Remove the forms from the paper path. Push down on either of the paper release lever tabs (Figure 5-2) while pulling as much of the paper out of the printer mechanism as possible.

2. Move the print head away from the jammed paper by rotating the jam relief gear shown in Figure 5-6.

    - Rotate the gear clockwise to move the print head to the left side.
    - Rotate the gear counterclockwise to move the print head to the right side.

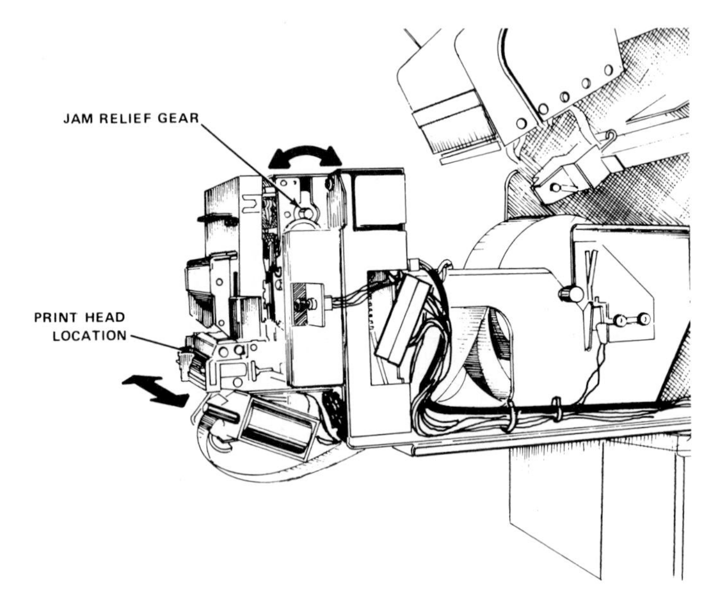

Figure 5-6 Moving the Print Head

3. Use a pincerlike tool (such as needle nose pliers) to grasp and pull the remaining paper remnants from the printer.

Paper remnants remaining in the Consumer Printer will cause further jamming.

4. Push the retaining clip away from the plastic dust collector and pull the dust collector from its position (Figure 5-7). (You may choose either the right or the left retaining clip.)

5. Press the cutter solenoid several times (Figure 5-7).

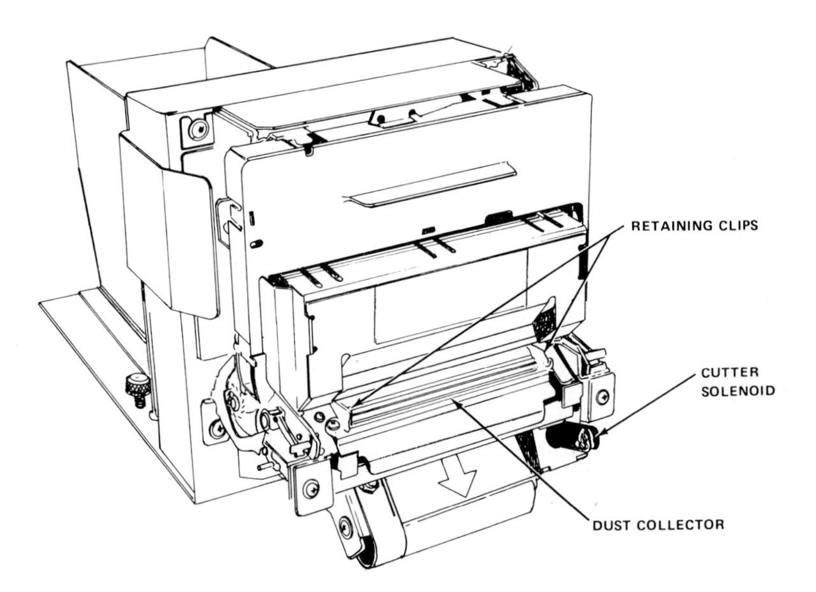

Figure 5-7 Removing the Dust Collector

6. Clear the cutter bar area of any paper fragments using a soft brush or cloth.

7. Replace the plastic dust collector.

   - Match one end of the dust collector with the appropriate end of the retaining clip.

   - Snap the dust collector in place matching the dust collector with the other retaining clip.

   - Press the cutter solenoid several times to ensure proper placement of the dust collector.

8. Restore power to the ETM.

    - *Turn on* the ETM by pressing the ON/OFF switch.

    - If the Consumer Printer is properly connected, the red and green lights on the Consumer Printer flash (Figure 7-5). Allow up to 8 minutes for the system to initialize and the status lights to flash.

      If the red and green lights are not flashing, check that the connector is properly interlocked with its receptacle. If the connection is secure but the lights are still not flashing, contact your Diebold service representative or your supervisor.

9. Place the ETM in the Maintenance mode (Section 3.2).

10. Select SUPPLY on the Main menu.

11. Reinstall the forms.

    - Guide the forms up through the paper feed opening at the front of the Consumer Printer.

    - Press the paper feed button (Figure 5-3) while guiding the forms supply up through the paper feed opening.

    - Press the paper feed button several more times until the forms supply feeds smoothly.

12. Perform the steps to return the ETM to the Operating mode (Section 3.4).

If your attempts to clear the paper jam using the procedure in this section have been unsuccessful, remove and replace the device.

Model 8-2 Commentary

These procedures were written by a student in response to Exercise 5 in this chapter. The student uses the playscript organization to present the system for handling the company tuition-refund program.

Headings call the readers' attention to the sequence of actions. Each step is a command, and the steps are numbered sequentially from beginning to end, no matter which employee is performing the step.

Discussion

1. Assume that you are an employee interested in the company's tuition-refund program. Is there any point in these procedures where you might be uncertain what to do next? If so, discuss how to revise.

2. The introduction discusses the company policy on tuition refunds and includes some additional information about eligibility. Is there any information in the numbered steps that you believe should be included in the introduction? If so, rewrite the introduction and incorporate all the information you believe readers should know about the procedures before starting the process of applying for a tuition refund.

Procedures for Employee Tuition Refund Program

The Whitstone Corporation offers an employee tuition-refund program to encourage employees to continue their technical training or complete advanced degrees. All full-time employees who have worked at Whitstone for one year may receive full reimbursement for courses taken for academic credit at fully accredited technical schools, colleges, or universities according to the following procedures:

BEFORE REGISTRATION

Supervisor:	1. Send a memo to all employees on August 1, December 1, and May 1, informing them of the company tuition program and inviting them to discuss education plans with you.
Employee:	2. See your supervisor at least two weeks before the start of classes to allow time to approve your courses for the tuition-refund program.
Supervisor:	3. Discuss employee's education plans and explain the application form.
Employee:	4. Complete the application form.

5. Submit the application form to supervisor.

Supervisor:

6. Review the application form for completeness and send it to Human Resources Clerk.

Human Resources Clerk:

7. Check the application for compliance with company regulations.
8. Approve or deny the application and return it to supervisor with comments.

Supervisor:

9. Inform employee of the decision.
10. If the decision is favorable, sign the Tuition-Refund Contract with employee and send it to Human Resources Clerk.
11. Be sure employee knows the required grade and employment status for a full refund.

AFTER REGISTRATION

Employee:

12. Submit fee invoice form to your supervisor after paying tuition, so your record is in order for reimbursement.

Supervisor:

13. Send employee's fee invoice to Human Resources Clerk with Form 26A.

Human Resources Clerk:

14. Record the fee invoice as a potential tuition refund in employee's file.

Supervisor:

15. Meet informally with employee once a month to ascertain progress in the course.
16. If the employee is having difficulty with the course, suggest a meeting with the Training Director to discuss study approaches.

AFTER COMPLETION

Employee:

17. Upon completion of the course, with a grade of at least B—, submit your final grade report to your supervisor.

Supervisor:

18. Send employee's final grade report with Form 26B to Human Resources Clerk.

Human Resources Clerk:

19. Check course identification and status and prepare a reimbursement check for the full course fee.

Supervisor:

20. Present the check to the employee and offer company congratulations.

Model 8-3 Commentary

These procedures for testing blood for the AIDS virus are from *The Complete Guide to Medical Tests* by H. W. Griffith (Fisher Books, Tucson, Ariz., 1988), pages 32–33. The procedures are written to acquaint patients with what to expect when they have a test for this disease.

The procedures are divided into major categories—"Before the test," "The test," "After the test," and "Test results." The steps listed under "Description of test" are not numbered but are in sequential order. Each test covered in this book uses the same format.

Discussion

1. Discuss the organization and headings used in these procedures. How helpful are they to the intended readers? Are any definitions needed for these readers? Notice that in the section "Equipment used" the writer assures readers that they cannot become infected from the test itself. Discuss whether this information would be useful placed anywhere else or repeated anywhere else.

2. Assume that you must write instructions for the nurse who will draw blood for the test. Using the information provided under "Description of test," rewrite the procedures so that they are suitable instructions for the nurse.

AIDS (Acquired Immune Deficiency Syndrome) Western Blot Test

Category: Immune response.
Subcategory: Autoantibody tests.
Material studied: Blood.
Estimated cost of test: $75.00.
Patient time for test: 3–5 minutes.
Reliability of test results: Good.
Available as home self-test? No.
Note: This is a more expensive and time-consuming laboratory test than the ELISA test for AIDS. See page 30. The ELISA test is commonly used for screening. The Western blot test checks "positive" ELISA tests.

BEFORE THE TEST

Purpose of test:

- Confirms presence of serum antibodies for human T-cell lymphotrophic virus type III (HTL-III) AIDS virus.

Where is test performed?

- Commercial laboratory, hospital, doctor's office.

Who performs test?

- Lab technician, nurse, doctor.

Risks and precautions:

- If tourniquet is applied on the arm too long (over 1 minute), it may cause an inaccurate test result. Request another sample to be collected to ensure accuracy.

Patient preparation:

- Activity—No changes necessary.
- Diet—No changes necessary.
- Medicines—No changes necessary.
- Disrobing—None required. Roll up sleeve only.

THE TEST

Sensory factors:

- Touching—You will feel mild discomfort when the needle is inserted into the vein or when the lancet pricks a finger, heel, or earlobe.
- Seeing—You will see the technician, nurse, or doctor, the basket or tray to hold the equipment, the needles, syringes, collecting tubes, and bandages.
- Feeling—Some degree of apprehension or fear is normal and should be expected. Discomfort disappears when the test is finished.
- Other senses (hearing, smell, taste)—Not affected.

Equipment used:

- Needles, syringes, and heparinized collecting tubes. Sterile, disposable equipment prevents contamination or spread of infection. There is no risk of becoming infected with the hepatitis virus, AIDS virus, or any other infecting germ.

Description of test:

- Technician, doctor, or nurse applies a tourniquet or blood pressure cuff to the upper arm if blood is collected from a vein.
- Skin over the vein to be stuck is cleaned with alcohol or other antiseptic on a piece of cotton.
- When blood is drawn from a vein, the operator feels the vein to be used and then punctures both the skin and vein in one quick stroke. The needle used is a sterile, disposable needle attached to a sterile, disposable syringe.

- Operator withdraws the needle and transfers sample from the collecting syringe into sterile tubes (identified with your name) before sending samples to the laboratory for analysis. Tubes are treated with an anticoagulant chemical to prevent clotting.
- If blood is collected from a finger, heel, or earlobe, skin over the selected site is cleaned with an antiseptic. The operator quickly pierces the skin to a shallow depth, using a sterile, disposable metal lancet. The drop or two of blood produced is collected into a capillary pipette.

AFTER THE TEST

Immediate posttest care:

- Apply pressure to the puncture site with cotton provided by the laboratory.
- If a vein has been punctured, raise your entire arm over your head while applying pressure.
- Some discoloration, soreness, or swelling may develop at the venepuncture site. This responds well to moist, warm compresses applied every 2 to 4 hours.

Activity after test:

- If test is reported as "abnormal" or "positive," consult your physician immediately! Although this test does *not prove* you have AIDS, it *does prove* you have become infected with the AIDS virus and you are now a carrier of the virus.
- If you are a man or a woman, it means you are likely to develop AIDS at some time in your future.
- Watch for the following symptoms:
  - Recurrent respiratory and skin infections.
  - Chronic fatigue.
  - Persistent diarrhea.
  - Unexplained weight loss.
  - Fever.
  - Swollen lymph glands throughout your body.
- If you are a woman, you may pass the virus to your unborn child.
- Practice "safe sex" or abstain from sexual intercourse.
- Do not donate blood.
- Don't share IV needles.

Time before test results available:

- 1 to 2 weeks.

TEST RESULTS

Test values:

- Test results are determined by immunoabsorbent assay to detect antibodies.

Normal values:

- Absence of AIDS virus.

What "abnormal" or "positive" may indicate:

- Presence of AIDS virus.

Taking these drugs may affect test results:

- None expected.

Other factors that may affect test results:

- Laboratory error.
- Immunity.
- Subclinical infection.
- Cross-reactivity with other viral agents.

Model 8-4 Commentary

This process explanation of kicking as it relates to football is from a description of football in *The Encyclopedia Americana,* Vol. 11 (Danbury, Conn.: Grolier, 1988), page 535. Kicking is one process included in the overall process explanation of how football is played. The writer identifies three types of kicks and then describes how they are made and their relevance to the game's progress.

Discussion

1. Read this process description and discuss how it does or does not reflect the style advice presented in this chapter.

2. Create a graphic aid that would effectively illustrate one element in this process explanation.

Kicking. Kicking—a punt, a placekick, or a dropkick—is a skill that only a few members on each team need master. The punt is generally used when a team, unable to make a first down in three plays, chooses to kick the ball out of its own territory on the fourth down. The punter sends the ball as far and as high as possible into his opponent's territory, or out of bounds when the distance to the goal is short of the punter's range. Speed and accuracy in punting are essential. The punter, awaiting the ball about 15 yards behind the scrimmage line, catches a direct pass from the center, positions the ball in his hands, lets it drop, and kicks it with a quick leg snap. To position the ball properly, a right-footed punter places his left hand on the front of the ball on its left side and his right hand to the rear of the ball on its right side. As the punter prepares to drop the ball he takes the first of several steps forward to gain momentum. When he drops it, the instep of his right foot should meet the ball just to the left of center before it touches the ground and sends it in a spiral flight toward the defensive team. (A left-footed kicker reverses this procedure.)

The placekick is used on the kickoff, on a free kick after a safety, or from scrimmage. It can also be used on a try for a field goal. For a placekick the ball must be set nearly upright on the field or on a tee of prescribed height (2 inches in the college game, 3 inches in the professional). For a field goal, a teammate kneeling by the spot where the ball is to be set receives the snap from the center and positions the ball for the kick. The kicker takes one forward step and, with a fast leg snap, meets the ball between the middle and bottom, lifting it high over charging defending players. The ball is hit with the toes, not the instep. In the NFL the ball must be placed on the ground; in the USFL a tee may be used.

The dropkick, archaic though still legal, is executed by letting the ball fall to the ground point foremost and kicking it quickly as it rises. The toe meets the ball slightly below the middle, resulting in an end-over-end flight.

Model 8-5 Commentary

This process explanation of how a smoke alarm works is from an information pamphlet published by the U.S. Department of Commerce, National Fire Prevention and Control Administration. The writer first describes the general operating principle and then the series of actions that result in an alarm. The drawing illustrates the interior of the alarm, showing how smoke enters the sensing chamber and reflects the light to the photocell, thus creating an alarm. The conclusion explains the effect smoke particle size has on the speed of the alarm.

Discussion

1. Identify the strategies the writer uses to make this process explanation easy for general readers.

2. In groups, draft a set of instructions that could accompany this explanation of the smoke alarm in which consumers are told what to do when the alarm sounds.

The Photoelectric Detector and How It Works

The photoelectric detector uses a small light source—either an incandescent bulb or a light-emitting diode (LED)—which shines its light into a dark sensing chamber. The sensing chamber also contains an electrical, light-sensitive component known as a photocell. The light source and photocell are arranged so that light from the source does not normally strike the photocell. When smoke particles enter the sensing chamber of the photoelectric detector, the light is reflected off the surface of the smoke particle, allowing it to strike the photocell and increase the voltage from the photocell. (This reflection of light is the same means by which we see smoke in the air. That is, light from the room strikes the smoke and reflects it to our eyes.) When the voltage reaches a predetermined level, the detector alarms.

Smoke particles that scatter visible light are larger in diameter than those which an ionization detector senses. Since smoldering fires produce these larger smoke particles in their greatest numbers, photoelectric detectors respond slightly faster to smoldering fires than ionization detectors.

In general, the size of the average smoke particle from a fire is inversely proportional to the combustion temperature. That is, the higher the temperature, the smaller the average particle size.

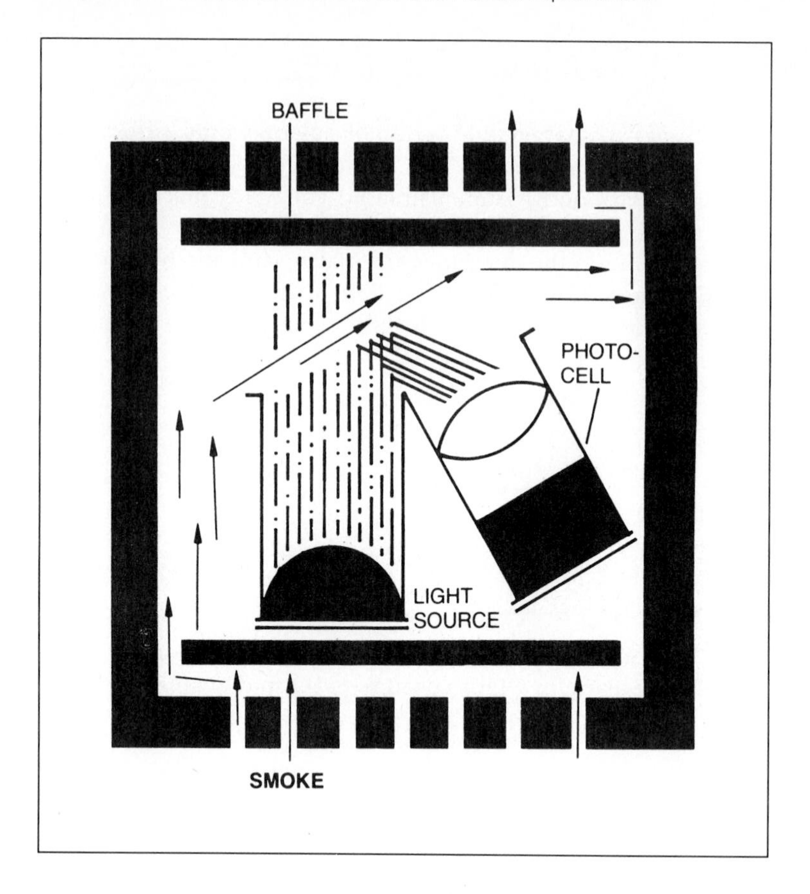
BAFFLE
PHOTO-
CELL
LIGHT
SOURCE
SMOKE

Chapter 8 EXERCISES

1. Write a set of instructions for one of the following activities. Identify your reader before you write. Include appropriate warnings and cautions and *one* graphic aid illustrating some aspect of the instructions.

Using a food processor
Changing a bicycle tire
Jump starting a car
Developing film
Installing a light fixture
Planting flowers
Stripping paint from furniture
Making pottery
Assembling a piece of equipment
Painting a house
Grilling hamburgers outdoors
Pouring concrete
Installing bathroom tiles
Parallel parking
Replacing an automobile fan belt
Cleaning an oven
Performing a particular laboratory test
Trimming hedges
Operating a particular machine
Performing a specific activity necessary to your field

2. Write a set of troubleshooting instructions that supplement the instructions you wrote for Exercise 1.

3. Write a set of safety tips for one of the following activities. Your reader is a consumer who has just purchased the product or who is just starting this activity. Include *one* appropriate graphic aid.

Riding a bicycle
Using a lawn mower
Installing an outside TV antenna
Playing a particular sport
Performing a particular laboratory procedure
Handling firearms

Riding a motorcycle

Sailing

Using a power saw

Cutting meat, fish, or vegetables

4. Using the process explanations of the "Walk-and-Turn Test" in Figure 8-5, write a set of instructions for a rookie police officer who is administering the test for the first time.

5. Write a set of procedures, using playscript organization, for a system involving several employees at your job. These procedures will be included in the new company procedures manual.

6. The following are some Buchanan Manufacturing Company safety rules regarding compressed gas, oxygen, and acetylene cylinders. Using this information, write a formal set of company procedures for handling these cylinders properly. These rules are not sequential steps.

- Never use cylinders as rollers or supports even if they are empty.
- Workers should store cylinders in the assigned areas and chain them in place to prevent upset or damage.
- Oxygen cylinders should be stored in the same area as gas cylinders.
- Avoid rough handling, dropping, or damaging cylinders.
- Do not use a hammer or wrench to open oxygen cylinder valves.
- Workers must close cylinder valves when work is finished.
- Cylinders that leak must not be used.
- If valves cannot be opened by hand, supervisors should be notified.
- Compressed gas cylinders must not be stored near furnaces, radiators, or any heat source.
- Cylinders stored outside must be protected from direct sun, moisture, and accumulated dirt.
- Acetylene cylinders must be stored, transported, and used with the valve up.
- Do not connect cylinders until contents have been identified from supplier invoices.
- Workers must use cylinders in the order they are shipped from the supplier.
- Keep empty cylinders separate from full ones.
- Cylinders should have a few pounds of pressure left even when considered empty.
- Acetylene should not be transferred from one cylinder to another nor mixed with any other gas.

- Do not let recessed tops of cylinders fill with water.
- Do not attempt to repair the cylinder valves on new acetylene cylinders.
- Never tamper with fusible safety plugs on acetylene cylinders.
- Cylinders should not come in contact with live wires.

7. Write a process explanation of one of the following actions. Your reader is a foreign student in the Orient who is studying English and wants to become familiar with American activities. Include a flowchart illustrating the process.

Trimming a Christmas tree

Baking cookies

Decorating Easter eggs

Making a pizza

Decorating a high school gym for the senior prom

Wrapping Christmas or birthday presents

8. Write a process explanation for a test or system you use in your major. Your reader is a student in another major who is interested in learning about the process used in your field.

9. Write a process explanation for a procedure used at your job. Your reader is your instructor who is interested in learning about your work. Include *one* graphic aid.

CHAPTER 9

Letters and Memos

Understanding Letters and Memos
- Letters
- Memos

Developing Effective Tone
- Natural Language
- Positive Language
- You-Attitude

Organizing Letters and Memos
- Direct Organization
- Indirect Organization
- Persuasive Organization

Writing Memos as a Manager

Selecting Letter Format
- Date Line
- Inside Address
- Salutation
  - Use Titles
  - Use Attention Lines
  - Omit Salutations
- Body
- Close
- Signature Block
- Notations
- Second Page

Selecting Memo Format

Writing Job-Application Letters

Writing Résumés

Chapter Summary

Supplemental Readings in Part Two

Models

Exercises

Understanding Letters and Memos

Correspondence includes both letters and memos. Although these frequently are addressed to one person, they often have multiple readers because the original reader passes along the correspondence to other parties or the writer sends copies to everyone involved. Letters and memos are the most versatile written documents and serve many purposes.

Letters

Letters are written primarily to people outside an organization and cover a variety of situations, such as (1) requests, (2) claims, (3) adjustments, (4) orders, (5) sales, (6) credit, (7) collections, (8) goodwill messages, (9) announcements, (10) records of agreements, (11) follow-ups to telephone conversations, (12) transmittals of other technical documents, and (13) job applications.

Memos

Memos are written primarily to people inside an organization. With the exception of job applications, memos cover the same topics as letters. In addition, many internal reports, such as trip reports, progress reports, and short proposals, may be in memo form.

Give the same careful attention to readers, purpose, and situation for letters and memos that you do in writing all technical documents. Because of its person-to-person style, however, correspondence may create emotional responses in readers that other technical documents do not. A reader may react positively to a set of new procedures in a company manual, but a memo from a superior announcing these same procedures may strike the reader as demanding and dictatorial. When you write letters and memos, consider tone and organizational strategy from the perspective of how your readers will respond emotionally as well as practically to the message.

Developing Effective Tone

Tone refers to the feelings created by the words in a message. Business letters and memos should have a tone that sounds natural and conveys cooperation, mutual respect, sincerity, and courtesy. Tone of this kind establishes open communication and reflects favorably on both the writer and the company. Because words on paper cannot be softened with a smile or a gesture, take special care in word choice to avoid sounding harsh or accusing. Remarks that were intended to be objective may strike your reader as overly strong commands or tactless insinuations, especially if the subject is at all controversial. Remember the following principles for creating a pleasant and cooperative tone in your letters and memos.

Natural Language

Write letters and memos in clear, natural language, and avoid falling into the habit of using old-fashioned phrases that sound artificial. A reader who has to struggle through a letter filled with out-of-date expressions probably will become annoyed with both you and your message, resulting in poor communication. This sentence is full of out-of-date language:

> Per yours of the tenth, please find enclosed the warranty.

No one really talks this way, and no one should write this way. Here is a revision in more natural language:

> I am enclosing the warranty you requested in your letter of March 10.

Keep your language simple and to the point. This list shows some stale business expressions that should be replaced by simpler, more natural phrases:

Old-Fashioned	*Natural*
Attached hereto . . .	Attached is . . .
We beg to advise . . .	We can say that . . .
Hoping for the favor of a reply . . .	I hope to hear from you . . .
As per your request . . .	As you requested . . .
It has come to my attention . . .	I understand that . . .
Prior to receipt of . . .	Before we received . . .
Pursuant to . . .	In regard to . . .
The undersigned will . . .	I will . . .

If you use such out-of-date expressions, your readers may believe you are as out of date in your information as you are in your language.

Positive Language

Keep the emphasis in your letters and memos on positive rather than negative images. Avoid writing letters or memos when you are angry, and never let anger creep into your writing. In addition, avoid using words that emphasize the negative aspects of a situation; emphasize the positive whenever you can, or at least choose neutral language. Shown below are sentences that contain words that emphasize a negative rather than a positive or neutral viewpoint. The revisions show how to eliminate the negative words.

Negative: When I received your complaint, I checked our records.

Positive: When I received your letter, I checked our records.

Negative: To avoid further misunderstanding and confusion, our sales representative will visit your office and try to straighten out your order.

Positive: To ensure that your order is handled properly, our sales representative will visit your office.

Negative: I am sending a replacement for the faulty coil.

Positive: I am sending a new coil that is guaranteed for one year.

Negative: The delay in your shipment because we lost your order should not be longer than four days.

Positive: Your complete order should reach you by September 20.

In these examples, a simple substitution of positive-sounding words for negative-sounding words improves the overall tone of each sentence, whereas the information in each sentence and its purpose remain the same.

No matter what your opinion of your reader, do not use language that implies that the reader is dishonest or stupid. Notice how the following revisions eliminate the accusing and insulting tone of the original sentences.

Insulting: Because you failed to connect the cable, the picture was blurred.

Neutral: The picture was blurred because the V-2 cable was not connected to the terminal.

Insulting: You claimed that the engine stalled.

Neutral: Your letter said that the engine stalled.

Insulting: Don't let carelessness cause accidents in the testing laboratory.

Neutral: Please be careful when handling explosive compounds.

Negative language, either about a situation or about your reader, will interfere with the cooperation you need from your reader. The emphasis in correspondence should be on solutions rather than on negative events.

You-Attitude

The *you-attitude* refers to the point of view a writer takes when looking at a situation as the reader would. In all correspondence, try to convey an appreciation of your reader's position. To do this, present information from the standpoint of how it will affect or interest your reader. In these sample sentences, the emphasis is shifted from the writer's point of view to the reader's by focusing on the benefits to the reader in the situation:

Writer emphasis: We are shipping your order on Friday.
Reader emphasis: You will receive your order by Monday, October 10.
Writer emphasis: To reduce our costs, we are changing the billing system.
Reader emphasis: To provide you with clear records, we are changing our billing system.
Writer emphasis: I was pleased to hear that the project was completed.
Reader emphasis: Congratulations on successfully completing the project!

By stressing your reader's point of view and the benefits to your reader in a situation, you can create a friendly, helpful tone in correspondence. Of course, readers will see through excessive praise or insincere compliments, but they will respond favorably to genuine concern about their opinions and needs. Here are guidelines for establishing the you-attitude in your correspondence:

1. Put yourself in your reader's place, and look at the situation from his or her point of view.
2. Emphasize your reader's actions or benefits in a situation.
3. Present information as pleasantly as possible.
4. Offer a helpful suggestion or appreciative comment when possible.
5. Choose words that do not insult or accuse your reader.
6. Choose words that are clear and natural, and avoid old-fashioned or legal-sounding phrases.

Organizing Letters and Memos

Most letters and memos are best organized in either a *direct* or an *indirect pattern,* depending on how the reader is likely to react emotionally to the message. If the news is good, or if the reader does not have an emotional stake in the subject, use the direct pattern with the main idea in the opening. If, however, the news is bad, the indirect pattern with the main idea after the explanation is often most effective because a reader may not read the explanation of the situation if the bad news appears in the opening. Most business correspondence uses the direct pattern. The indirect pattern, however, is an important strategy whenever a writer has to announce bad news in a sensitive situation and wants to retain as much goodwill from the reader as possible. A third letter pattern is the *persuasive pattern*, which also places the main idea in the middle portion or even the closing of a message. Writers use this pattern for sales messages and when the reader needs to be convinced about the importance of a situation before taking action.

Memos present a special writing challenge in that very often they have mixed purposes. A memo announcing the installation of new telephone equipment is a message to inform, but if the memo also contains instructions for using the equipment, the purpose is both to announce and to direct. Moreover, a memo can be either good or bad news, and if the memo is addressed to a group of employees, it can contain *both* good and bad news, depending on each individual's reaction to the topic. A memo announcing a plant relocation can be good news because of the expanded production area and more modern facility; however, it also can be bad news because employees will have to uproot their lives and move to the new location. You will have to judge each situation carefully to determine the most effective approach and pattern to use.

Direct Organization

Figure 9-1 shows a letter that conveys good news and is written in the direct pattern. In this situation, a customer has complained about excessive wear on the tile installed in a hospital lobby. Because the tile was guaranteed by the company that sold and installed it, the customer wants the tile replaced. The company will stand behind its guarantee, so the direct pattern is appropriate in this letter because the message to the customer is good news. The direct pattern generally has three sections:

1. The opening establishes the reason for writing the letter and presents the main idea.
2. The middle paragraphs explain all relevant details about the situation.
3. The closing reminds the reader of deadlines, calls for an action, or looks to future interaction between the reader and the writer.

In Figure 9-1, the opening contains the main point, that is, the good news that the company will grant the customer's request for new tile. The middle paragraph reaffirms the guarantee, describes the tile, and explains why the replacement tile should be satisfactory. In the closing, the manager tells the customer to expect a call from the sales representative and also thanks the customer for her information about the previous tile order. The writer uses a

PRESCOTT TILE COMPANY
444 N. Main Street
Ottumwa, Iowa 52555

April 13, 1990

Ms. Sonia Smithfield
General Manager
City Hospital
62 Prairie Road
Fort Madison, IA 52666

Dear Ms. Smithfield:

We will be more than happy to replace the Durafinish tile in front of the elevators and in the lobby area of City Hospital as you requested in your letter of April 6.

When we installed the tile—model 672—in August 1989, we guaranteed the nonfade finish. The tile you selected is imported from Paloma Ceramic Products in Italy and is one of our bestselling tiles. Recently, the manufacturer added a special sealing compound to the tile, making it more durable. This extrahard finish should withstand even the busy traffic in a hospital lobby.

Our sales representative, Dawn Truhart, will call on you in the next few days to inspect the tile and make arrangements for replacement at no cost to you. I appreciate your calling this situation to my attention because I always want to know how our products are performing. We guarantee our customers' satisfaction.

Sincerely yours,

Michael Allen
Product Installation Manager

MA:tk

c: Dawn Truhart

Figure 9-1 A Sample Letter Using Direct Organization (in Semiblock Style)

pleasant, cooperative tone and avoids repeating any negative words that may have appeared in the customer's original request; therefore, the letter establishes goodwill and helps maintain friendly relations with the reader.

Indirect Organization

The basic situation in the letter in Figure 9-2 is the same as in Figure 9-1, but in this case the letter is a refusal because the guarantee does not apply. Therefore, indirect organization is used.

The indirect pattern generally contains these elements:

1. The opening is a "buffer" paragraph that establishes a friendly, positive tone and introduces the general topic in a way that will later support the refusal or negative information and help the reader understand it. Use these strategies when appropriate:
   - Agree with the reader in some way:

     You are right when you say that. . . .
   - State appreciation for past efforts or business:

     Thank you for all your help in the recent. . . .
   - State good news if there is any:

     The photographs you asked for were shipped this morning under separate cover.
   - Assure the reader the situation has been considered carefully:

     When I received your letter, I immediately checked. . . .
   - Express a sincere compliment:

     Your work at the Curative Workshop for the Handicapped has been outstanding.
   - Indicate understanding of the reader's position:

     We understand your concern about the Barnet paint shipment.
   - Anticipate a pleasant future:

     The prospects for your new business venture in Center City look excellent.

   Do not use negative words in the buffer or remind your reader about the unpleasantness of the situation. Buffers should establish a pleasant tone and a spirit of cooperation, but do not give the reader the impression that the request will be granted or that the main point of the message is good news.
2. The middle section carefully explains the background of the situation, reminding the reader of all the details that are important to the main point.

PRESCOTT TILE COMPANY
444 N. Main Street
Ottumwa, Iowa 52555

April 13, 1990

Ms. Sonia Smithfield
General Manager
City Hospital
62 Prairie Road
Fort Madison, IA 52666

Dear Ms. Smithfield:

You are certainly correct that we guarantee our tile for 20 years after installation. We always stand behind our products when they are used according to the manufacturer's recommendations and the recommendations of our design consultant.

When I received your letter, I immediately got out your sales contract and checked the reports of the design consultant. Our records indicate that the consultant did explain on March 6, 1989, that the Paloma tile--model 672--was not recommended for heavy traffic. Although another tile was suggested, you preferred to order the Paloma tile, and you signed a waiver of guarantee. For your information, I'm enclosing a copy of that page of the contract. Because our recommendation was to use another tile, our usual 20-year guarantee is not in force in this situation.

For your needs, we do recommend the Watermark tile, which is specially sealed to withstand heavy traffic. The Watermark tile is available in a design that would complement the Paloma tile already in place. Our design consultant, Trisha Lyndon, would be happy to visit City Hospital and recommend a floor pattern that could incorporate new Watermark tile without sacrificing the Paloma tile that does not show wear. Enclosed is a brochure showing the Watermark designs. Ms. Lyndon will call you for an appointment this week, and because you are a past customer, we will be happy to schedule rush service for you.

Sincerely,

Michael Allen
Product Installation Manager

MA/dc
Encs.: Watermark brochure
contract page
cc: Ms. Trisha Lyndon

Figure 9-2 A Sample Letter Using Indirect Organization (in Full Block Style)

3. The bad news follows immediately after the explanation and in the same paragraph. Do not emphasize bad news by placing it in a separate paragraph.
4. The closing maintains a pleasant tone and, if appropriate, may suggest alternatives for the reader, resell the value of the product or service, or indicate that the situation can be reconsidered in the future.

The emphasis in an indirect pattern for bad news should be to assure the reader that the negative answer results from careful consideration of the issue and from facts that cannot be altered by the writer. As the writer of a bad news letter, you do not want to sound arbitrary and unreasonable.

In Figure 9-2, the opening "buffer" paragraph does confirm the guarantee but mentions restrictions. In the next section the manager carefully explains the original order and reminds the customer that she did not follow the company's recommendation—thus the guarantee is not in effect. The manager encloses a copy of the waiver that the customer signed. The refusal sentence comes as a natural result of the events that the manager has already described. The final section represents a movement away from the bad news and suggests a possible solution—ordering different tile. The manager also suggests how some of the original tile can be saved. The final sentences look to the future by promising a call from the design consultant and noting the enclosure of a brochure that illustrates the suggested new tile. In this letter the manager does not use such phrases as "I deeply regret" or "We are sorry for the inconvenience" because these expressions may imply some fault on the company's part, where there is none. Instead, the manager emphasizes the facts and maintains a pleasant tone through his suggestion for a replacement tile and his offer to find a way to save some of the old tile.

Although you cannot give readers good news if there is none, use the indirect pattern of organization to help your readers understand the reasons behind bad news, and emphasize goodwill by suggesting possible alternatives.

Persuasive Organization

Figure 9-3 shows a letter written in the persuasive pattern used for sales messages. In this letter the writer urges the reader to become a museum supporter and purchase reproductions of eighteenth- and nineteenth-century toys. The persuasive pattern generally includes these elements:

1. The opening in a persuasive letter catches the reader's attention through one of these strategies:
   - A startling or interesting fact:

     Every night, over 2000 children in our city go to bed hungry.

THE MUSEUM OF AMERICAN TOYS
One Michigan Avenue Plaza
Chicago, Illinois 60001

Dear Friend:

The Museum of American Toys is unique among our nation's museums because it is the only one devoted exclusively to preserving over three hundred years of delightful mementos of childhood.

Since its opening at the 1893 World's Columbian Exposition in Chicago, the Museum has been the major source of information about American toys and their creators as well as serving as the repository of over 15,000 toys made in the United States from the Pilgrims' landing to the present. Now the most appealing of these enchanting, one-of-a-kind toys made before World War I are available as Museum reproductions for your own pleasure or as gifts to those young in years or young at heart.

The high quality of our reproductions results from a two-year development plan for each toy, beginning with its selection for our catalogue and ending when the final decorative touches are applied. Each toy is carefully reproduced, reflecting not only its original appearance but also the fashion and interests of its era. Every advance in science left its mark on American toys, and the Museum's 300 reproductions reflect the changes in science and culture that mark our nation's history.

The collection includes the colonial broomstick doll, beloved by all little girls, with its colorful straw face and perky blue and white colonial dress . . . the Great Western Railroad with tiny iron boxcars, fiery red engine, and jaunty red caboose . . . a stately, white Mississippi riverboat with revolving paddle wheel . . . San Francisco painted tin cable car, holding little tin passengers and a conductor . . . painted wooden cowboys and Indians astride prancing horses . . . all these as fresh and exciting as when they were first created by skilled craftsmen.

If you become a Museum member now, you will receive our catalogue illustrating and describing every toy available in reproduction with plenty of time to order gifts for the holidays at the member discount of 25% off catalogue price. The proceeds from the sale of these fascinating toys are used to support the Museum, and we unconditionally guarantee

Figure 9-3 A Sample Letter Using Persuasive Organization

your satisfaction with these works of art. Please return the enclosed card now or call our toll-free number 1-800-555-6678 to receive your membership card and copy of the latest Museum catalogue within ten days.

Experience the best of our national heritage through these exquisite toys, charming reminders of America's exciting past.

Sincerely,

Duncan Farrady
Curator

DF:ss

P.S. Become a Museum member before the end of the year and receive a miniature replica of a turn-of-the-century toy automobile.

Figure 9-3 cont.

- A solution to a problem:

  At last, a health insurance plan that fits your needs!

- A story:

  Our company was founded 100 years ago when Mrs. Clementine Smith began baking. . . .

- An intriguing question:

  Would you like to enjoy a ski weekend at a fabulous resort for only a few dollars?

- A special product feature:

  Our whirlpool bath has a unique power jet system like no other.

- A sample:

  The sandpaper you are holding is our latest. . . .

2. The middle paragraphs of a persuasive letter build the reader's interest by describing the product, service, or situation. Use these strategies when appropriate:
   - Describe the physical details of the product or service to impress the reader with its usefulness and quality.
   - Explain why the reader needs this product or service both from a practical standpoint and from an enjoyment standpoint.
   - Explain why the situation is important to the reader.
   - Describe the benefits to the reader that will result from this product or service or from handling the situation or problem.
3. After arousing the reader's interest or concern, request action, such as purchasing the product, using the service, or responding favorably to the persuasive request.
4. The closing reminds the reader of the special benefits to be gained from responding as requested and urges action immediately or by a relevant deadline.

In Figure 9-3 the first two paragraphs catch the reader's attention through interesting facts about the museum's history, purpose, and special service—reproductions of historical toys. In the middle paragraphs, the writer emphasizes the care that goes into each reproduction and describes some of the toys available for purchase. The closing urges the reader to become a member and benefit from a discount on the toys, while the return card and toll-free number make it convenient for the reader to respond. The final sentence in the letter reminds the reader about the special qualities of the museum. Notice that a persuasive sales letter may differ from other letters in these ways:

- It may not be dated. Companies often use the same sales letters for several months, so the date of an individual sales letter is not significant.

- It may be a form letter without a personal salutation. The museum letter is addressed to "Dear Friend," setting a pleasant tone that implies the reader is already interested in the museum.
- It may include a P.S. In most letters, using a P.S. implies that the writer neglected to organize the information before writing. In persuasive sales letters, however, the P.S. is often used to urge the reader to immediate action or to offer a new incentive for action. The P.S. in the museum letter announces a special gift if the reader becomes a member before the end of the year.
- It may not include prices if other enclosed materials explain the costs. Sometimes the product cost is one of its exciting features and appears in the opening or in the product description. Usually, however, price is not a particularly strong selling feature. The museum letter did not mention the membership fee because it is on the enclosed membership card. Most sales letters, such as those for industrial products, magazine subscriptions, or mail-order items, include supplemental materials that list the prices.

Persuasive memos rarely follow this persuasive sales pattern completely, but they often do begin with an opening designed to arouse reader interest, such as

> With a little extra effort from all of us, South Atlantic Realty will have the biggest sales quarter in the history of the company.

Writing Memos as a Manager

A mutually cooperative communication atmosphere is just as important between managers and employees as it is between employees and people outside a company. The tone and managerial attitude in a manager's memos often have a major impact on employee morale. Messages with a harsh, demanding tone that do little more than give orders and disregard the reader's emotional response will produce an atmosphere of distrust and hostility within a company. For this reason, as a manager, you need to remember these principles when writing memos:

1. *Provide adequate information.* Do not assume that everyone in your company has the same knowledge about a subject. Explain procedures fully, and be very specific about details.
2. *Explain the causes of problems or reasons for changes.* Readers want more than a bare-bones announcement. They want to know *why* something is happening, so be sure to include enough explanation to make the situation clear.

3. *Be clear.* Use natural language, and avoid loading a memo with jargon. Often employees in different divisions of the same company do not understand the same jargon. If your memo is going beyond your unit, be sure to fully identify people, equipment, products, and locations.
4. *Be pleasant.* Avoid blunt commands or implications of employee incompetence. A pleasant tone will go a long way toward creating a cooperative environment on the job.
5. *Motivate rather than order.* When writing to subordinates, remember to explain how a change in procedure will benefit them in their work, or discuss how an event will affect department goals. Employees are more likely to cooperate if they understand the expected benefits and implications of a situation.
6. *Ask for feedback.* Be sure to ask the reader for suggestions or responses to your memo. No one person has all the answers; other employees can often make valuable suggestions.

Here is a short memo sent by a supervisor in a testing laboratory. The memo violates nearly every principle of effective communication.

> Every Friday afternoon there will be a department cleanup beginning this Friday. All employees must participate. This means cleaning your own area and then cleaning the complete department. Thank you for your cooperation.

The tone of this memo indicates that the supervisor distrusts the employees, and the underlying implication is that the employees may try to get out of doing this new task. The final sentence seems to be an attempt to create a good working relationship, but the overall tone is already so negative that most employees will not respond positively. The final sentence is also a cliché closing because so many writers put it at the end of letters and memos whether it is appropriate or not. Although the memo's purpose is to inform readers about a new policy, the supervisor does not explain the reason for the policy and does not provide adequate instructions. Employees will not know exactly what to do on Friday afternoon. Finally, the memo does not ask for feedback, implying that the employees' opinions are not important. Here is a revision:

> As many of you know, we have had some minor accidents recently because the laboratory equipment was left out on the benches overnight and chemicals were not stored in sealed containers. Preventing such accidents is important to all of us, and therefore, a few minutes every Friday afternoon will be set aside for cleanup and storage.
>
> Beginning on Friday, October 14, and every Friday thereafter, we'll take time at 4:00 p.m. to clean equipment, store chemicals, and straighten up the work areas. If we all pitch in and help each other, the department should be in good order within a half hour. Please let me know if you have any ideas about what needs special attention or how to handle the cleanup.

This memo, in the direct organizational pattern, explains the reason for the new policy, outlines specifically what the cleanup will include, mentions safety as a reader benefit, and concludes by asking for suggestions. The overall tone emphasizes mutual cooperation. This memo is more likely to get a positive response from employees than the first version is. Since a manager's success is closely linked to employee morale and cooperation, it is important to take the time to write memos that will promote a cooperative, tension-free environment.

Selecting Letter Format

The two most common letter formats are illustrated in Figure 9-1, the semiblock style, and Figure 9-2, the full block style. In the *full block style,* every line—date, address, salutation, text, close, signature block, and notations—begins at the left margin. In the *semiblock style,* however, the date, close, and signature block start just to the right of the center of the page. Business letters have several conventions in format that most companies follow.

Date Line

Since most company stationery includes an address, or letterhead, the date line consists only of the date of the letter. Place the date two lines below the company letterhead. If company letterhead is not used, put your address directly above the date:

1612 W. Fairway Street
Dayton, OH 45444
May 12, 1989

Spell out words such as *street, avenue,* and *boulevard* in addresses.

Inside Address

Place the reader's full name, title, company, and address two to eight lines below the date and flush with the left margin. Spell out the city name, but use the Postal Service two-letter abbreviations for states. Put one space between the state and the ZIP code. The number of lines between the date and the inside address varies so that the letter can be attractively centered on the page.

Salutation

The salutation, or greeting, appears two lines below the inside address and flush with the left margin. In business letters, the salutation is always followed by a colon. Address men as Mr. and women as Ms., unless a woman specifically indicates that she prefers Miss or Mrs. Professional titles, such as

Dr., Judge, or Colonel, may be used as well. Here are some strategies to use if you are unsure exactly who your reader is.

Use Titles

When writing to a group or to a particular company position, use descriptive titles in the salutations:

Dear Members of Com-Action:

Dear Project Director:

Dear Customer:

Dear Contributor:

Use Attention Lines

When writing to a company department, use an attention line with no salutation. Begin the letter two lines below the attention line:

Standard Electric Corporation
Plaza Tower
Oshkosh, WI 54911

Attention: Marketing Department

According to our records for 1988. . . .

Use an attention line also if the reader has not been identified as a man or a woman:

Standard Electric Corporation
Plaza Tower
Oshkosh, WI 54911

Attention: J. Hunter

According to our records for 1988. . . .

Omit Salutations

When writing to a company without directing the letter to a particular person or position, omit the salutation and begin the letter three lines below the inside address:

Standard Electric Corporation
Plaza Tower
Oshkosh, WI 54911

According to our records for 1988. . . .

General salutations, such as "Dear Sir" or "Gentlemen," are not used much anymore because they might imply an old-fashioned, sexist attitude.

Body

The body of a letter is single-spaced, and double-spaced between paragraphs. If the letter is very short, it can be double-spaced with the paragraphs indented five spaces. In this case, the semiblock style is more likely to look balanced on the page. Although word processors can justify (end the lines evenly) the right margins, most people are used to seeing correspondence with an uneven right margin. Justified margins imply a mass printing and mailing, while unjustified margins imply an individual message to a specific person.

Close

The close appears two lines below the last sentence of the body and consists of a standard expression of goodwill. In semiblock style, the close appears just to the right of the center of the page (see Fig. 9-1). In full block style, the close is at the left margin (see Fig. 9-2). The most common closing expressions are "Very truly yours," "Sincerely," and "Sincerely yours." As shown in Figure 9-1, the first word of the close is capitalized, but the second word is not. The close is always followed by a comma.

Signature Block

The signature block begins four lines below the close and consists of the writer's name with any title directly underneath. The signature appears in the four-line space between the close and the signature block.

Notations

Notations begin two lines below the signature block and flush with the left margin. In Figure 9-1, the capital initials "MA" represent the writer, and the lowercase initials "tk" represent the typist. A colon or slash always appears between the two sets of initials.

If materials are enclosed with the letter, indicate this with either the abbreviation *Enc.* or the word *Enclosure*. Some writers show only the number of enclosures as "Encs. (3)"; others list the items separately, as shown in Figure 9-2.

If other people receive copies of the letter, be sure to indicate this. The notation *cc* means carbon copy, and although companies use photocopies now, *cc* is still the most popular way to indicate copies. Some companies now, however, use *c* alone or *copy* to signal the reader that other people received copies of this correspondence (see Figs. 9-1 and 9-2). For copy notations, use full names alone (see Fig. 9-1) or include courtesy titles (see Fig. 9-2). List people receiving copies according to company rank, the highest first. If space permits, type the copy notation two lines below any other notations, as shown in Figure 9-1. If the letter is long, the copy notation should follow other notations without the extra line, as in Figure 9-2.

Second Page

If your letter has a second page, type the name of the addressee, page number, and date across the top of the page:

Ms. Sonia Smithfield 2 April 13, 1990

This heading also may be typed in the upper left-hand corner:

Ms. Sonia Smithfield
page 2
April 13, 1990

Selecting Memo Format

Many companies have printed memo forms so that all internal messages have a consistent format. If there is no printed form, the memo format shown in Model 9-4 is often used.

The subject line should be brief but clearly indicate a specific topic. Use key words so that readers can recognize the subject quickly and so that the memo can be filed easily. Capitalize the major words.

Memos do not have a close, so you may initial your typed name in the opening or sign your name at the bottom of the page. The writer's initials and the typist's initials appear two lines below the last line of the memo, followed by enclosure or copy notations. If a second page is required, use the same heading as for a second page of a letter.

Writing Job-Application Letters

One of the most important letters you will write is a letter applying for a job. The application letter functions not only as a request for an interview, but also as the first demonstration to a potential employer of your communication skills. Model 9-7 illustrates an application letter from a new college graduate. An application letter should do three things:

1. *Identify what you want.* In the opening paragraph, identify specifically the position you are applying for and how you heard about it—through an advertisement or someone you know. In some cases, you may write an application letter without knowing if the company has an opening. Use the first paragraph to state what kind of position you

are qualified for and why you are interested in that particular company.

2. *Explain why you are qualified for the position.* Do not repeat your résumé line for line, but do summarize your work experience or education and point out the specific items especially relevant to the position you are applying for. In discussing your education, mention significant courses or special projects that have enhanced your preparation for the position you want. If you have extracurricular activities that show leadership qualities or are related to your education, mention these in this section. Explain how your work experience is related to the position for which you are applying.
3. *Ask for an interview.* Offer to come for an interview at the employer's convenience; however, you also may suggest a time that may be suitable. Tell the reader how to reach you easily by giving a telephone number or specifying what time of the day you are available.

In Model 9-7, the writer begins her application letter by explaining how she heard about the position, and she identifies her connection with the person who told her to apply for the job. In discussing her education, she points out her training in computers and mentions her laboratory duties that gave her relevant experience in product chemistry. Her work experience has been in research laboratories, so she explains the kind of testing she has done. In closing, she suggests a convenient time for an interview and includes a phone number for certain daytime hours. Overall, the letter emphasizes the writer's qualifications in chemistry and points out that the company could use her experience (reader benefits). An application letter should be more than a brief cover letter accompanying a résumé; it should be a fully developed message that provides enough information to help the reader make a decision about offering an interview.

Writing Résumés

The résumé for a new college graduate is usually one page long and summarizes the applicant's career objective, education, work experience, and activities. Model 9-8 shows the résumé that would accompany the application letter in Model 9-7. Remember that anyone reading a résumé wants to find information easily and quickly, so the headings that direct the reader to specific information should stand out clearly. Because the résumé in Model 9-8 is for a new college graduate, education has a prominent position. There are many résumé styles; this one is a basic chronological format:

Heading. The heading should consist of name, address, and telephone number. A business telephone number also can be listed so that the reader can contact the applicant easily.

Objective. List a specific position that matches your education and experience, because employers want to see a clear, practical objective. Avoid vague descriptions, such as, "I am looking for a challenging position where I can use my skills." Availability to travel is important in some companies, so the phrase "willing to travel" can be used under the objective. If you are not willing to travel, say nothing.

Education. List education in reverse chronological order, your most recent degree first. Once you have a college degree, you can omit high school. Be sure to list any special certificates or short-term training done in addition to college work. Include courses or skills that are especially important to the type of position for which you are applying. List your grade-point average if it is significantly high, and indicate the grade-point scale. Some people list their grade-point average in their majors only, since this is likely to be higher than the overall average.

Work Experience. List your past jobs in reverse chronological order. Include the job title, the name of the company, the city, and the state. Describe your responsibilities for each job, particularly those which provided practical experience connected with your career goals. In describing responsibilities, use action words, such as *coordinated, directed, prepared, supervised,* and *developed.* Dates of employment need not include month and day. Terms such as *vacation* or *summer* with the relevant years are sufficient.

Honors and Awards. List scholarships, prizes, and awards received in college. Include any community honors or professional prizes as well. If there is only one honor, list it under "Activities."

Activities. In this section list recent activities, primarily those in college. Be sure to indicate any leadership positions, such as president or chairperson of a group. Hobbies, if included, should indicate both group and individual interests. In Model 9-8, the applicant lists piano (an individual activity) and tennis (a group activity).

References. Model 9-8 shows the most common way to mention references. If you do list references, include the person's business address and telephone number. Be sure to ask permission before listing someone as a reference.

It is best to omit personal information about age, height, weight, marital status, and religion. Employers are not allowed to consider such information in the employment process, and most prefer that it not appear on a résumé.

Model 9-9 shows a résumé for a job applicant who is more advanced in his chemistry career than the applicant in Model 9-8. This résumé lists experience first, in the most prominent position. Other significant differences from the résumé for the new college graduate include the following:

Major Qualifications. Instead of an objective, this applicant wants to highlight his experience, so he uses the *capsule résumé* technique to call attention to his years of experience and his specialty. He can discuss his specific career goals in his application letter.

Professional Experience. This applicant has had more than one position with the same employer because he has been promoted within the company. Dates of the two positions are given but are not emphasized, as are the dates in the résumé in Model 9-8. The applicant highlights his experience by stacking his responsibilities into impressive-looking lists.

Education. This applicant includes his date of graduation but does not emphasize it, and he omits his grade-point average because his experience is now more important than his college work. He lists his computer knowledge because such knowledge is useful in his work.

Activities and Memberships. After several years of full-time work, most people no longer list college activities. Instead, they stress community and professional service. This applicant lists his professional membership in the American Chemical Society to indicate that he is keeping current in his field.

The purpose of the application letter and résumé is to present an interest-attracting package that will result in an interview. Because the letter and résumé are often the first contact a new college graduate has with a potential employer, the initial impression these documents make may have a decisive impact on career opportunities. Table 9-1 lists the 35 résumé items considered most important for new college graduates by personnel administrators at Fortune 500 companies.

Table 9-1 Résumé Content Ranked by Importance: The 35 Top-Ranked Items

Résumé Content Items in Order of Importance	***Percent Ranking Items Important (Fortune 500 Personnel Administrators)***
1 Name	100.0
2 Telephone no.	100.0
3 Degree	100.0
4 Name of college	100.0
5 Jobs held (titles)	100.0
6 Address	99.3
7 Major	99.3
8 Employing company(s)	99.3
9 Dates of employment	99.3
10 Date of college graduation	95.3
11 Duties—work experience	94.7
12 Special aptitudes/skills	91.9
13 Job objective	91.7
14 Awards, honors—college achievements	91.4
15 Grade-point average	90.8
16 Willingness to relocate	89.4
17 Achievements—work experience (learning, contributions, accomplishments)	89.2
18 Professional organizations—college extracurricular activities	84.7
19 Scholarships—college achievements	84.1
20 Career objective	83.3
21 Years attended—college	83.0
22 Previous employers—references	81.8
23 Minor	80.8
24 Military experience	80.1
25 Student government activities—college	78.7
26 Publications	77.9
27 Combined job and career objective	76.6
28 References supplied only on request	75.5
29 Current organization memberships	75.5
30 Summary of qualifications	70.8
31 Community involvement	70.2
32 Yearbook editor, etc.—college	68.2
33 Professors/teachers—references	65.3
34 Athletic involvement—college	63.3
35 Social organizations—college	62.9

CHAPTER SUMMARY

This chapter discusses how to write effective letters, memos, job-application letters, and résumés. Remember:

- Letters are written primarily to people outside an organization, and memos are written primarily to people inside an organization.
- Effective tone in letters and memos requires natural language, positive language, and the you-attitude.
- Letters and memos can be organized in the direct, indirect, or persuasive pattern.
- Direct organization presents the main idea in the opening.
- Indirect organization delays the main idea until details have been explained.
- Persuasive organization begins with an attention-getting opening.
- Managers need to create a cooperative atmosphere in their memos to employees.
- The two most common letter formats are full block style and semiblock style.
- A job-application letter presents a writer's qualifications for a position and requests an interview.
- Résumés list a writer's most significant achievements relative to a position.

SUPPLEMENTAL READINGS IN PART TWO

Boe, E. "The Art of the Interview," *Microwaves and RF.*

Calabrese, R. "Designing and Delivering Presentations and Workshops," *The Bulletin of the Association for Business Communication.*

Davidson, J. P. "Astute DP Professionals Pay Attention to Business Etiquette," *Data Management.*

Erdlen, J. D. "A Good Résumé Counts Most," *Machine Design.*

Kirtz, M. K., and Reep, D. C. "A Survey of the Frequency, Types, and Importance of Writing Tasks in Four Career Areas," *The Bulletin of the Association for Business Communication.*

"The Practical Writer," *The Royal Bank of Canada.*

Model 9-1 Commentary

This model is a typical order letter. The writer lists the items in a three-column table—the number of items ordered, the product number, and a description of the items. The letter concludes with information about how to ship the order.

Discussion

1. Discuss why the information in this letter is necessary for the reader. What is the advantage of listing the ordered items rather than discussing them in a paragraph?

2. In groups, make a list of foods you would want for a party of 25 friends. Draft an order letter to Special Occasion Catering in your city and order the foods. Ask the company to bill you.

Heritage Americana Interiors, Inc.
62 West Mitchell Parkway
Geneva City, Ohio 44439

November 1, 1989

Hilltop Office Products
950 South Kaye Street
Fort Nichols, IN 46408

Attention: Order Department

Please send me the following office supplies as listed and priced in your fall catalog, No. 79:

6 pkgs.	B11C-688	Color-code files with heavyweight manila folders--letter size
4 boxes	B11C-672	Square-bottom box files--2 in. capacity
24	B11C-5223	Large capacity 3-ring binders--letter size
2	B11C-1781	Hole-punches with movable heads
2	B11C-60155	Mobile file cabinets--chocolate brown

Please charge these to the Heritage Americana Interiors, Inc., account. I would appreciate rush service on these items. Please ship by the fastest freight available.

Sincerely,

Melanie Svenson
Office Manager

MS:rb

Model 9-2 Commentary

This model is the response to the order letter in Model 9-1. The writer uses the direct organization pattern and opens by telling the reader which items were shipped. The writer explains the delay in shipping one item by describing its popularity and assuring the reader that it will be shipped soon. The writer also encloses a new catalog to encourage future orders.

Discussion

1. Identify the organizational strategies used in this letter, and discuss why the writer probably chose these strategies.

2. In what ways does the writer imply that the delay is insignificant and the hole punches are worth waiting for?

3. Assume that you are Melanie Svenson, and draft a letter to Robin Hammel canceling the order for hole punches. You do not want to wait, and you think you can get the items from another dealer.

HILLTOP OFFICE PRODUCTS
950 South Kaye Street
Fort Nichols, Indiana 46408

November 6, 1989

Ms. Melanie Svenson
Office Manager
Heritage Americana Interiors, Inc.
62 West Mitchell Parkway
Geneva City, OH 44439

Dear Ms. Svenson:

The following items were shipped to you today on Worldwide Express, rush service:

Color-code files--letter size (6 pkgs.)

Square-bottom box files, 2 in. capacity (4 boxes)

3-ring binders, letter size (24)

Mobile file cabinets, chocolate brown (2)

Enclosed is the invoice for $410.59, including sales tax. Your order for two hole punches with movable heads (B11C-1781) should reach you within 14 days, also by Worldwide Express. Because of the great popularity of these durable, high-capacity hole punches, we are temporarily out of stock. A new shipment from the supplier is due shortly, and when it arrives, we will fill your order immediately.

Thank you for your order, and please let me know if I can be of service in the future. For your convenience, I am enclosing a preview copy of the winter catalog, No. 80, that will be mailed to our customers in early December.

Sincerely,

Robin Hammel
Supervisor

RH:rb

Enc.: Catalog No. 80

Model 9-3 Commentary

This persuasive letter is individually addressed but is actually a form letter that the bank sends to all new city residents. The purpose is to introduce the bank and its services and gain new business.

Discussion

1. Identify the organizational strategies used in this letter, and discuss why the writer probably chose these strategies.

2. Find all the places in this letter where the writer uses language that will increase the reader's interest in the bank and its services.

3. In groups, draft a persuasive letter to incoming freshmen to encourage them to join a campus activity at your school. Do not emphasize one particular group; you want the readers to consider all campus activities before joining one.

Community Bank and Trust
Marine Plaza
Milwaukee, Wisconsin 53201

February 4, 1990

Mr. Jonathan Quigley
4237 N. 63rd Street
Milwaukee, WI 53212

Dear Mr. Quigley:

Community Bank and Trust welcomes you to the Milwaukee area. We know you will enjoy living in this fast-paced, but friendly community.

As you know, financial planning is becoming increasingly complicated in today's rapidly changing economy where interest rates rise and fall overnight and government tax laws require careful monitoring of investments and retirement plans. Community Bank and Trust can relieve you of this complicated long-range planning.

Your personal banker in our office will be pleased to review with you the wide range of savings and investment plans we offer to meet your financial needs. For example, our Super Money Market Deposit account is a variable-rate savings account that also offers checking privileges with certain minimum deposits. Your CBT Investment Certificates allow you to choose maturity dates that support your plans for investments or important expenditures, such as college expenses or a home purchase. Our certificates offer maturity dates from 30 days to 2 years, with the rate guaranteed for the full period of the deposit. Both the Super Money Market Deposit account and the CBT Investment Certificate can be opened for as little as $1000, and they are protected by FDIC insurance.

We look forward to the opportunity to serve you in all your banking needs. Please call 555-4471 today and ask for Suzanne Cabot, our Customer Service Manager. She will assign a personal banker to your accounts immediately. And you can begin financial planning in your new home with the professional advice of investment experts at Community Bank and Trust.

Sincerely yours,

Victor Neumann
Vice President

VN:aa

Model 9-4 Commentary

This memo is from a supervisor to a division head, asking for maintenance support.

Discussion

1. Identify the organizational strategies the writer uses, and discuss why the writer probably chose these strategies.

2. Consider the tone of this memo. How does the writer place blame for the situation? How does the writer convey the seriousness of the problems?

3. In what ways does the writer show the reader that there are benefits in solving these problems? Is there any advantage to addressing the reader by name in the memo?

4. Write a brief memo to D. P. Paget from T. L. Coles announcing that R. Fleming, the maintenance supervisor, has been told to report on the problems within two days. Assure Paget that the problems are being taken seriously.

April 13, 1989

To: T. L. Coles, Division Chief

From: D. P. Paget, Supervisor

Subject: Problems in Cost Center 22, Paint Line

As I've discussed with Bill Martling, our department has had problems with proper maintenance for six weeks. Although the technicians in the Maintenance Department respond promptly to our calls, they do not seem to be able to solve the problems. As a result, our maintenance calls focus on the same three problems week after week. I've listed below the areas we need to deal with.

1. For the past two weeks, our heat control has not consistently reached the correct temperature for the different mixes. The Maintenance technician made six visits in an attempt to provide consistent temperature ranges. Supervisor R. Fleming investigated and decided that the gas mixture was not operating properly. He did not indicate when we might get complete repairs. In the last ten working days, we've experienced nearly five hours of total downtime.

2. The soap material we currently use in the water fall booth does not break up the paint mixtures, and as a result, the cleanup crew cannot remove all the excess paint. Without a complete cleanup every twenty-four hours, the booth malfunctions. Again, we have downtime. We need to investigate a different soap composition for the water fall booth.

3. The parking lever control has not been in full operation since January 14. Maintenance technicians manage to fix the lever for only short periods before it breaks down again. I asked for a complete replacement rather than continued repairs, but was told it wasn't malfunctioning enough to warrant being replaced. This lever is crucial to the operation because when it isn't functioning the specified range slips and a lot of paint literally goes down the drain. As a result, we use more paint than necessary and still some parts are not coated correctly.

I'd appreciate your help in getting these problems handled, Tom, because our cost overruns are threatening to become serious. Since all maintenance operations come under your jurisdiction, perhaps you could consider options and let me know what we can do. I'd be glad to discuss the problems with you in more detail.

Model 9-5 Commentary

This model shows two drafts of a supervisor's memo to employees in a chemical laboratory announcing changes in the way they have been preparing reports. She is changing procedures because the chemical engineers who rely on the tests have complained about incomplete reports, and the laboratory's only purpose is to support the work of the engineers. Delays of hours and sometimes days are occurring because the reports are incomplete or in the wrong format. The laboratory technicians, however, will not like these changes because they will add about 15 minutes preparation time to each report.

Discussion

1. Discuss the revisions from the first draft to the second and whether or not these changes have improved the memo.

2. Discuss how to further improve the tone and clarity of this memo. How could the writer add reader benefits to the second draft?

December 16, 1989

To: Lab Personnel

From: Corinne Desmond

Subject: Final Report Requirements

Beginning Monday, December 19, all our final test reports must include the following:

1. Test results
2. Dimensions in metric terms
3. Photos in proper order--also identify each one on the back
4. Include the distribution list
5. Write the report immediately after the test
6. Be sure all terms are spelled correctly
7. Complete formulas

December 16, 1989

To: Laboratory D-66 Personnel

From: Corinne Desmond, Supervisor

Subject: Final Test Report Requirements

I've received some requests for changes in our test reports from the chemical engineers who use them. Therefore, beginning Monday, December 19, all final test reports must include the following:

1. Full test results at each stage of the testing process
2. Dimensions stated in metric terms
3. Photos in proper order and each identified on the back
4. The distribution list
5. Correctly spelled terms
6. Full formulas

Please write your reports immediately after completing the test while the data are fresh in your mind. I'm sure with these minor adjustments in report style, we can give the engineers what they need.

Model 9-6 Commentary

This persuasive memo is from a supervisor in an insurance company to the district managers of the agents in the field. The supervisor asks the district managers to encourage the insurance agents to prepare routine proposals themselves rather than sending all proposals to the home office for preparation.

Discussion

1. Identify and discuss the persuasive strategies used in this memo. What are the advantages to using this organization in this situation?

2. Underline the places in the memo where the writer emphasizes the you-attitude and reader benefits. If you were the reader, how would you respond?

April 13, 1989

To: District Sales Directors

From: Julie Kelly
Manager, Sales Proposal Division

Subject: Requests for Proposal Service

Thanks to the terrific job your agents are doing in the field, our division is experiencing a large increase in the number of proposal requests each week. Because we want to continue to offer high-quality service on complicated proposals, we need some assistance on the routine ones.

As you know, the company has an established format for Major Medical, Franchise, and Disability Income proposals. I'm enclosing sample forms as well as a checklist of items that must be included on each form. By using these checklists, agents can quickly prepare the simple proposals for presentation to clients. The usual 12 to 14 days our office needs to handle these requests can be eliminated, allowing agents to make follow-up calls as soon as they complete the proposals.

Please ask agents to use the samples and checklists to prepare the Major Medical, Franchise, and Disability Income proposals in their field offices. The complicated plans, such as those for company pension agreements, should be sent to this office. We will be able to prepare them within 4 to 6 days if the simpler plans are done in the field.

Can we have your help in asking agents to forward the complex proposals to us and to finish the easy-to-do proposals themselves?

JK:td
Enc.: Forms (3)
Checklists (3)

cc: J. F. Cooper
W. G. Simms

Model 9-7 Commentary

This job-application letter is from a recent college graduate inquiring about a specific company opening.

Discussion

1. Identify the organizational strategies the writer uses.

2. Read this letter in conjunction with the résumé in Model 9-8. Consider the specific details about education and experience in the letter. How do the letter and résumé together make a stronger case for the applicant than either would alone?

3. How does the writer manage to put the you-attitude into her letter?

1766 Wildwood Drive
Chicago, Illinois 60666
July 12, 1989

Mr. Eric Blakemore
Senior Vice President
Alden-Chandler Industries, Inc.
72 Plaza Drive
Milwaukee, WI 53211

Dear Mr. Blakemore:

Professor Julia Hedwig suggested that I write to you about an opening for a product chemist in your chemical division. Professor Hedwig was my senior adviser this past year.

I have just completed my B.S. in chemistry at Midwest University with a 3.9 GPA in my major. In addition to chemistry courses, I took three courses in computer applications and developed a computer program on chemical compounds. As a laboratory assistant to Professor Hedwig, I entered and ran the analyses of her research data. My senior project, which I completed under Professor Hedwig's guidance, was an analysis of retardant film products. The project was given the Senior Chemistry Award, granted by a panel of chemistry faculty.

My work experience would be especially appropriate for Alden-Chandler Industries. At both Ryan Laboratories and Century Concrete Corporation, I have worked extensively in compound analysis, and I am familiar with standard test procedures. My position at Century was part-time during the academic year and full-time during summers, so I have been able to participate in the full testing cycle for several projects from beginning to end.

I would appreciate the opportunity to discuss my qualifications for the position of product chemist. I am available for an interview every afternoon after three o'clock, but I could arrange to drive to Milwaukee any time convenient to you. My telephone number during the day between 10:00 a.m. and 3:00 p.m. is (312) 555-1212.

Sincerely,

Kimberly J. Oliver

Enc.

Model 9-8 Commentary

This résumé is for the job applicant in Model 9-7 and is in the chronological format most recent college graduates use.

Discussion

Identify the types of information the writer includes, and discuss how this information helps create an image of the writer.

Kimberly J. Oliver
1766 Wildwood Drive
Chicago, Illinois 60666
(312) 555-6644

Objective: Chemist with responsibility in product development and production. Willing to travel.

Education

1986-1990 Midwest University, Chicago, Illinois
B.S. in Chemistry; GPA: 3.6 (4.0 scale)

Computer Languages: BASIC, FORTRAN, PASCAL

Work Experience

1988-present: Century Concrete Corp., Chicago, Illinois

Research Assistant: Set up chemical laboratories, including purchasing equipment and materials. Perform ingredient/compound analysis. Produce experimental samples according to quality standards. (part-time)

Summers, 1985-1987: Ryan Laboratories, Chicago, Illinois

Laboratory Assistant: Assisted in testing compounds. Prepared standard test solutions; recorded test data; operated standard laboratory equipment; wrote operating procedures.

Honors and Awards

Chemical Society of Illinois Four-Year Scholarship, 1986-1990.
Senior Chemistry Award, Midwest University, 1989.
Outstanding Chemistry Major, Midwest University, 1988.

Activities

Member, American Chemistry Association, 1988-present.
President, American Chemistry Association (student chapter), 1987-1988.
Member, Toastmaster Association, 1985-1988.
Council Member, Women in Science Student Association, 1986-1988.
Hobbies: sailing, piano, tennis

References

Available on request.

Model 9-9 Commentary

This résumé is for a job applicant with significant professional experience.

Discussion

1. Compare this résumé with the one in Model 9-8. Discuss the differences in format and the ways those differences reflect the differences in the qualifications of the two job applicants.

2. Discuss why this résumé does not include the college GPA and scholastic honors, deemphasizing the applicant's education.

Jack Montgomery
21 Camelot Court
Skokie, IL 60622
(312) 555-5620

Major Qualifications: Seven years of experience in product development chemistry. Specialty: Adhesives

PROFESSIONAL EXPERIENCE

CDX TIRE AND RUBBER CORPORATION, Chicago, IL

Formulation Chemist, 1987-present.
- Develop new hot-melt adhesives, hot-melt pressure sensitives, and hot-melt sealants
- Perform analysis of competitors' compounds
- Determine new product specification
- Calculate raw materials pricing
- Conduct laboratory programs for product improvement

Analytical Chemist, 1985-1987.
- Performed ingredient/compound analysis
- Developed and tested compounds (cured/uncured)
- Wrote standard operating procedures
- Developed improved mixing design
- Supervised processing technicians

STERLING ANALYTICAL LABORATORIES, Pittsburgh, PA

Research Laboratory Assistant, 1983-1985 (part-time)
- Prepared solutions
- Drafted laboratory reports
- Supervised students during testing procedures
- Inventoried laboratory equipment

EDUCATION

B.S. in Chemistry, Minor in Biology. University of Pittsburgh, Pittsburgh, May, 1985.

COMPUTER SKILLS

COBOL, C, FORTRAN, PASCAL

ACTIVITIES AND MEMBERSHIPS

District Chairman, United Way Campaign, 1988
Member, Illinois Consumer Protection Commission, 1988
Member, American Chemical Society, 1985-present

References furnished on request.

Chapter 9 Exercises

1. Doris Caswell wrote to Mutual Insurance to cancel her automobile policy because she is selling her car. Doris is spending the next 6 months in Spain working on a project for Imperial Oil, Inc., the company she works for. After the project, she will return to her present position in the company's headquarters in Denver. After she canceled her policy, she received the following letter from Mutual:

> Dear Ms. Caswell:
> As per your request, we are canceling herewith your policy #B-167-A. Enclosed please find a check in the amount of $42.36, the refund owing to you.
>
> Sincerely yours,

Rewrite the letter for a more positive, natural tone. Also, try to conclude the business relationship on a friendly note. Doris will, after all, return to the United States within the next year and probably will buy a new car.

2. The following letter uses overly negative language and emphasizes the customer's problem. Rewrite the letter with more you-attitude and less negative language.

> Dear Mr. Schultz:
> I am sorry to hear that you have experienced problems with the Mark-7 Sander. You claim that when you use the sander on plastics, the heat causes melting and the sander belt becomes gummed up. I don't understand how this problem occurred because the Mark-7 is our newly developed four-belt sander, designed for increased versatility, with a low-speed setting of 150 feet per minute. This speed should prevent damage to the plastic. Because your complaint is so serious, I am asking Rick Cansino, our dealer in Center City, to check the defective sander. Thank you for reporting this problem, and we regret the inconvenience you have suffered.
>
> Sincerely,

3. Select a "real world" memo or letter of at least one page in length. Make enough copies for class members. Present a critique of the memo or letter to the class, pointing out both effective and ineffective aspects of the sample. *In addition,* submit a revision of the sample. The revision should reflect your ideas about how to improve the original memo or letter for greatest effectiveness. Do not change the original situation in the writing sample. Your job is to improve and develop the message. You may add any logical material to your revision in order to improve it.

4. Write to JTD Computers, Inc., to lease three Hewlett-Packard laserjet printers. You want one printer shipped to each of the three branch offices of the Wisconsin Plastics Company. The three addresses are 1753 W. Market Street, Chicago, Illinois 60622; One Industrial Plaza, Cleveland, Ohio 45207; 9350 Forest Reserve Drive, Fort Mason, Indiana 47967. Remind JTD Computers that you need proper protocol interfaces and cables to attach to a System 38. You would like font cartridges A and B. The branch offices need these quickly, so ask for special rush service. You would prefer a 3-year lease but will take whatever is standard.

5. As supervisor of the painting and finishing department of a major farm machinery manufacturer, you are faced with a safety problem. Two weeks ago, the company changed suppliers and is using new paints and rust coatings. The ventilation system apparently is not adequate for these new compounds. Fumes are building up. Yesterday two employees went home early, saying they were sick from the fumes. The union foreman suggested in a friendly manner that employees would not tolerate the situation more than a few days more. Write a memo to the division manager stating the problem, stressing the urgency, and asking for immediate repairs or replacement of the ventilation system.

6. You are director of the publications department of Porter Housewares. Professor Joseph Cicinni has written to you asking you to speak to his technical writing class. He would like you to describe the writing and publication process of the consumer manuals for your company's many electrical housewares. As he says in his letter, "Cooperation between education and industry is extremely important if students are to be well trained for future jobs." You have to be out of town for the next week, and when you get back, the department workload will be too great to allow you any free time. You may have time next semester after the manuals for the new product line are finished. Write Professor Cicinni and decline his invitation, but try to maintain good relations.

7. As the executive director of the local chapter of Tomorrow's Leaders, a national organization for girls ages 12 to 16, you are pleased at the great success of the annual chocolate brownie sale. Shogun International, a large manufacturer headquartered in your city, has ordered 1000 boxes of the rich, chewy brownies. Delivery has been promised for January 20. Shogun International plans to give each employee a box of brownies at the company's celebration marking the 100th anniversary of its founding on January 21. On January 14, you receive word that the Renfield Bakery has had a terrible fire and cannot deliver the brownies on time. You make arrangements for the Cromwell Cookie Company to take over the order, but there will be a delay of at least 10 days. There will be no brownies on January 20 for Shogun International, a company that has always supported your organization. Write to J. G. Shelby, Director of Special Events at Shogun, and explain the situation. You may have to cancel the order, but you would rather persuade Shelby to wait and accept the brownies later than planned.

8. As the manager of Quick and Easy Car Wash, write a letter to Ms. Lynda Bradley, an irate customer. Ms. Bradley is upset because when she sent her 1990 Korean import through the car wash, the side mirror snapped off during the washing process. You are willing to make repairs, but you've encountered a problem getting a replacement mirror. The dealer you ordered the mirror from told you it would be in "next week." That was three weeks ago. The next closest dealer is 62 miles away and would also have to order the part from Korea. Write Ms. Bradley and explain the delay, assuring her that you will replace the mirror when you can. Send her some complimentary car wash coupons.

9. Find a technical advertisement in a trade magazine. Assume that you work for a company that uses a product similar to the one in the advertisement. Write a persuasive memo to your supervisor suggesting that the company switch to the brand in the advertisement. Point out the unique features of the product, and attach a copy of the advertisement to your memo.

10. Prepare a résumé based on the formats discussed in class or using a sample from this chapter. Select a company in your area, and apply for a summer job (or a part-time job during the semester) in your career area. Graduating seniors should apply for a full-time job in their career area. Write an application letter that realistically discusses your interest in and qualifications for a position based on your *present* situation. Key your remarks to your résumé, and attach your résumé to the letter.

11. Select a company you would like to work for upon graduation. Prepare a résumé that reflects the qualifications you expect to have when you graduate. Then write an application letter asking for an entry position in your field. Key your remarks to your résumé, and attach your résumé to the letter.

12. *Oral report project.* Your instructor will divide the class into groups, and each group will work on one of the topics listed below. Each group member will be responsible for gathering information from outside sources for one specific area of the panel topic. Consult Chapter 10 for a list of information sources. Read Chapter 13 and the article by Calabrese in Part Two for guidance in organizing your oral presentation. Plan your oral report as a 10-minute presentation or for whatever length of time your instructor assigns, and practice it so you know it fits the required time. Remember that your audience consists of your classmates who are interested in information that will help them begin their careers. Tailor your oral report to this audience. After giving your oral report, submit an executive summary of your talk to your instructor. Write a transmittal memo to accompany the executive summary. Guidelines for executive summaries and transmittal memos are included in Chapter 11.

a. *Women in careers*—Report on women in business and industry today—promotion chances, statistics on numbers of women in specific career areas, sexual harassment problems and laws, mentors, and networks for career information.

b. *Job prospects for this year's college graduate*—Report on the opportunities for new graduates in specific career areas that students in your class are studying, such as engineering, nursing, or marketing. Also cover types of entry jobs available in these fields, promotion possibilities, starting salaries, and how to find information about potential jobs.
c. *Body language and appearance as factors in the job search*—Report on body language (turn-ons and turn-offs), dress for success (men and women), and effective body language in job interviews.
d. *Effective job interviewing*—Report on questions to expect during a job interview, difficult or hostile questions, and information sources for learning about the company with which you are interviewing. Also explain how to effectively present your experience and education.
e. *Business etiquette*—Report on how to evaluate your position in the company hierarchy, etiquette points the ambitious beginner *must* know, international business and culture/etiquette differences Americans need to be aware of, how to take part in a meeting, and symbols of status and power.

CHAPTER 10

Short and Long Reports

Understanding Reports

Developing Short Reports

- Opening-Summary Organization
  - Opening Summary
  - Data Sections
  - Closing
- Delayed-Summary Organization
  - Introductory Paragraph
  - Data Sections
  - Closing Summary

Developing Long Reports

- Planning Long Reports
- Gathering Information
  - Secondary Sources
  - Primary Sources
- Taking Notes
- Interpreting Data
- Drafting Long Reports
  - Introductory Section
  - Data Sections
  - Concluding Section

Chapter Summary

Supplemental Readings in Part Two

Models

Exercises

Understanding Reports

Next to correspondence, reports are the most frequently written documents on the job. Reports usually have several purposes, most commonly these:

- To inform readers about company activities, problems, and plans so that readers are up to date on the current status and can make decisions —for example, a progress report on the construction of an office building or a report outlining mining costs.
- To record events for future reference in decision making—for example, a report about events that occurred during an inspection trip or a report on the agreements made at a conference.
- To recommend specific actions—for example, a report analyzing two production systems and recommending adoption of one of them or a report suggesting a change in a procedure.
- To justify and persuade readers about the need for action in controversial situations—for example, a report arguing the need to sell off certain company holdings or a report analyzing company operations that are hazardous to the environment and urging corrective action.

Reports may vary in length from one page to several hundred, and they may be informal memos, formal bound manuscripts, rigidly defined form reports, such as an accident report at a particular company, or documents for which neither reader nor writer has any preconceived notion of format and organization. Reports may have only one reader or, more frequently, multiple readers with very different purposes.

Whether a report is short or long depends on how much information the reader needs for the specific purpose in the specific situation, not on the subject or the format of the report. The Kirtz and Reep article in Part II shows that people on the job tend to write more short reports (five pages or less) than long reports. Busy people on the job do not want to read reports any longer than necessary to meet their needs. Furthermore, for most company projects, writers supply several reports, or even a regular series of reports, to keep those concerned with the subject up to date rather than waiting to write one lengthy report at the end of a project. Long reports are usually necessary in these situations: (1) scientists reporting on the results of major investigations, (2) consulting firms providing clients with evaluations of their operations and recommendations for future actions, (3) company analysts predicting future trends and making forecasts on how well the company will perform in the future, and (4) writers reporting on any complex company situation with large amounts of data needed by multiple readers. These long reports usually include summaries for readers who do not need all the detailed information in the reports and are concerned only with conclusions and recommendations.

For all reports, analyze reader and purpose before writing, as discussed in Chapter 2, and organize your information, as outlined in Chapter 3. In addition, use the guidelines for planning, gathering information, taking notes, interpreting data, and drafting discussed later in this chapter. Reports for specific purposes are discussed in Chapter 12. The rest of this chapter focuses on the general structure of short and long reports and on ways to develop long reports that deal with complex subjects and require information from many sources.

Developing Short Reports

Most short reports within a company are written as memos, whereas short reports to people outside an organization are usually written as letters or as formal reports with the elements described in Chapter 11. Because short reports provide so much information essential to conducting business and readers on the job are always pressed for time, most short reports are organized in the opening-summary, or front-loaded, pattern. The delayed-summary organization is used primarily when readers are expected to be hostile to the report contents.

Opening-Summary Organization

The *opening-summary organization* is often called *front-loaded* because this pattern supplies the reader with the most important information—the conclusions or results—before the specific details that lead to those results or conclusions. The pattern has two or three main sections, depending on the writer's wish to emphasize certain points.

Opening Summary

The opening is a summary of the essential points covered in the report. Through this summary, the reader has an immediate overall grasp of the situation and understands the main point of the report. Include these elements:

- The subject or purpose of the report
- Special circumstances, such as deadlines or cost constraints
- Special sources of information, such as an expert in the subject
- Main issues central to the subject
- Conclusions, results, recommendations

Readers who are not directly involved in the subject often rely solely on the opening summary to keep them informed about the situation. Readers who are directly involved prefer the opening-summary pattern because it previews the report for them so that when they read the data sections, they already know how the information is related to the report conclusions. Here is an opening from a short report sent by a supervisor of a company testing laboratory to his boss:

> The D120 laboratory conducted a full temperature range test on the 972-K alternator. The test specifications came from the vendor and the Society of Automotive Engineers and are designed to check durability while the alternator is functioning normally. Results indicate that the alternator is acceptable, based on temperature ranges, response times, and pressure loads.

In this opening summary, the supervisor identifies the test he is reporting, cites the standards on which the test is based, and reports the general result—the alternator is acceptable. The reader who needs to know only whether the alternator passed the test does not have to read further.

Data Sections

The middle sections of a short report provide the specific facts relevant to the conclusions or recommendations announced in the opening. The data sections for the report about the alternator would provide figures and specific details about the three tested elements—temperature ranges, response times, and pressure loads—for readers who need this information.

Closing

Reports written in the opening-summary pattern often do not have closings because the reader knows the conclusions from the beginning and reads the data sections for details, making any further conclusions unnecessary. However, in some cases you may wish to reemphasize results or recommendations by repeating them in a brief closing. Depending on the needs of your readers, your closing should:

- Repeat significant results/recommendations/conclusions
- Stress the importance of the matter
- Suggest future actions
- Offer assistance or ask for a decision

Here is the closing paragraph from the short report written by the supervisor in the testing laboratory:

> In conclusion, the 972-K alternator fits our needs and does not require a high initial cost. The quality meets our standards as well as those set by the

> Society of Automotive Engineers. Because we need to increase our production to meet end-of-the-year demands, we should make a purchase decision by November 10. May I order the 972-K for a production run starting November 20?

In this closing, the supervisor repeats the test results—that the alternator meets company standards. He also reminds his reader about production deadlines and asks for a decision about purchasing the alternators. Because of a pressing deadline, the supervisor believes he needs a closing that urges the reader to reach a quick decision.

Figure 10-1 presents a short report in the opening-summary pattern. In this report situation, the writer was told to investigate a new scheduling system for working hours and report back with information. The opening summary reminds the reader of the previous discussions leading to the report. The writer also explains how he gathered information and states his general conclusion—the advantages outweigh the disadvantages in the new system. Although the reader of this opening summary does not know the details, he does have the answer to his original inquiry about the effectiveness of the new scheduling system.

The middle sections provide details about how the system works and about the advantages and disadvantages reported by those already using the system. The writer uses a conclusion section to emphasize how successful the system appears to be, and he offers further assistance. The organization of this short report gives the reader a quick overall understanding of the situation, followed by easy access to the details.

Delayed-Summary Organization

The *delayed-summary organization,* as its name suggests, does not reveal the main point or result in the opening. This organizational pattern slows down rather than speeds up the reader's understanding of the situation and the details. Use the delayed-summary pattern when you believe that your reader is not likely to agree with your conclusions readily and that reading those conclusions at the start of the report will trigger resistance to the information rather than acceptance. In some cases, too, you may know that your reader usually prefers to read the data before the conclusions or recommendations or that your reader needs to understand the data in order to understand your conclusions or recommendations. The delayed-summary organization has three main sections.

Introductory Paragraph

The introductory paragraph provides the same information as the opening summary, except that it does not include conclusions, results, or recommendations.

To: Lee Eastland Date: January 23, 1990

From: Aaron Berger

Subject: Flex-Time Advantages and Disadvantages

As you recall, at the last division meeting we discussed the possibility of instituting flex-time scheduling in all company departments, and I offered to check on the flex-time concept. Since Best Hospital Products and American Cabinetry in our city have operations similar to ours, I consulted with the personnel managers at both companies. In addition, I talked to the personnel director at our San Antonio branch, which uses flex-time scheduling now. Overall, the personnel managers report that the advantages appear to outweigh any disadvantages in the system.

Flex-Time Scheduling System

Flex-time scheduling is a system that allows individual employees to arrange their own work schedules, within defined limits. The work day usually extends 10 to 12 hours, five days a week, and certain hours are designated as core hours during which all employees must be on the job. Arrival and departure times are flexible for employees.

Best Hospital Products and American Cabinetry, for example, both use the same schedule. The companies open at 6:30 a.m. when one-third of the employees arrive. The core hours are between 9:00 a.m. and 3:30 p.m., and the closing hour is 6:00 p.m., with about one-third of the employees still on the job. Our San Antonio branch opens at 7:00 a.m., closes at 7:00 p.m., and the core hours are 9:30 to 4:30 p.m.

All three personnel managers reported that about 45% of the employees prefer to work the traditional hours of 8:00 a.m. to 5:00 p.m. Female employees with children in daycare centers and single, unmarried employees under age 35 were the most likely to arrange flex-time hours. No one reports any disputes among employees over scheduling.

Advantages of Flex-Time Scheduling

All three personnel managers cited significant advantages in the flex-time scheduling:

1. Employees have lower rates of absence and tardiness because they pick hours that complement their personal lives.

2. Employees have higher productivity when on the job because they take care of personal business outside the office.

3. Employee morale is high because employees perceive the management as interested in their needs and problems.

Figure 10-1 A Short Report Organized in an Opening-Summary Pattern

4. Company customers have greater access to departments since business can begin earlier, accommodating customers on the East Coast, and end later, accommodating customers on the West Coast.

5. Use of office equipment, such as computers and copying machines, is more efficient because the use is spread over 12 hours instead of 8 hours.

With our nationwide network of dealers and customers, the convenience of expanded hours could contribute substantially to our growth.

Disadvantages of Flex-Time Scheduling

Personnel managers all cited the following disadvantages:

1. Key employees in one department may not be on hand when key employees in another department or customers need them.

2. Total utility bills are higher because the office is open longer hours and facilities are in use longer.

3. Cleaning services must accommodate the longer hours, and the San Antonio branch reports having to pay a premium rate because the service must start later than it does at other companies.

4. Direct employee supervision is not possible for a full day because supervisors will either start later than some employees or leave earlier.

Careful scheduling might compensate for the possible problems resulting if key employees are not on the job at the same time.

Conclusion

In spite of some disadvantages, the personnel managers report that there is general satisfaction with the flex-time scheduling system. Best Hospital Products conducted a formal survey after six months on flex-time, and results indicated employees were comfortable with the system and did not want to return to the traditional working hours. American Cabinetry and our San Antonio division did not formally survey employees, but both personnel managers reported employees appeared to be satisfied. If you would like to discuss this matter further, please let me know. I would be happy to give a presentation of my study of flex-time at the next division meeting.

AB:ss

Figure 10-1 cont.

Data Sections

The data sections cover details relative to the main subject just as the data sections in the opening-summary pattern do.

Closing Summary

Because the introductory paragraph does not include results, conclusions, or recommendations, a closing that summarizes the main points, results, general observations, and recommendations is essential. The closing also may include offers of assistance, reminders about deadlines or other constraints, and requests for meetings or decisions.

The report in Model 10-2 illustrates the delayed-summary organization. Because the writer is asking for expensive equipment and he suspects that the reader will react unfavorably, he uses the delayed-summary organization so that he can present the rationale before the request. In his closing, the writer stresses the need for the new equipment and asks for a meeting to discuss the matter. In this situation, the opening-summary organization would be too direct and might result in a quick denial of the writer's request.

Developing Long Reports

Long reports (over five pages) may be written as memos, but frequently they are written as formal documents with such elements as appendixes and a table of contents, as explained in Chapter 11. Long reports tend to deal with complex subjects that involve large amounts of data, such as an analysis of how well a company's 12 branch offices are performing with recommendations for shutting down or merging some offices. In addition to dealing with complicated information, a long report nearly always has multiple readers who are concerned with different aspects of the subject and have different purposes for using the information in the report.

Planning Long Reports

Consider these questions when you begin planning a long report:

- *What is the central issue?* In clarifying the main subject of a long report, consider the questions your readers need answered: (1) Which choices are best? (2) What is the status of the situation, and how does it affect the company's future? (3) What changes must we make and why? (4) What results will various actions produce? (5) What are the solutions to certain problems? (6) How well or how badly does something work? Not all your readers are seeking answers to the same questions, so in planning a long report you should consider all possible aspects that might concern your readers.

- *Who are your readers, and what are their different purposes?* Because long reports tend to have multiple readers, such reports usually cover several aspects of a subject. You must consider your readers' purposes and how to effectively provide enough information for each purpose. Remember that long reports generally need to include more background information than short reports because (1) multiple readers rarely have an identical understanding of the subject and (2) long reports may remain in company files for years and serve as an information source to future employees dealing with a similar situation. Chapter 2 explains specific strategies for identifying readers' purpose.
- *How much information do I need to include in this report?* The relevant data for all readers and all purposes sometimes constitute masses of information. Determine what kinds of facts you need before you begin to gather information, so you are sure you are covering all areas.

Gathering Information

When you gather information for a long report, consult secondary and primary sources for relevant facts.

Secondary Sources

Secondary sources of information consist of documents or materials already prepared in print, on tape, or on film. For a long project, consult secondary sources first because they are readily available and easy to use. Also, the information in secondary sources reflects the work of others on the same subject and may help direct your plans for further research or trigger ideas about types of information your readers need to know. These are your major secondary sources:

Internal Company Documents. Company records, reports, and correspondence contain information about past operations, trends in production and sales, decisions and their rationales, and agreements with people inside and outside the company. Most company projects have their beginnings in past decisions or events, and readers usually need to know background information about the main subject of a long report.

Government Agencies and Foundations. Many government agencies and private foundations support research in science and technology. The scientists conducting research must write a final project report on results, test methods, and conclusions. If the information is not classified, you can obtain copies of the final research reports from most agencies and foundations.

Library. The library offers a rich collection of facts, figures, and research articles representing both historical and current information. Librarians are available to help you locate specific information. Here are the most frequently used sources in the library for science, technology, and business subjects:

1. *Library catalog.* As either a computer catalog or file of bibliography cards, the library catalog of holdings lists all books, films, tapes, records, and any materials on microfilm or microfiche in the library. The catalog information will tell you the call number of each item, so you can send for it or find it on the shelf. The computer catalog also usually indicates whether the item is currently in the library or checked out and due back on a certain date.
2. *Periodical indexes.* Published articles about science, technology, and business subjects are listed in periodical indexes, most of which are published monthly. The most useful for people in professional areas include *Engineering Index, Business Periodicals Index, Applied Science and Technology Index, Index Medicus, General Science Index, Biological and Agricultural Index,* and *Business Index.*
3. *Newspaper indexes.* Newspaper indexes list the contents of a newspaper by subject, so you can find, for example, articles a newspaper may have run on toxic waste. The two indexes available in most college libraries are *The New York Times Index* and *The Wall Street Journal Index.*
4. *Abstract indexes.* Abstract indexes provide summaries of published articles about science, technology, and business topics. Read the abstract to decide whether the article contains information you need for your report. If it does, look for the article in the journal listed in the abstract. Abstract indexes cover most fields and include such indexes as *Biology Digest, Chemical Abstracts, Electrical and Electronic Abstracts, Geological Abstracts, Mathematical Reviews, Microbiology Abstracts, Nuclear Science Abstracts, Water Resources Abstracts,* and *Psychological Abstracts.*
5. *Annual reports.* Libraries usually collect the annual reports of the Fortune 500 companies as well as those of local companies in the immediate geographic area. The reports provide financial information for the previous year.
6. *Government documents.* The federal government publishes hundreds of scientific and technical reports, newsletters, and brochures every year. These documents are not listed in the periodical indexes, but the

library should have available an index such as the *Monthly Catalog of United States Government Publications.*

7. *Encyclopedias, directories, business guides, and almanacs.* Specialized encyclopedias on science and technology cover such subjects as computer science, space, energy, construction technology, and wildlife. Business guides and directories provide information about corporations, their officers, products, stock offerings, and sales trends. Almanacs give specific dates and facts about past events.

8. *Computer databases.* Your library probably has direct on-line access to hundreds of databases concerning science, technology, and business, allowing you to search for articles and reports more quickly than you can manually. Since databases are regularly updated and do not have to be printed, as most periodical indexes are, they provide the most current listings of information. Databases rely on key words to find information. After you select key words for your subject, the system searches for those words in titles and abstracts of articles. One limitation of this method is that titles of important articles may not include the key words you are using and the articles will be missed. For a comprehensive search, you should consult other sources as well as use databases. Information-retrieval services charge a fee for every search through a database, so you should consult a librarian to assess costs and arrange for a search.

Unless you have a specific author or title of an article in mind when you use these secondary sources, you must search for information by subject. For effective searches, check under other key topics as well as the main subject. If you are looking for information about diamonds, check under the key word *diamond,* as well as under such topics as mining, precious gems, minerals, South Africa, and carbon. A librarian can help you select the best key words for database searches.

Primary Sources

Primary sources of information involve research strategies to gather unpublished or unrecorded facts. Scientific research reports, for example, are usually based on original experiments, a primary source. A marketing research report may be based on a consumer survey, also a primary source. Secondary sources can provide background, but they cannot constitute a full research project. Only primary sources can provide new information to influence decision making and scientific advances. Here are the major primary sources of information:

Personal knowledge. When you are assigned to a report project on the job, it is usually because you are already deeply involved in the subject. Therefore, much of the information needed for the report may be in your head. Do not rely on remembering everything you need, however—make notes. Writing down what you know about a subject will clarify in your own mind whether you know enough about a particular aspect of the subject or you need to gather more facts.

Observation. Gathering information through personal observation is time-consuming, but it can be essential for some subjects. Experienced writers collect information from other sources before conducting observations so that they know which aspects they want to focus on in observation. Scientists use observation to check, for example, how bacteria are growing under certain controlled conditions. Social scientists may use observation to assess how people interact or respond to certain stimuli. Technology experts may use observation to check on how well machinery performs after certain adjustments. Do not interfere with or assist in any process you are observing. Observation is useful only when the observer remains on the sidelines, watching but not participating.

Interviews. Interviews with experts in a subject can yield valuable information for long reports. Writers usually gather information from other sources first and then prepare themselves for an interview by listing the questions they need answered. Know exactly what you need from the person you are interviewing so that you can cover pertinent information without wasting time. Use the guidelines for conducting interviews in Chapter 2.

Tests. Tests can be useful in yielding information about new theories, systems, or equipment. A scientist, for example, must test a new drug to see its effects. A market researcher may test consumer reaction to new products. The tests themselves should be conducted by an expert to ensure a reliable study.

Surveys. Surveys collect responses from individuals who represent groups of people. Results can be analyzed in various ways, based on demographic data and responses to specific questions. A social scientist who wants to find out about voting patterns for people in a specific geographic area may select a sample of the population that represents all voting age groups, men and women, income groups, and any other categories thought to be significant. Designing an effective questionnaire is complicated because questions must be tested for reliability. If you are not trained in survey research, consult survey research handbooks and ask the advice of experts before attempting even an informal survey on the job.

Taking Notes

Taking useful notes from the multiple sources you consult for a long report is necessary if you are to write a well-developed report. No one can remember all the facts gathered from multiple sources, so whenever you find relevant information, take notes.

Although putting notes in a spiral notebook may seem more convenient than shuffling note cards while you are gathering information, notes in a notebook will be more difficult to use when you start to write. Because such notes reflect the order in which you found the information and not necessarily the order relevant to the final report organization, the information will seem "fixed" and nearly impossible to rearrange easily, thus interfering with effective organization for the reader. You would have to flip pages back and forth, wasting time searching for particular facts as you try to organize and write. A more efficient way to record facts is to use note cards, because such cards are easy to rearrange in any order and are durable enough to stand up to frequent handling. Here are some guidelines for using note cards to keep track of the information you gather for a long report:

1. Fill out a separate note card for each source of information you use at the time you use it. Put complete bibliographic information on the card for a library source. For other sources, such as interviews, record the names of people, dates, subjects discussed, and any other identifying material. Figure 10-2 shows a note card that contains a record of a source used for a long report. Notice that the authors, book title, publisher, city of publication, and date of publication are on the card.

Figure 10-2 A Sample Note Card Indicating a Source

If any of this information is missing, you would have to return to the library to get it before citing this source in your report. Recording all the publishing information when you use the source will avoid delays later in the writing process.

2. Put only one item of information from only one source on each card. Putting multiple notes or multiple sources on one card will make it difficult to sort and reorder your notes when you begin to write.
3. Condense the information you find, but do not change the meaning. If the original source states, "The responses from consumers appear to indicate . . . ," do not write on your note card, "Consumer responses indicate. . . ." The original sentence implied some uncertainty by using the word *appears*. This meaning should be clear in your notes.
4. If you plan to quote directly from a source, put quotation marks on your note card so you are certain later exactly which words are direct quotations.
5. Record precise figures and doublecheck your facts, so you do not write "1947" when you want to write "1974."
6. Record enough information so that you can recall the full meaning from the original. One or two key words are not usually enough to help you remember detailed information.

Figure 10-3 shows the notes taken for a student science report from this paragraph:

> The gorilla is a true knuckle walker. It is the one ape that has returned to the ground almost completely. All its food—coarse vegetable matter, roots, bamboo shoots, berries—is found there. It has given up running in favor of large size and enormous strength. Gorillas depend for survival on being big and powerful—up to four hundred pounds for an adult male—and by looking fierce. Young gorillas hang and play in trees, but adults are too large and lethargic for that. Sometimes they will squat in groups on very low, very broad branches, but they really prefer the ground. They are extremely sedentary, stay in the same places for long periods of time, and almost never run at all.[1]

Note cards should contain three basic elements: (1) a brief statement of the information you need from the paragraph, (2) the authors' last names and the page number from which the information was taken in one upper corner, (3) a topic heading in the other upper corner for convenient sorting and organizing. Notice that the notes in Figure 10-3 are not copies of the sentences in the original paragraph. Put notes in your own words to avoid inadvertently copying the original.

Figure 10-3 Sample Notes for a Student Science Report

Interpreting Data

Readers need the facts you present in a long report, but they also need help in interpreting those facts, particularly as to how the information affects their decision making and company projects. When you report information, tell readers what it means. Explain why the information is relevant to the central issue and how it supports, alters, or dismisses previous decisions or how it calls for new directions. Your objective in interpreting data for readers is to help them use the facts efficiently. Whatever your personal bias about the subject, be as objective as you can, and remember that several interpretations of any set of facts may be possible. Alert your readers to all major possibilities. Declining sales trends in a company's southwest sales region may indicate that the company's marketing efforts are inadequate in the region, but the trends also may indicate that consumers are dissatisfied with the products or that the products are not suitable for the geographic region because of some other factors. Give the readers all the options when you interpret facts.

After you have gathered information for your report and before you write, ask yourself these questions:

- Which facts are most significant for which of my readers?
- How do these facts answer my readers' questions?
- What decisions do these facts support?
- What changes do these facts indicate are needed?
- What option seems the best, based on these facts?
- What trends that affect the company do these facts reveal?
- What solutions to problems do these facts support?
- What actions may be necessary because of these facts?
- What conditions do these facts reveal?
- What changes will be useful to the company, based on these facts?
- How does one set of facts affect another set?
- When facts have several interpretations, which interpretations are the most logical and most useful for the situation?
- Is any information needed to further clarify my understanding of these facts?

Presenting data is one step, and interpreting their meaning is the second step in helping readers. If you report to your readers that a certain drainage system has two-sided flow channels, tell them whether this fact makes the system appropriate or inappropriate for the company project. Your report should include the answers to questions relevant to your subject and to your readers' purposes.

Drafting Long Reports

Like short reports, long reports may be organized in either the opening-summary or delayed-summary pattern, depending on your analysis of your readers' probable attitudes toward the subject. Whatever the pattern, long reports generally have three major sections.

Introductory Section

The introductory section in a long report usually contains several types of information. If this information is lengthy, it may appear as subsections of the introduction or even as independent sections. Depending on your subject and your readers, these elements may be appropriate in your introduction:

Purpose. Define the purpose of your report—for example, to analyze results of a marketing test of two colognes in Chicago. Include any secondary purposes—for example, to recommend marketing strategies for the colognes or to define the target consumer groups for each cologne. For a report on a scientific research project, tell your readers what questions the research was intended to answer. Also clarify in this section why you are writing the report. Did someone assign the writing task to you, or is the report subject your usual responsibility? Be sure to say if one person told you to write the report and address it to a third person.

Methods. Explain the methods you used to obtain information for your report. In a research report, the methods section is essential for readers who may wish to duplicate the research and for readers who need to understand the methods in order to understand the results. Include in this section specific types of tests used, who or what was tested, the length of time involved, and the test conditions. If appropriate, also describe how you conducted observations, interviews, or surveys to gather information and under what circumstances and with which people. You need not explain going to the library or searching through company files for information.

Background. If the report is about a subject with a long history, summarize the background for your readers. The background may include past research on the same subject, previous decisions, or historical trends and developments. Some readers may know nothing about the history of the subject; others may know only one aspect of the situation. To effectively use the report information, readers need to know about previous discussions and decisions.

Report Limitations. If the information in the report is limited by what was available from specific sources or by certain time frames or conditions, clarify this for your readers so that they understand the scope of the information and do not assume that you neglected some subject areas.

Report Contents. For long reports, include a section telling readers the specific topics covered in the report. Some readers will be interested only in certain information and can look for that at once. Other readers will want to know what is ahead before they begin reading the data. Indicate also why these topics are included in the report and why they are presented in the way they are to help your readers anticipate information.

Recommendations/Results/Conclusions. If you are using the opening-summary pattern of organization, include a section in which you provide an overview of the report conclusions, research results, or recommendations for action.

Data Sections

The data sections include the facts you have gathered and your interpretations of what they mean relative to your readers. Present the facts in ways best suited to helping your readers use the information. The patterns of organization explained in Chapter 3 will help you present information effectively. Chapter 12 explains data sections in specific types of reports. For a report of a research study, the data sections cover all the results obtained from the research.

Concluding Section

Long reports generally have a concluding section even if there is an opening summary because some readers prefer to rely on introductory and concluding sections. If the report is in the opening-summary pattern, the concluding section should summarize the facts presented in the report and the major conclusions stemming from those facts. Also explain recommendations based on the conclusions. Some readers prefer that recommendations always appear in a separate section, and if you have a number of recommendations, a separate section will help readers find and remember them.

If the report is in the delayed-summary pattern, the concluding section is very important because your readers need to understand how the facts lead to specific conclusions. Present conclusions first; then present any recommendations that stem from those conclusions. Include any suggestions for future studies of the subject or future consideration of the subject. Do not, however, as you may do in short reports, include specific requests for meetings or remind readers about deadlines. The multiple readers of a long report and the usual formal presentation of long reports make these remarks inappropriate. You may include such matters in the transmittal letter that accompanies your long report, as explained in Chapter 11.

CHAPTER SUMMARY

This chapter discusses the usual organization of short and long reports and provides guidelines for report planning, information gathering, note taking, data interpretation, and drafting. Remember:

- Reports generally have one or more of these purposes: to inform, to record, to recommend, to justify, and to persuade.
- Most short internal reports are written as memos, and most short external reports are written as letters or as formal documents.
- Most short reports are organized in the opening-summary pattern.

- Long reports may be in the opening-summary pattern or the delayed-summary pattern and are written as formal documents.
- Planning reports includes identifying the central issue, analyzing readers and purpose, and deciding how much information is necessary for your readers.
- Writers gather information for reports from both primary and secondary sources.
- Notes are best collected on note cards, one item per card.
- For effective reports, writers must interpret the data for their readers.
- Long reports usually have three major divisions: an introduction, information sections, and a concluding section.

SUPPLEMENTAL READINGS IN PART TWO

Benson, P. J. "Writing Visually: Design Considerations in Technical Publications," *Technical Communication.*

ENDNOTE

1. From D. C. Johanson and M. A. Edey, *Lucy: The Beginnings of Humankind* (New York: Simon and Schuster, 1981), p. 318.

Model 10-1 Commentary

This short report is organized in the opening-summary pattern. The reader and the writer have previously discussed the need for computer equipment, but the reader delayed any purchases because of costs. Now cost is no longer a problem, and the writer has decided to recommend the equipment again. He uses the opening summary to review the problem and recommend purchasing equipment.

Because the reader needs to carefully consider how the computers will benefit the office, the writer describes the features in detail. For the reader's convenience, the writer presents costs in a table. The closing repeats the recommendation for new computers and points out the expected benefits from such a purchase.

Discussion

1. Discuss the headings used in this report. What subheadings might be helpful to the reader in the features section?

2. Discuss why the writer probably decided to close the report by restating the recommendation and pointing out the expected benefits.

June 21, 1990

TO: B. A. Neilsen, Manager

FROM: E. A. Anderson

SUBJECT: Purchase of Laptop Computers

As you know, our department is handicapped by not being able to perform spreadsheet analysis on out-of-town audits. The lack of word-processing capabilities requires staff members to prepare all communications by hand. In addition, the lack of ability to use spreadsheet analysis techniques in our statistical tests hampers the effectiveness of a professional audit. To solve this difficulty, I recommend that we purchase two laptop computers to be used by staff on all out-of-town audits.

Background

We have discussed purchasing laptop computers several times during the past year. Each time, the subject has been delayed for further study and review. Budget considerations earlier this year also argued against further purchases of computers. As you indicated at our last staff meeting, these budget constraints are no longer an issue. Therefore, with the help of Janice Allread in the Data Processing Department, I researched the features and costs of the NEC Multispeed HD System.

Features

The NEC Multispeed HD System has a number of features that are not found in most laptop computers. Among them are

20-megabyte hard disk
battery operation
built-in 2400-baud modem
enhanced color screen

Unlike most laptop computers, the NEC Multispeed HD System contains a 20-megabyte hard disk and one floppy disk drive instead of the standard two floppy disk drives. The inclusion of the hard disk will eliminate the need to continually use program disks and floppy disks to access and maintain the software. All the software applications and individual programs can be permanently stored on the hard disk, and the floppy disks can be used as backup.

The NEC Multispeed HD System can run off either an AC power source or batteries. Battery operation is important for us because an auditor may wish to operate the computer on a plane, in a hotel room, or anywhere an outlet is not readily available.

The built-in 2400-baud modem is also an attractive feature because an auditor can dial access to the mainframe computer system in order to access important information that may otherwise be unavailable. With the dial access to the mainframe, files can be accessed and downloaded to the laptop computer for integration with the detailed testing techniques. Delays in data transmission will be reduced. The enhanced color screen is standard equipment on the NEC Multispeed HD System. It provides approximately 25% higher color resolution than the standard color screen, and it also supports the graphics software packages we use.

Cost

I have asked several area dealers for a price list and investigated possible documents. Following are the costs for various components:

NEC Multispeed (20MB)	$2700
Carrying case	68
RS-232 cable (10 ft.)	45
Diconex model 150 printer	600
2400-baud modem	500
Hardware Total	$ 3913
Crosstalk XXV	$ 190
Flowcharting III Plus	350
Lotus 1-2-3 Version 6	550
DisplayWrite 5 Version 4	600
Software Total	$ 1690
Total System Cost	$ 5603

The cost of two full units would be $11,206. Rayburn Electronics has offered a 6% discount, available until September. The discounted cost would be approximately 20% less than other models without a discount.

Recommendation

The increased efficiency of our out-of-town audits with this laptop equipment will result in savings of both time and money during business trips. The company will benefit in two major ways. First, we can prepare more complete audit reports for local management during the visit. The time spent waiting in airports, etc., will be used more effectively to complete the final reports and follow-up material. Second, better testing techniques will result in substantial cost savings in the full audit. I recommend, therefore, that we purchase the two full units immediately. If you wish to discuss the features and costs of the system in more detail, please let me know.

Model 10-2 Commentary

This short report is organized in the delayed-summary pattern because the writer believes he must persuade the reader that the current ultrasound equipment is inadequate. The writer begins with a review of the current hospital situation—increasing requests for ultrasound examinations. He then presents two alternatives that could solve the problem to some degree and includes the cost of each alternative. The closing recommends one of the alternatives and suggests a deadline for action.

Discussion

1. Notice that the writer organizes the two information sections differently. In one he provides cost figures before the discussion. In the other he discusses the alternative before providing cost figures. Discuss why the writer probably chose these organization patterns.

2. Discuss how effectively the writer presents the problem of increased demand in the opening.

3. Assume that you had decided to write this report in the opening-summary pattern. Rewrite the opening appropriately.

TO: T. R. Lougani
Vice President, Support Services

January 16, 1990

FROM: C. S. Chen
Asst. Director of Radiology

SUBJECT: Ultrasound Equipment

Since ultrasound is a noninvasive procedure and presents no radiation risk to the patient, it has become increasingly popular with most physicians. In our department, ultrasound use over the past five years has increased 32%. In 1985, we performed 1479 procedures, and last year we performed 2186 procedures. We expect this increased demand to continue. To prepare for the demand on our equipment (estimated 2297 this year), we need to consider ways to increase our capability.

Updating Current Equipment. One possibility is to update the current ultrasound unit with new software. Costs for updating the present equipment are as follows:

Current unit update	$28,000
Maintenance contract (first year)	8,500
Total	$36,500

The maintenance contracts will increase slightly in each succeeding year. The maintenance charge is rather high because as the equipment ages, more replacement parts will be needed. The basic hardware is now seven years old. Although updating the present equipment will improve the performance of the equipment, it will only marginally increase the number of procedures that can be performed over a 24-hour period. We cannot reasonably expect to make up the cost of updating our equipment because our volume will not increase significantly.

Purchasing New Equipment. Purchasing new ultrasound equipment will provide Columbia Hospital with the ability to keep up with the expected growth in requests for the procedure. A new unit with increased capability will allow us to expand by 15% in the first year. The unit alone will enable this expansion; we will not need new staff or floor space. Increased capability will also increase our patient referrals. At an average cost of $120 per examination, an additional $41,000 will result in the first year, and the full cost of equipment should be recovered within the first two years. Patients may also use other services at the hospital once they

T. R. Lougani - 2 - January 16, 1990

come here for ultrasound, resulting in further income. Cost of new equipment is as follows:

New unit	$74,000
Maintenance contract	0
Total	$74,000

Conclusion

Updating old equipment is merely a holding technique, even though such a decision would save money initially. To increase our capability and provide the latest technology for our patients, I recommend that we purchase a new unit to coincide with the opening of the new community clinic on our ground floor, March 1. I would appreciate the opportunity to discuss this matter further. If you wish to see sales and service literature, I can supply it.

Model 10-3 Commentary

This court report was written by a social worker for the county children's services agency. The report will be read by the judge hearing the case, the prosecutor working for the county, and the defense attorney if the parents contest the agency's recommendation. The social worker is concerned with presenting an effective summary of the case file (containing possibly dozens of pages) so that the judge will grant the agency's request.

Social workers make notes in a chronological log for the files. For the court report, however, the writer must organize by topic so that the judge and prosecutor can refer quickly to specific information.

The writer opens with a summary that includes the agency request for custody and emphasizes the danger the children are in if they are not in agency custody. Headings call attention to the children and the specific conditions affecting their welfare. The closing repeats the agency request and specifies future actions.

Discussion

1. Decide which information sections begin with a summary or topic sentence. Write an opening-summary sentence for any section that doesn't have one.

2. Assume that you are the prosecutor in this case. How would this report aid you in presenting a case to the judge?

REQUEST FOR TEMPORARY CUSTODY

Winthrop County Children Services Agency requests Temporary Custody of Betsy White, age 6, and David White, age 8. This agency was granted Emergency Custody on April 4, 1989, which has continued to date. The mother, Susan White, has a history of violent, unpredictable behavior, problems with alcohol, and inability to provide medical care for the children. The father, Martin White, is facing criminal charges for injuries to both children. Because of the mother's hostile reaction to Agency contact and the father's pending criminal charges, Temporary Custody is necessary to protect the children.

Betsy White

Betsy White has suffered from emotional outbursts at Harland Elementary School. Mrs. Doris Winters, her teacher, reports that Betsy frequently cries and causes a disturbance during class time. Betsy had several fading bruises when the social worker visited the White home, but no recent injury. Records at City Hospital show that Betsy was treated May 17, 1988 for third-degree burns on both hands. At that time, Mrs. White told the doctor that Betsy had been playing near the stove. Betsy, however, told the doctor that she had "been bad" and was punished. Investigation at the time did not reveal sufficient evidence of child abuse.

David White

David White was taken to City Hospital by his mother on April 1, 1989. He suffered from a large gash on the right side of his face and various welts and bruises on his body. Dr. Raymond Weaver treated David and ordered chest x-rays because of the severity of the bruises. The x-rays revealed no new fractures, but they did show two healed breaks in the rib cage. David told hospital personnel and Detective Roslyn Lovett of the Winthrop County Sheriff's Department that his father had beaten him because he broke a dish. David expressed fear of both his parents to Detective Lovett.

Living Conditions

The White home is in poor condition and unhealthy for the family. Garbage is strewn everywhere, and the kitchen counters are piled with dirty dishes. There are no towels in the bathroom, no sheets on the beds, and no clean clothes for the children. The bathroom sink is stopped up.

The gas heat has been turned off, although other utilities are still on. Mrs. White said she would clean up the apartment when the Emergency Custody order went into effect, but she had not done so as of November 12, 1989.

Parental Behavior

Both parents in previous agency contact had promised to work on improving their home environment, but didn't. Mrs. White refused to admit the social worker to the home on three occasions and refuses to attend alcohol treatment sessions. She appears to have been drinking each time the social worker visited the home. Although Mrs. White took David to City Hospital for treatment on April 1, 1989, she indicated that she did it primarily to annoy her husband. She did not seek treatment for older injuries. She stated that the children fell a lot and were clumsy, thus accounting for the multiple injuries. She remains hostile and uncooperative with the Agency and is angry that her husband is facing criminal charges. She insists she does not need counseling. Mr. White threatened the social worker during the investigation of David's injuries by stating that he would "get" her. Neighbors report that he has a violent temper and once killed a neighbor's dog when the animal strayed into the White yard. He has not shown remorse for the children's injuries.

Summary

Since the Criminal charges against the father are still pending, and since the mother is uncooperative and refuses counseling, Winthrop County Children Services Agency requests Temporary Custody of both Betsy White and David White. The children should not return home until the mother attends parental counseling regularly and receives alcohol abuse counseling. The Agency also requests a 90-day review to evaluate the cooperation of the mother.

Model 10-4 Commentary

This model consists of excerpts from a 58-page formal report written by a research team for NASA on the subject of communications practices of aeronautical engineers and scientists.

The introductory section explains the purpose of the study, describes the survey method the writers used, provides background on the reasons for conducting the study, and lists the primary research assumptions or questions the researchers wanted to answer.

The data sections begin with an explanation of the statistical methods used to analyze the survey results. The survey results relative to each research assumption appear in a separate section, such as the first one shown here entitled, "Survey Objective 1: The Importance of Technical Communications." Tables illustrating survey results are included in the text.

The conclusion consists of two parts: (1) "Validity of the Assumption," a section discussing what the results reveal about each research assumption, and (2) "Concluding Remarks," a summary of the major conclusions reached in the report.

Discussion

1. Read the data section headed "Survey Objective 1: The Importance of Technical Communications." Then read the concluding section connected with this topic, "Validity of the Assumptions" (the first item in the section). Discuss how well the concluding section summarizes the results and their meaning for Assumption 1. If you were reading this report out of general interest in the subject, would the analysis be sufficient, or would you have more questions?

2. Discuss the format of the introduction. In groups, draft a revision of the introduction to make it more helpful for a busy reader.

3. Discuss how the tables support the writers' explanation of results. Why did the writers probably decide to include the tables in the text rather than collect them in an appendix?

Technical Communications in Aeronautics: Results of an Exploratory Study

An Analysis of Managers' and Nonmanagers' Responses

INTRODUCTION

This exploratory study investigated the technical communications practices of aeronautical engineers and scientists. The study, which utilized survey research in the form of a self-administered mail questionnaire, had a twofold purpose: (1) to gather baseline data regarding several aspects of technical communications in aeronautics and (2) to develop and validate questions that could be used in a future study concerning the role of the U.S. government technical report in aeronautics.

The study had five specific objectives: first, to solicit the opinions of aeronautical engineers and scientists regarding the importance of technical communications to their profession; second, to determine the use and production of technical communications by aeronautical engineers and scientists; third, to seek their views about the appropriate content of an undergraduate course in technical communications; fourth, to determine aeronautical engineers' and scientists' use of libraries, technical information centers, and on-line databases; and fifth, to determine the use and importance of computer and information technology to them. The study, which spanned the period from July 1988 to November 1988, was conducted in conjunction with Old Dominion University under Contract NAS1-18584, Task 28, to help ensure the objectivity and confidentiality of the data and to obtain research skills not readily available to the project.

Research Design and Methodology for the Exploratory Study

Data were collected by means of the self-administered mail questionnaire shown in the Appendix. The questionnaire was developed within the project team, circulated to selected technical communicators for review and comment, and pretested at the NASA Ames Research Center, the NASA Langley Research Center, and the McDonnell Douglas Corporation in St. Louis.

Members of the American Institute of Aeronautics and Astronautics (AIAA) comprised the study population. The sample frame consisted of approximately 25,000 AIAA members in the United States with either academic, government, or industry affiliations. Simple random sampling was used to select 2000 individuals from the sample frame to participate in the exploratory study. Six hundred and six (606) usable questionnaires (30.3% response rate) were received by the established cutoff date.

The questionnaire used in the study contained 35 questions: 25 questions concerned technical communications in aeronautics, 8 questions concerned demographic information about the survey respondents, and 2 open-ended questions allowed survey respondents to comment on the topics covered in the questionnaire and to offer suggestions for improving technical communications in aeronautics.

The data were analyzed by using the Statistical Package for the Social Sciences-X (SPSS-X) designed for use with a personal computer. Cross-tabulations were prepared to explore the relationships between the responses to the 25 questions and the respondents' organizational affiliations. Affiliations included "academic" (both academic and not-for-profit organizations), government (NASA and non-NASA), and industry. The Chi-Square and one-way ANOVA (Analysis of Variance) tests at the 0.05 level of statistical significance were used as the nonparametric and parametric tests for relationships between the responses to the 25 questions and the organizational affiliations of the respondents. The results of the exploratory study are presented in NASA Technical Memorandum 101534, Parts 1 and 2 (Pinelli et al., 1989).

Background for the Analysis of Managers' and Nonmanagers' Responses

This report represents an analysis of the management and nonmanagement responses to the data collected in the exploratory study. These responses were analyzed to test the primary assumption that aerospace managers and nonmanagers have different technical communications practices.

Many technical communicators believe that managers and nonmanagers have different technical communications practices. This assumption of differences is based on the presumption that the duties of managers and nonmanagers are fundamentally different. Consequently, these two groups would develop different information use and production strategies that would, in turn, manifest themselves as distinctive technical communications practices.

There is, however, little empirical evidence to support the presumption that managers and nonmanagers, in particular, have different technical communications practices. For example, Pinelli et al. (1984) found little difference in the choice of report components used by aerospace managers and nonmanagers to decide to read a NASA technical report. Additionally, there was little difference in the order in which the components of a NASA technical report were read. Furthermore, aerospace managers and nonmanagers expressed little difference in their preferences regarding the production (i.e., format and layout) of NASA technical reports (Pinelli et al., 1982).

The assumption of differences is stated as a research question, "Do aerospace managers and nonmanagers have different technical communications practices?" rather than as a research hypothesis for the following reasons:

1. The study is exploratory in nature and, as such, has certain limitations.
2. The low response rate of 30.3%, which is fairly typical for mail surveys, prohibits generalizing the findings to the "nonrespondents" and the population being studied.
3. The available related research and literature regarding the technical communications practices of managers and nonmanagers does not provide a sufficient research foundation.

Assumptions

Five secondary assumptions were made regarding the five study objectives. These assumptions, which are given below, were tested and were used to answer the research question.

1. The importance of communicating technical information effectively is equally significant to aerospace managers and nonmanagers. A significant difference in the reported responses of aerospace managers and nonmanagers regarding "importance" would support the presumption

of different technical communications practices between the two groups.

2. The use and production of technical information and technical information products are different for aerospace managers and nonmanagers because of the different duties performed by the two groups. A significant difference in the reported responses of aerospace managers and nonmanagers regarding "use and production" would support the presumption of different technical communications practices between the two groups.
3. The content for an undergraduate course in technical communications should be viewed differently by aerospace managers and nonmanagers. A significant difference in the reported responses of aerospace managers and nonmanagers regarding "content" would support the presumption of different technical communications practices between the two groups.
4. The use of libraries, technical information centers, and on-line (electronic) databases differs for aerospace managers and nonmanagers because of the different duties performed by the two groups. A significant difference in the reported responses of aerospace managers and nonmanagers regarding "usage" would support the presumption of different technical communications practices between the two groups.
5. The use and importance of computer and information technology differs for aerospace managers and nonmanagers because of the different duties performed by the two groups. A significant difference in the reported responses of aerospace managers and nonmanagers regarding "use and importance" would support the presumption of different technical communications practices between the two groups.

Presentation and Discussion of Managers' and Nonmanagers' Responses

The data in this report are presented for each survey objective and are discussed in terms of management/nonmanagement responses. Background data collected as part of the survey revealed that approximately 76% of the respondents held nonmanagement positions and approximately 24% held administrative/managerial positions.

The Chi-Square and *t*-test for a difference between two independent means were used as the nonparametric and parametric tests for relationships between the responses to the 25 questions and the management and nonmanagement respondents. Attempts were made to establish the extent to which the characteristics of the population may reasonably be inferred from the attributes of the sample. Such inference is then subject to various conventions regarding statistical significance. The appropriate application of such conventions to the primary effort ($n = 606$) is called "estimate of parameters." The population parameter, in this case a population proportion (P), is estimated from a sample proportion (p). Such estimates are dependent in part on sample size, the overall response rate, and the sample size (response) for each question.

Given the general range of sample sizes and the nature of the sampling distribution of proportions, it can be stated that at the 95% confidence level, the true population proportion (P) for managers lies within $\pm 8.4\%$ of the sample proportion (p) and the true population proportion (P) for nonmanagers lies within $\pm 4.8\%$ of the sample proportion (p).

Although a confidence and tolerance level can be established, readers are cautioned that while a random sample of AIAA members were sent questionnaires, no assurances of randomness can be made regarding the questionnaires that were returned. Because the overall response rate was less than 50%, which is traditionally considered to be "representative," the figures given above should be used with caution when making generalizations about the population.

Survey Objective 1: The Importance of Technical Communications

To determine the importance of technical communications in aeronautics, survey respondents were asked to indicate the importance of communicating technical information effectively, the number of hours spent each week communicating technical information to others, the number of hours spent each week working with technical communications received from others, and how professional advancement has affected the amount of time they spend communicating technical information to others and working with technical communications from others.

Approximately 99% of the managers and nonmanagers surveyed (Table 1) indicate that the ability to communicate technical information effectively is important. Fewer than 1.0% indicate that this ability is not at all important.

Table 1 Importance of Technical Communications

	Managers		***Nonmanagers***	
How Important	***No.***	***%***	***No.***	***%***
Very	129	89.6	411	89.8
Somewhat	14	9.7	45	9.8
Not at all	1	.7	2	.4
Total	144	100.0	458	100.0

Managers spend an average of 13.6 hours per week communicating technical information to others (Table 2), and nonmanagers spend an average of 14.0 hours per week. Based on a 40-hour work week, both groups spend approximately 35% of their work week communicating technical information to others.

Managers and nonmanagers spend approximately 13 hours a week working with technical communications received from others (Table 3), which is approximately 31% of their 40-hour work week.

Considering both the time spent working on the preparation of technical information and the time spent working with technical information received

Table 2 Time Spent Communicating Technical Information to Others

	Managers		Nonmanagers	
Time Spent Per Week, Hour	**No.**	**%**	**No.**	**%**
5 or less	22	15.6	79	17.7
6 to 10	48	34.1	140	30.9
11 to 20	58	41.1	179	39.5
21 or more	13	9.2	55	11.9
Total	141	100.0	453	100.0
Mean	13.6		14.0	

Table 3 Time Spent Working With Technical Information Received from Others

	Managers		Nonmanagers	
Time Spent Per Week, Hour	**No.**	**%**	**No.**	**%**
5 or less	14	9.9	111	24.6
6 to 10	65	46.2	156	34.3
11 to 20	54	38.3	143	31.5
21 or more	8	5.6	44	9.6
Total	141	100.0	454	100.0
Mean	13.0		12.5	

from others, technical communications takes up approximately 66% of the manager's and nonmanager's 40-hour work week.

Approximately 59% of the managers and 76% of the nonmanagers indicate that as they advanced professionally, the amount of time they spent communicating technical information to others increased (Table 4). Approximately 11% of the managers and 17% of the nonmanagers indicate that the amount of time spent communicating technical information to others stayed the same. Approximately 31% of the managers and 7% of the nonmanagers indicate that the amount of time they spent communicating technical information to others decreased as they advanced professionally. In terms of the amount of time spent communicating technical information to others, nonmanagers were more likely to say that the amount of time has increased and managers were more likely to say it has decreased.

Approximately 63% of the managers and 61% of the nonmanagers indicate that as they advanced professionally, the amount of time they spent working with technical communications received from others increased (Table 5). Approximately 18% of the managers and 28% of the nonmanagers indi-

Table 4 Professional Advancement and Amount of Time Spent Communicating Technical Information to Others

	Managers		***Nonmanagers***	
Time Spent Communicating	***No.***	***%***	***No.***	***%***
Increased	84	58.7	349	*76.0
Stayed the same	15	10.5	76	16.6
Decreased	44	*30.8	34	7.4
Total	143	100.0	459	100.0

*Differences between managers and nonmanagers are significant at $p > 0.05$.

Table 5 Professional Advancement and Amount of Time Spent Using Technical Information Received from Others

	Managers		***Nonmanagers***	
Time Spent Using	***No.***	***%***	***No.***	***%***
Increased	89	62.7	278	61.0
Stayed the same	25	17.6	129	*28.3
Decreased	28	*19.7	49	10.7
Total	142	100.0	456	100.0

* Differences between managers and nonmanagers are significant at $p > 0.05$.

cate that the amount of time they spent working with technical communications received from others stayed the same as they advanced professionally. Approximately 20% of the managers and 11% of the nonmanagers indicate that the amount of time they spent working with technical communications received from others decreased as they advanced professionally. Nonmanagers were more likely than managers to say that the amount of time they spent working with technical communications received from others had stayed the same, and managers were more likely than nonmanagers to say that it had decreased.

Survey Objective 2: The Use and Production of Technical Communications

Survey respondents were asked to indicate the amount and type of technical information products they produced and used as well as the sources of help they sought in producing technical information and in solving technical problems.

Memos, letters, and audiovisual (A/V) materials are the technical information products most frequently produced by both managers and nonmanagers (Table 6). On the average, managers. . . .

Table 6 Production of Technical Information Products

Products	***6-Month Average***	
	Managers	***Nonmanagers***
Letters	*30.5	19.6
Memos	*49.0	22.6
Technical reports—government	*2.1	1.4
Technical reports—other	1.8	1.9
Proposals	*2.1	1.6
Technical manuals	0.3	0.3
Computer program documentation	0.5	*1.6
Journal articles	0.3	0.4
Conference/meeting papers	*1.5	0.9
Trade/promotional literature	*1.5	0.9
Press releases	*0.4	0.2
Drawings/specifications	2.1	3.6
Speeches	*3.6	1.8
Audiovisual materials	*9.6	5.6

* Differences between managers and nonmanagers are significant at $p > 0.05$.

VALIDITY OF THE ASSUMPTIONS

The following conclusions are presented concerning the validity of the five study assumptions.

Assumption 1: The Importance of Communicating Technical Information Effectively Is Equally Significant to Aerospace Managers and Nonmanagers

The responses of managers and nonmanagers to the five questions associated with this assumption were very similar. The importance of communicating technical information effectively is significant to aerospace managers and nonmanagers alike. There is very little difference in the average amount of time the two groups spend communicating technical information to others and working with technical communications received from others. Nonmanagers were more likely than managers to say that the amount of time spent communicating technical information to others has increased, whereas managers were more likely than nonmanagers to say it has decreased. Nonmanagers were more likely than managers to say that the amount of time spent working with technical communications from others has stayed the same, whereas managers were more likely than nonmanagers to say that the amount of time spent working with technical communications from others has decreased. However, based on the overall responses to questions dealing with this assumption, the conclusion of no difference in technical communications practices is reached for Assumption 1.

Assumption 2: The Use and Production of Technical Information and Technical Information Products Are Different for Aerospace Managers and Nonmanagers

The responses of managers and nonmanagers to the seven questions associated with this assumption were very different. Significant differences were found for 10 of the 14 types of technical information products produced and used. The magnitudes of difference were greatest for the numbers of memos, letters, drawings/specifications, and A/V materials produced and used. Significant differences existed for how managers and nonmanagers produce artwork and the sources they consult for help in preparing technical communications.

Significant differences also exist in the types of technical information produced and used by managers and nonmanagers in the performance of their duties and in the sources of technical information used to solve technical problems. Nonmanagers were more likely than managers to use experimental techniques and computer programs, whereas managers were more likely than nonmanagers to use government rules and regulations and economic information. Nonmanagers were more likely than managers to produce scientific and technical information, experimental techniques, and computer programs, whereas managers were more likely than nonmanagers to produce economic information. When solving a technical problem, nonmanagers were more likely than managers to use discussions with supervisors, government technical reports, other technical reports, journal articles, conference/meeting papers, textbooks, and handbooks/standards, whereas managers were more likely than nonmanagers to use experts outside the organization. Therefore, the conclusion of difference in technical communications practices is reached for Assumption 2.

Assumption 3: The Content for an Undergraduate Course in Technical Communications Should Be Viewed Differently by Aerospace Managers and Nonmanagers

The responses of managers and nonmanagers to the six questions associated with this assumption were very similar. There is very little difference in the percentage of managers and nonmanagers who had taken technical communications coursework and in the percentages of managers and nonmanagers who indicated that such coursework had helped them to better communicate technical information. Further, there were very few differences in the types of principles, mechanics, on-the-job communications, and types of technical reports to be included in an undergraduate technical communications curriculum for aeronautical engineers and scientists. Therefore, the conclusion of no difference in technical communications practices is reached for Assumption 3.

Assumption 4: The Use of Libraries, Technical Information Centers, and On-Line (Electronic) Databases Differs for Aerospace Managers and Nonmanagers

The responses of managers and nonmanagers to the three questions associated with this assumption were different. Nonmanagers were more likely than managers to use a library or technical information center and were more likely to use on-line (electronic) databases than managers. Nonmanagers were more likely than managers to do all or most of their own searches. Therefore, the conclusion of difference in technical communications practices is reached for Assumption 4.

Assumption 5: The Use and Importance of Computer and Information Technology Differs for Aerospace Managers and Nonmanagers

The responses of managers and nonmanagers to three of the six questions associated with this assumption were different. Nonmanagers were more likely than managers to use computer technology for preparing technical communications and were more likely to say that the use of computer technology has increased their ability to communicate technical information "a lot." Nonmanagers were more likely than managers to use scientific graphics software and managers were more likely than nonmanagers to use business graphics software.

Managers were more likely than nonmanagers to "already use" audiotapes and cassettes, whereas nonmanagers were more likely than managers to say that they "doubt if they will" use this technology. Managers were more likely than nonmanagers to "already use" video tape whereas nonmanagers were more likely than managers to say that they "doubt if they will" use it. Managers were more likely than nonmanagers to "already use" electronic mail, whereas nonmanagers were more likely than nonmanagers to say they "don't but may" use it in the future. Therefore, the conclusion of difference in technical communications practices is reached for Assumption 5.

Concluding Remarks

Aerospace managers and nonmanagers have different technical communications practices for three of the five assumptions tested. Therefore, in response to the study's research question, it is concluded that aerospace managers and nonmanagers do have different technical communications practices.

However, while the results of this study provide empirical evidence regarding the technical communications practices of aerospace managers and nonmanagers, data supporting the presumption that the "difference" is attributable to the duties performed by aerospace managers and nonmanagers are neither conclusive nor compelling. The limitations of this exploratory study and the study's research design prohibit reaching that conclusion. Nevertheless, the implication that these differences arise from differing professional duties is hard to resist.

There are perhaps several explanations for both the similarities and the differences in the findings regarding the technical communications practices of aerospace managers and nonmanagers. One possible reason for the similarities is that the managers in this study have risen through the ranks and have retained many of the technical communications practices formed while they were nonmanagers. Another possible explanation is that many of the managers included in this study are actually working supervisors and, consequently, utilize technical communications practices common to both managers and nonmanagers.

The differences may be variously explained. One explanation can be attributed to a difference in the duties performed by the two groups. For example, it seems logical that managers would produce more economic information than nonmanagers and that managers would use more economic information and government rules and regulations than nonmanagers. Likewise, it seems logical that different duties would explain why nonmanagers produce and use significantly more experimental techniques and computer programs than do managers. Could other factors or variables (e.g., organizational affiliation) account for the different technical communications practices?

Accessibility or availability of support help may also explain certain technical communications practices among aerospace managers and nonmanagers. Managers are more likely than nonmanagers to seek the help of a secretary to prepare written technical communications. Likewise, managers are more likely than nonmanagers to use a secretary to help prepare their artwork. Does accessibility or availability explain why neither managers nor nonmanagers make extensive use of technical writers and editors? Could familiarity, experience, ease of use, or expense account for this finding?

Managers make greater use of experts outside of the organization to solve technical problems. One possible explanation is that managers have greater access to outside experts. Another is that the use of outside experts to solve problems is a fairly common practice among managers. On the other hand, nonmanagers are far more likely than managers to use a variety of information sources when seeking solutions to technical problems. Is the use of various information sources by nonmanagers more an indication of the different type(s) of problems being solved? Both groups, however, display a preference for personalized, informal information sources when solving technical problems. This similarity may be more attributable to social/professional enculturation than to any other possible factor or variable.

Both managers and nonmanagers prefer personalized, informal information sources to libraries, technical information centers, and on-line electronic databases. This similarity may also be attributable to social/professional enculturation. On the other hand, the finding that nonmanagers are more likely than managers to use libraries, technical information centers, and on-line electronic databases may be attributed to a difference in the duties performed by the two groups.

Nonmanagers are more likely than managers to use computer technology for preparing written technical communications, a distinction that may be more dependent upon the lack of secretarial support for nonmanagers than differences in duties. Furthermore, the fact that managers are more likely than nonmanagers to use certain information technology may be dependent upon managers' access to the technology because of their position within the organization rather than because of differences in duties.

Although the results of this study add to a rather limited empirical knowledge base, more research regarding the technical communications practices of aerospace managers and nonmanagers is clearly needed. The data reported here offer limited but useful insight into the technical communications practices of aerospace managers and nonmanagers. Technical communications educators may find the results useful in curriculum planning, technical information managers may find the results useful when planning and providing for information policy and services, and researchers may find the results useful for planning a more indepth investigation of the topic.

Chapter 10 Exercises

1. The company you work for now has hired a consultant to prepare a communications audit of all aspects of communication, both internal and external. As a preliminary step, your supervisor, Peter Harris, has asked you to write a report in which you explain the major communications aspects of your job. Communications may include written or oral duties. Include in your report not only a description of what you do, but also explanations of why the communications function is important, and explain any problems connected with this communications duty. You know that your supervisor will pass this report on to a consultant, so you need to keep this reader in mind as you prepare your report. Decide whether to use the opening-summary or delayed-summary pattern. *Note:* If you do not have a current job, you may use a previous one or write your report about your communications duties in one of your classes.

2. You are a summer intern for Investors Information, Inc. Your supervisor, Monica Quarles, has asked you to go to the library and look up the latest annual reports of three major manufacturing companies. She wants to know the latest sales figures, whether the company is doing better or worse than in previous years, how optimistic or pessimistic the tone of the president's letter is, and significant company developments in new products, plant expansion, or acquisitions. Ms. Quarles would like your opinion as to which of these three companies would be a good investment for her client J. Danforth Mitchell, who likes to invest in companies with good potential for growth. Ms. Quarles prefers the opening-summary report pattern.

3. Your company is expanding its library, which contains important professional periodicals such as *Civil Engineering* or *Journal of Accountancy.* The librarian, Clarence Tillman, has asked you to recommend two professional periodicals from your field. You will need to examine several issues before writing your report. When you make your recommendations, describe the typical content and format of the periodicals, mention special features and regular columns, explain who the readers are, and include the subscription information so Tillman can order the periodicals easily. Be sure to explain *why* these would be good additions to the company library. Write your report in the opening-summary pattern.

4. As plant safety engineer, you have been asked by your supervisor, Jeff Steinberg, division head, to check on regulations for fire hose equipment and report back on what is needed in the plant to comply with standards. Upon checking, you find that the National Fire Protection Association has issued a bulletin, *Regulations for the Installation of Fire Hose Equipment,* No. 316D. The bulletin indicates that all areas in a Class II facility (your company) should be within 20 ft of a nozzle. This limitation seems to indicate that the company will have to buy 100-ft lengths for the plant because the hoses have to reach from the reel locations, across plant aisles, and around racks and other equipment. All hoses should carry either the Underwriters' Laboratories, Inc., approval or a seal from Factory Mutual Laboratories. Hose-coupling threads have to be the same as those of the public fire department and have to be rocker lug-type. Hose nozzles are to be of the full-range adjustable type. Some of the current hoses do not have this feature, but according to the bulletin, they need not be replaced until they are in poor condition. The plant will need two sizes of hose—the hand hose (1½ in.) for use in plant areas in 100-ft lengths, and the large hose (2½ in.) in 50-ft lengths for motorized hose carts. The hose carts must be near outside exits. Regulations call for at least two carts for small plants and at least four carts for large assembly plants. Your plant could comply with the regulations by using only two carts, but because the plant uses significant quantities of hazardous materials, you decide to recommend four carts. The hand hose for inside the plant must be mounted on reels and attached to walls or columns so the hose is about 6½ ft above the floor. All hose should be rubber lined with a single cotton or synthetic fabric jacket. Unlinen lined hose or rubber-covered hose is not acceptable. Write this information report to Jeff Steinberg, division head. He will need the information as a guide in ordering appropriate equipment.

5. Assume that you are already employed in your field. Your supervisor, Bonnie Bromlet, has decided to update your department with the latest technological equipment, systems, testing methods, or anything else that would put your department and work in a "state of the art" condition. As part of her plans, she asks you to check on one topic (you select it) and report to her, comparing the latest options. Be sure to describe each option thoroughly enough for her to understand it, and then make a strong recommendation for one option. Write an opening-summary report to Bromlet. *Or,* if your instructor prefers, prepare a long report in which you present options for several changes and make recommendations for each specific area. If your instructor asks for a formal report, follow the guidelines in Chapter 11.

CHAPTER 11

Formal Report Elements

Selecting Formal Report Elements

Writing Front Matter

- Title Page
- Transmittal Letter or Memo
- Table of Contents
- List of Figures
- List of Tables
- Abstract and Executive Summary
  - Descriptive Abstract
  - Informative Abstract
  - Executive Summary

Writing Back Matter

- References
- Glossary/List of Symbols
- Appendixes

Documenting Sources

- APA System: Citations in the Text
- APA System: List of References
- Number-Reference System

Chapter Summary

Supplemental Readings in Part Two

Models

Exercises

Selecting Formal Report Elements

Most in-house reports, whatever their purpose, are in memo format, as illustrated in Chapter 10. Long reports, however, whether internal or external, more often include the formal report elements discussed in this chapter.

Management or company policy usually dictates when formal report elements are appropriate. In some companies, certain report types, such as proposals or feasibility studies, always include formal report elements. In addition, long reports addressed to multiple readers often require formal report elements, such as glossaries and appendixes, to effectively serve all reader purposes.

The formal report is distinguished from the informal report by the inclusion of some or all of the special elements described in this chapter.

Writing Front Matter

Front matter includes all the elements that precede the text of a report. Front-matter elements help a reader (1) locate specific information and (2) become familiar with the general content and organization of the report.

Title Page

A *title page* records the report title, writer, reader, and date. It is usually the first page of a long formal report. In some companies, the format is standard If it is not, include these items:

- Title of the report, centered in the top third of the page.
- Name, title, and company of the primary reader or readers, centered in the middle of the page
- Name, title, and company of the writer, centered in the bottom third of the page
- Date of the report, centered directly below the writer's name.

The title should accurately reflect the contents of the report. Use key words that identify the subject quickly and inform readers about the purpose of the report.

Title: Parking
Revision: Feasibility Study of Expanding Parking Facilities

Title: Office Equipment
Revision: Proposal for Computer Purchases

Transmittal Letter or Memo

A *transmittal letter* or *memo* sends the report to the reader. In addition to establishing the title and purpose of the report, the transmittal letter provides a place for the writer to add comments about procedures, recommendations, or other matters that do not fit easily into the report itself. Here, too, the writer may offer to do further work or credit others who assisted with the report. The transmittal letter is placed either immediately after the title page or immediately before it, as company custom dictates. In some companies, writers use a transmittal memo, but the content and placement are the same as for a transmittal letter. If the transmittal letter simply sends the report, it often is before the title page. However, if the transmittal letter contains supplementary information about the report or recommends more study or action, some writers place it immediately after the title page, where it functions as a part of the report. In your transmittal letter,

- State the report title, and indicate that the report is attached.
- Establish the purpose of the report.
- Explain why, when, and by whom the report was authorized.
- Summarize very briefly the main subject of the report.
- Point out especially relevant facts or details.
- Explain any unusual features or organization.
- Acknowledge those who offered valuable assistance in gathering information, preparing appendixes, and so on.
- Mention any planned future reports.
- Thank readers for the opportunity to prepare the report, or offer to do more study on the subject.
- Recommend further action, if needed.

Table of Contents

The *table of contents* alerts the reader to (1) pages that contain specific topics, (2) the overall organization and content of the report, and (3) specific and supplemental materials, such as appendixes. The table of contents is placed directly after the transmittal letter or title page, if that is after the transmittal letter, and before all other elements. The title page is not listed in the table of contents, but it is counted as page i. All front matter for a report is numbered in small Roman numerals. The first page of the report proper is page 1 in Arabic numbers, and all pages after that have Arabic numbers. In your table of contents,

- List all major headings with the same wording used in the report.
- List subsections, indented under major headings, if the subheadings contain topics that readers are likely to need.
- List all formal report elements, such as the abstract and appendixes, except for the title page.
- Include the titles of appendixes, for example:

  APPENDIX A: PROJECTED COSTS

- Do not underline headings in the table of contents even if they are underlined in the text.

List of Figures

Any graphic aid, such as a bar graph, map, or flowchart that is not a table with numbers or words in columns is called a *figure*. The list of figures follows the table of contents. List each figure by both number and title, and indicate page numbers:

Figure 1. Map of Mining Surveys 6
Figure 2. Wall Excavation 8
Figure 3. Light Diffusion 9

List of Tables

The list of tables appears directly after the list of figures. List each table by number and title, and indicate page numbers:

Table 1. Equipment Downtime 3
Table 2. Production by Region 5

If your report contains only two or three figures and tables, you may combine them into one "List of Illustrations." In this case, list figures first and then tables.

Abstract and Executive Summary

An *abstract* is a synopsis of the most important points in a report and provides readers with a preview of the full contents. An abstract, which can be either descriptive or informative, is usually one paragraph of no more than 200 words. An *executive summary* is a longer synopsis of one to two pages that provides a more comprehensive overview than the abstract does. An executive summary covers a report's main points, conclusions, recommendations, and the impact of the subject on company planning. In some cases, readers may rely completely on such a synopsis, as, for instance, when a nonexpert reader must read a report written for experts. In other cases, readers use these synopses to orient themselves to the main topics in a report before reading it completely.

The abstract or executive summary in a formal report usually follows the list of tables if there is one. In some companies, the style is to place an abstract on the title page or immediately after the title page. Whether you decide to include the longer executive summary or the shorter abstract depends on company custom and the expectations of your readers. In some companies, executives prefer to see both an abstract and an executive summary with long reports. If a report is very long, an executive summary allows a fuller synopsis and provides readers with a better understanding of the report contents and importance than an abstract will. Even though they are all synopses, the two types of abstracts and the executive summary provide readers with different emphases.

Descriptive Abstract

A *descriptive abstract* names the topics covered in a report without revealing details about those topics. Here is a descriptive abstract for a proposal to modify and redecorate a hotel restaurant:

> This proposal recommends a complete redesign of the Bronze Room in the Ambassador Hotel to increase our appeal to hotel guests and local customers. A description of the suggested changes, as well as costs, suggested contractors, and management reorganization, is included.

Because a descriptive abstract does not include details, readers must read the full report to learn about specifics, such as cost and planned decor. Descriptive abstracts have become less popular in recent years because they do not provide enough information for busy readers who do not want to read full reports.

Informative Abstract

An *informative abstract,* the one frequently used for formal reports and technical articles, describes the major subjects in a report and summarizes the conclusions and recommendations. This informative abstract is for the same report covered by the preceding descriptive abstract:

> This proposal discusses the 5-year decline in occupancy rates and local customer traffic at the Ambassador Hotel. The recommendation is to convert

the Bronze Room into a turn-of-the-century supper club. This motif has been successful in hotels in other markets, and research shows that it creates an upscale atmosphere for the entire hotel. The construction and decorating will take about 10 weeks and cost $206,000. Finishing the Bronze Room by the end of the year will allow the Ambassador to compete for the International Homebuilders Convention, which attracts more than 2000 members.

This informative abstract includes more details and gives a more complete synopsis of the report contents than the descriptive abstract does.

Executive Summary

An *executive summary* includes (1) background of the situation, (2) major topics, (3) significant details, (4) major conclusions or results, (5) recommendations, and (6) a discussion of how the subject can affect the company. Writers sometimes use headings in an executive summary to make it look like a miniature report. Here is an executive summary for the hotel proposal:

The Ambassador Hotel has experienced declining occupancy rates and declining local customer traffic for the past 5 years. During that period, three new hotels, all part of well-known chains, have opened in the downtown area. The Ambassador's image as an upscale, sophisticated place to visit has been eroded by the competition of the larger and more modern hotels. Beryl Whitman of the Whitman Design Studio, a consultant to the hotel, has recommended complete redecoration of the Bronze Room. The suggested changes will convert the present restaurant into a turn-of-the-century supper club.

Research indicates that the image of the main dining room in a hotel has a direct impact on the public perception of the hotel in general. A turn-of-the-century motif for the main dining room has been successful in hotels in Indianapolis, Milwaukee, and St. Louis. These markets are comparable to ours, and our survey, conducted in May by Hathaway Management Consultants, indicated that local customers would be interested in a dining atmosphere that reflects sophistication and elegance. Redesign of the Bronze Room has several advantages:

1. A turn-of-the-century atmosphere will create the sophisticated, upscale image we want to project.
2. By not copying the ultramodern look of the competing hotels, we will achieve distinction based on a perceived return to elegant and gracious hotel service in the past.
3. Other hotel facilities will not be disturbed, and construction can be completed by November 1, in time for the holiday season.

Renovation will take 10 weeks, and the total projected cost of $206,000 could be regained within 18 months or sooner if we can immediately attract conventions, such as the International Homebuilders Association. Even without such a convention, the Ambassador could expect an immediate 10%

increase in occupancy rates and a 26% increase in local customers, based on the reports from Milwaukee, St. Louis, and Indianapolis. The recommendation is to redesign the Bronze Room with a target date of November 1.

Abstracts and executive summaries, important elements in formal reports, are difficult to write because they require summarizing in a few words what a report covers in many pages. Remember that an informative abstract or executive summary should stand alone for readers who do not intend to read the full report immediately. In preparing abstracts and executive summaries, follow these guidelines:

1. Write the abstract or executive summary after you complete the report.
2. Identify which topics are essential to a synopsis of the report by checking major headings and subheadings.
3. Rewrite the original sentences into a coherent summary. Simply linking sentences taken out of the original report will not produce a smooth style.
4. Write full sentences, and include articles *a, an,* and *the.*
5. Avoid overly technical language or complicated statistics, which should be in the data sections or an appendix.
6. Do not refer readers to tables or other sections of the original report.
7. If conclusions are tentative, indicate this clearly.
8. Do not add information or opinions not in the original report.
9. Edit the final draft for clarity and coherence.

Writing Back Matter

Back matter includes supplemental elements that some readers need to understand the report information or that provide additional specialized information for some readers.

References

A *reference list* records the sources of information in the report and follows the final section of the report body. Preparation of a reference list is explained later in this chapter under "Documenting Sources."

Glossary/List of Symbols

A *glossary* defines technical terms, such as *volumetric efficiency,* and a *list of symbols* defines scientific symbols, such as *Au.* Include a glossary or list of symbols in a long report if your readers are not familiar with the terms and symbols you use in the report. Also include informal definitions of key terms or symbols in the text to aid readers who do not want to flip pages back and forth looking for definitions with every sentence. If you do use a glossary or list of symbols, say so in the introduction of the report to alert readers to them. The glossary or list of symbols usually follows immediately after the references. In some companies, however, custom may dictate that the two lists follow the table of contents. In either case, remember these guidelines:

- Arrange the glossary or list of symbols alphabetically.
- Do not number the terms or symbols.
- Include also any terms or symbols that you are using in a nonstandard or limited way.
- List the terms or symbols on the left side of the page, and put the definition on the right side on the same line:

bugseed an annual herb in northern temperate regions
Pt platinum

Appendixes

Appendixes are the final elements in formal reports that contain supplemental information or information that is too detailed and technical to fit well into the body of the report or that some readers need and others do not. Appendixes can include documents, interviews, statistical results, case histories, lists of pertinent items, specifications, or lists of legal references. The recent trend in formal reports has been to place highly technical or statistical information in appendixes for those readers who are interested in such material. Remember these guidelines:

- Label appendixes with letters, such as "Appendix A" and "Appendix B," if you have more than one.
- Provide a title for each appendix, such as "Appendix A. Questionnaire Sample."
- Indicate in the body of the report that an appendix provides supplemental information on a particular topic, such as "See Appendix C for cost figures."

Documenting Sources

Documenting sources refers to the practice of citing original sources of information used in formal reports, journal articles, books, or any document that includes evidence from published works. Cite your information sources for the following reasons:

- Readers can locate the original sources and read them if they wish.
- You are not personally responsible for every fact in the document.
- You will avoid charges of plagiarism. *Plagiarism* is the unacknowledged use of information discovered and reported by others or the use of their exact words, copied verbatim.

In writing a report that relies somewhat on material from other sources, remember to document information when you are doing either of the following:

- Using a direct quotation from another source
- Paraphrasing information from another source

If, however, the information you are presenting is generally known and readily available in general reference sources, such as dictionaries and encyclopedias, you need not document it. You would not have to document a statement that water is made up of two parts hydrogen and one part oxygen or that it boils at 212°F and freezes at 32°F. In addition, if your readers are experts in a particular field, you need not document basic facts or theories that all such specialists would know.

The documentation system frequently used in the natural sciences, social sciences, and technical fields is the *American Psychological Association (APA) system,* also called the *author/date system.* Another system often used in the sciences is the *number-reference system.* Some fields use reference systems that differ slightly from the APA and number-reference systems. Following is a list of style guides for particular associations or specific fields. Check for the latest edition in the library or in *Books in Print.* Always use the reference system that is appropriate for your situation and audience or requested by your editor if you are writing for a technical journal.

American National Standard for Bibliographic References

CBE Style Manual: A Guide for Authors, Editors, and Publishers in the Biological Sciences

Guide for Preparation of Air Force Publications

Handbook for Authors of Papers in American Chemical Society Publications

A Manual for Authors of Mathematical Papers

NASA Publications Manual

Style Manual for Engineering Authors and Editors

Suggestions to Authors of the Reports of the United States Geological Survey
Uniform Requirements of Manuscripts Submitted to Biomedical Journals

APA System: Citations in the Text

In the APA system, when you refer to a source in the text, include the author and the date of publication. This paragraph from a psychologist's report illustrates in-text citation:

> Television's portrayal of women has been called "the best of recent years" (Steenland, 1986, p. 17). Since reader-response theory (Allen, 1987) suggests that viewers interpret images individually, research is now focusing on testing viewer perceptions. Atwood, Zahn, and Webber (1986) used random telephone dialing to test viewer response to women on television and found no differences between male and female viewers. A British study (Gunter & Wober, 1982) found that heavy viewers of crime action shows had lowered perceptions of women as traditionally feminine. Durkin (1985) provides a comprehensive survey of all relevant studies of audience perception involving children.

Notice these conventions of the APA citation system:

1. If an author's name begins a sentence, place the date of the work in parentheses immediately after.
2. If an author is not referred to directly in a sentence, place both the author's last name and the year of publication, separated by a comma, in parentheses at the end of the sentence.
3. If there are multiple authors, cite all the names up to six. If there are more than six authors, cite only the first author and follow with *et al.*:

   Jones et al. (1988) reported. . . .
4. If multiple authors are cited in parentheses, separate names with commas, and use an ampersand (&) between the last two names.
5. If you are using a direct quotation, give author, year, and page number.
6. If you are citing two works by the same author, give the year of publication of both works:

   Young (1979, 1984) disagrees. . . .
7. If you are citing two works by the same author, published in the same year, distinguish them by *a, b, c,* and so on:

   Lucas (1987a, 1987b) refers to. . . .
8. If you are citing several works by different authors in the same parentheses, list the works alphabetically by first author:

Several studies tested for side effects but found no significant results (Bowman & Johnson, 1980; Mullins, 1979; Roberts & Allen, 1975; Townsend, 1988).

9. Use last names only unless two authors have the same surname; then include initials to avoid confusion:

   W. S. Caldwell (1987) and R. D. Caldwell (1988) reported varied effects. . . .

   Heightened effects were noted (W. S. Caldwell, 1987; R. D. Caldwell, 1988).

10. Use an *ellipsis* (three spaced periods) to indicate omissions from direct quotations:

    Bagwell (1987) commented, "This handbook will not fulfill most needs of a statistician, but . . . model formulas are excellent."

11. If the omission in a quotation comes at the end of the sentence, use four periods to close the quotation:

    According to Martin (1984), "Researchers should inquire further into effects of repeated exposure. . . ."

12. Use brackets to enclose any information you insert into a quotation:

    Krueger (1988) cited "continued criticism from the NCWW [National Commission on Working Women] regarding salary differences between the sexes."

APA System: List of References

The list of references for a report or an article includes each source cited in the document. Follow these conventions:

1. Do not number the list.
2. Indent the second line of the reference three spaces.
3. List items alphabetically according to the last name of the first author.
4. Alphabetize letter by letter:

   Bach, J. K.
   Bachman, D. F.
   DeJong, R. T.
   DuVerme, S. G.
   MacArthur, K. O.
   Martin, T. R.
   McDouglas, T. P.
   Sebastian, J. K.
   St. John, J. L.
   Szartzar, P. O.

5. Place single-author works ahead of multiple-author works if the first author is the same:

   Fromming, W. R.
   Fromming, W. R., Brown, P. K., & Smith, S. J.

6. Alphabetize by the last name of the second author if the first author is the same in several references:

   Coles, T. L., James, R. E., & Wilson, R. P.
   Coles, T. L., Wilson, R. P., & Allen, D. R.

7. List several works by one author according to the year of publication. Repeat the author's name in each reference. If two works have the same publication date, distinguish them by using *a, b, c,* and so on, and alphabetize by title:

   Deland, M. W. (1978).
   Deland, M. W. (1987a). Major differences . . .
   Deland, M. W. (1987b). Separate testing . . .

Remember that each reference should include author, year of publication, title, and publication data. Here are some sample references for typical situations:

1. *Journal article—one author:*

   Payne, R. (1974). Songs of humpback whales. *Science, 173,* 587–597.
   Note:
   - Use initials, not first names, for authors.
   - Capitalize only the first word in a title, except for proper names.
   - Underline journal names and volume numbers.

2. *Journal article—two authors:*

   Hawkins, R. P., & Pingree, S. (1981). Uniform messages and habitual viewing: Unnecessary assumptions in social reality effects. *Human Communication Research, 7,* 291–301.
   Note:
   - Use a comma and ampersand between author names.
   - Capitalize the first word after a colon in an article title.
   - Capitalize all important words in journal names.

3. *Journal article—more than two authors:*

   Geis, F. L., Brown, V., Jennings, J., & Corrado-Taylor, D. (1984). Sex vs. status in sex-associated stereotypes. *Sex Roles, 11,* 771–785.

Note:

- Use an ampersand between the names of the last two authors.
- List all authors even if *et al.* was used in text references.

4. *Journal article—issues separately paginated:*

Battison, J. (1988). Using effective antenna height to determine coverage. *The LPTV Report, 3*(1), 11.

Note: If each issue begins on page 1, place the issue number in parentheses directly after the volume number without a space between them.

5. *Magazine article:*

Elmer-DeWitt, P. (1988, March 14). When the dead are revived. *Time,* pp. 80–81.

Note:

- Include day, month, and year for weekly or daily publications.
- Do not include volume number.
- Use *pp.* before page numbers for magazines and newspapers, but not for professional and technical journals published monthly or less frequently.

6. *Newspaper article—no author:*

Proposed diversion of Great Lakes. (1987, December 7). *Cleveland Plain Dealer,* C, p. 14.

Note:

- If an article has no author, alphabetize by the first word of the title, excluding *a, an,* and *the.*
- If a newspaper has several sections, include the section identification as well as page numbers.
- Capitalize all proper names in a title.

7. *Newspaper article—author:*

Sydney, F. H. (1986, January 7). Microcomputer graphics for water pollution control data. *The Milwaukee Journal,* D, pp. 6, 12.

Note: If an article appears on several pages, give all page numbers, separated by commas.

8. *Article in an edited collection—one editor:*

Allen, R. C. (1987). Reader-oriented criticism and television. In R. C. Allen (Ed.), *Channels of discourse* (pp. 74–112). Chapel Hill, NC: Univ. of North Carolina Press.

Note:

- Capitalize only the first word in a book title, except for proper names.

- Underline book titles.
- Use initials and the last name of the editor in standard order.
- Use *pp.* with page numbers of the article in parentheses after the book title.
- Use the Postal Service ZIP code abbreviations for states in publication information.

9. *Article in an edited collection—two or more editors:*

   Zappen, J. P. (1983). A rhetoric for research in sciences and technologies. In P. V. Anderson, R. J. Brockmann, & C. R. Miller (Eds.), *New essays in technical and scientific communications: Research theory, practice* (pp. 123–138). Farmingdale, NY: Baywood.

   Note:

   - List all editors' names in standard order.
   - Use *Eds.* in parentheses if there is more than one editor.

10. *One book—one author:*

    Sagan, C. (1980). *Cosmos.* New York: Random House.

    Note: Do not include the state abbreviation when the city of publication is New York.

11. *Book—more than one author:*

    Johanson, D. C., & Edey, M. A. (1981). *Lucy: The beginnings of humankind.* New York: Simon and Schuster.

    Note: Capitalize the first word after a colon in a book title.

12. *Edited book:*

    Ceram, C. W. (Ed.). (1966). *Hands on the past.* New York: Alfred A. Knopf.

    Note: Place a period after the *Ed.* in parentheses.

13. *Book edition after first edition:*

    Mills, G. H., & Walter, J. A. (1986). *Technical writing* (5th ed.). New York: Holt, Rinehart and Winston.

    Note: Do not place a period between the title and the edition in parentheses.

14. *Article in a proceedings:*

    Glover, R. W. (1975). Apprenticeship in America: An assessment. *Proceedings of the twenty-seventh annual winter meeting, December 28–29, 1974—San Francisco* (pp. 33–45). Madison, WI: Industrial Relations Research Association.

    Note: If the proceedings are published annually with a volume number, treat them the same as a journal.

15. *Unpublished conference paper:*

    Marshall, R. W. (1986, April). *The role of apprenticeship in an internationalized information world.* Paper presented at the meeting of the Technology Abroad Association, Chicago, IL.

    Note: Include the month of the meeting if available.

16. *Report in a document deposit service:*

    Haynes, P. N. (1976). *Clinical practice and nursing education.* Houston, TX: Houston University, Texas Teacher Center. (ERIC Document Reproduction Service No. ED 123 209)

    Note: Include document number at the end of reference in parentheses.

17. *Report—corporate author, author as publisher:*

    American Association of Junior Colleges. (1968). *Extending campus resources: Guide to selecting clinical facilities for health technology programs* (Rep. No. 67). Washington, DC: Author.

    Note:

    - If the report has a number, insert it between the title of the report and the city of publication, in parentheses, followed by a period.
    - If the report was written by a department staff, use that as the author, such as "Staff of Accounting Unit." Then give the corporation or association in full as the publisher.

18. *Dissertation—obtained on microfilm:*

    Jones, D. J. (1986). Programming as theory foundation: Microprocessing and microprogramming. *Dissertation Abstracts International, 46,* 4785B–4786B. (University Microfilms No. 86-08, 134)

    Note: Include the microfilm number in parentheses at the end of reference.

19. *Dissertation—obtained from a university:*

    Heintz, P. D. (1985). Television and psychology: Testing for frequency effects on children (Doctoral Dissertation, The University of Akron, 1985). *Dissertation Abstracts International, 45,* 4644A.

    Note: When using the printed copy of a dissertation, include the degree-granting university and the year of dissertation in parentheses after the title.

20. *Computer program:*

    Lippard, G. L. (1986). *Torsion analysis and design of steel beams* [Computer program]. Astoria, NY: Structural Software, Inc. (AX-P 34–47)

    Note:

    - Underline the title of the program.

- Identify the work as a computer program in brackets following the title without a period between the title and the brackets.
- If the program has an identification number, place that in parentheses following the name of the software publisher.

Number-Reference System

In the number-reference system, the references are written as shown above, but the reference list is numbered. The list may be organized in one of two ways:

1. List the items alphabetically by last name of author.
2. List the items in the order they are cited in the text.

The in-text citations use only the number of the reference in parentheses. Following is the same paragraph shown in APA style earlier to illustrate in-text citations. Here the number-reference system is used; the reference list is in the order the references appear in the text.

> Television's portrayal of women has been called "the best of recent years" (1:17). Since reader-response theory (2) suggests that viewers interpret images individually, research is now focusing on testing viewer perceptions. Reference 3 used random telephone dialing to test viewer response to women on television and found no differences between male and female viewers. A British study (4) found that heavy viewers of crime action shows had lowered perceptions of women as traditionally feminine. A comprehensive survey (5) of all relevant studies of audience perception involving children was completed in 1985.

Notice these conventions for the number-reference system:

1. Page numbers are included in the parentheses, separated from the reference number by a colon.
2. Sentences may begin with "Reference 6 states. . . ." However, for readability, rewrite as often as possible to include the reference later in the sentence.

For readers, the APA system is easier to use because it provides dates for information and the names of authors that readers may recognize as experts on the subject. The number-reference system requires readers to flip back and forth to the reference list to find dates and authors. Whichever system you use, be consistent throughout your report.

CHAPTER SUMMARY

This chapter explains how to prepare front and back matter for formal reports and how to document sources of information. Remember:

- Front matter consists of elements that help readers locate specific information and become familiar with report organization and content before reading the text. Included are the title page, transmittal letter or memo, table of contents, list of figures, list of tables, abstract, and executive summary.
- Back matter consists of elements that some readers need to understand the report information or that provide additional information for readers. Included are a glossary, list of symbols, and appendixes.
- Documenting sources is the practice of citing the original sources of information used in formal reports and other documents.
- Sources of information should be documented whenever a direct quotation is used or whenever information is paraphrased and used.
- The APA system of documentation is used frequently in the natural sciences, social sciences, and technical fields for citations in the text and for references.
- The number-reference system, similar to APA style, is also used in the sciences and technical fields.

SUPPLEMENTAL READINGS IN PART TWO

Ridgway, L. "The Writer as Market Researcher," *Technical Communication.*

Model 11-1 Commentary

This title page and the samples in Models 11-2 through 11-7 are from a student report.

Discussion

Discuss how helpful the title is in reflecting the contents and purpose of the report. Read the rest of the models for this report, and discuss possible titles that would more fully identify the report content and purpose.

A TYPING SERVICE AT CENTRAL UNIVERSITY

Prepared for
Ms. Erica Hunter
Director of Student Services
Central University

By
Kimberly James
Assistant Director of Student Services
Central University

April 27, 1987

Model 11-2 Commentary

This transmittal memo identifies the authorization for the report and the topic—a preliminary study. Notice that the pagination appears at center bottom. This memo was placed after the title page in the report.

The writer identifies others who assisted with gathering information for the report and mentions the primary recommendation in the report. The writer ends courteously by offering to discuss the report further.

Discussion

Compare this transmittal memo with the transmittal letter in Model 11-8. Which features of each seem most useful to readers?

TO: Ms. Erica Hunter
Director of Student Services

FROM: Ms. Kimberly James
Assistant Director of Student Services

DATE: April 27, 1987

SUBJECT: Proposed On-Campus Typing Service

As you requested, a preliminary study was conducted at Central University to determine if there is sufficient student interest in an on-campus typing service.

After surveying 55 full-time students, our research team (which included Mike Jones, David Fletcher, Elise McCaffrey, and me) determined that there is sufficient student interest in an on-campus typing service. The results of this preliminary study indicate that a more extensive study is justified.

Attached is an abstract and report on the study. I would like to thank my fellow researchers for all their hard work and cooperation, which made this report possible.

If you have any questions after reading the report, please contact me at extension 301.

Model 11-3 Commentary

In this table of contents, the writer uses all capital letter for main headings and capital and lowercase letters for subheadings.

Notice that the pagination is in small Roman numerals for front matter and Arabic numbers for the report sections and all other items.

Discussion

Discuss the visual design of the table of contents. What features help readers locate information?

TABLE OF CONTENTS

TRANSMITTAL MEMO ii
LIST OF FIGURES iv
ABSTRACT v
INTRODUCTION 1
  Background 1
  Purpose 3
  Method 4
  Limitations 5
RECOMMENDATIONS 5
QUESTIONNAIRE RESULTS 6
  Freshmen 7
  Sophomores 8
  Juniors 10
  Seniors 11
FORECAST OF USE OVER 10 YEARS 12
  Equipment 15
  Service Volume 18
SUMMARY OF FINDINGS 21
REFERENCES 22
APPENDIX: QUESTIONNAIRE 24

Model 11-4 Commentary

This list of figures directs readers to graphic aids that illustrate specific information.

Discussion

Discuss the advantages and disadvantages of including a list of figures separate from the table of contents.

LIST OF FIGURES

Figure 1. Age of Students 6
Figure 2. Class Rank 6
Figure 3. Students Needing a Typing Service 12

Model 11-5 Commentary

This abstract for the student report on a campus typing service begins with a statement about the project. The writer then briefly mentions the research method and cites the major recommendation.

Notice that the word *abstract,* in capital letters, is centered above the report title and the pagination appears centered at the bottom of the page.

Discussion

1. Identify the type of abstract this student uses for her report.
2. Draft an abstract of the other type using the information in this model.

ABSTRACT

A Typing Service at Central University

A preliminary study at Central University investigated student interest in an on-campus typing service. Research methods for determining this interest consisted of surveying 55 full-time students. Results indicated that 54% of the respondents were interested in a typing service. Based on the preliminary study results, the conclusion is that a second, more extensive study is needed to determine the services Central University students would want from an on-campus typing service.

Model 11-6 Commentary

This reference list for the student report on a typing service shows the sources of information she used for her report.

Discussion

Check the accuracy of the format of these references with the guidelines in this chapter.

Reference List

Blakemore, M. (1986). Projected use of campus services. Bulletin of University Planning, 6, 16–20.

MacRae, R. L. (1986). Rising demand for student services. Association of Student Directors Newsletter, 16(4), 6–7.

Rowland, E. (1984). Developing campus services. New York: Oxford.

Model 11-7 Commentary

This excerpt from the Appendix in the student report shows some of the questions used in the study. Notice that the writer has filled in total responses on the questionnaire for the reader's information. The student conducted the survey by having respondents fill out the questionnaire themselves.

Discussion

1. Discuss the visual design of this page of the questionnaire. How easy or difficult would it be for a respondent to quickly complete the questionnaire?

2. In groups, edit the introduction and the questions according to guidelines in Chapter 5.

APPENDIX: QUESTIONNAIRE

The purpose of this study is to determine student interest in an on-campus typing service. This service will provide experienced typists to complete any typing task with professional quality. Your confidential responses will be analyzed, and if a sufficient need is expressed, a more extensive study will be conducted to confirm these results. Your participation will be appreciated.

1. What is your rank?
   a. Freshman 13%
   b. Sophomore 14%
   c. Junior 32%
   d. Senior 41%

2. Do you type?
   a. Yes 78%
   b. No 22%

3. Approximately how many typed assignments do you turn in in a semester?
   a. 0 5%
   b. 1–3 42%
   c. 4–7 27%
   d. 8–10 16%
   e. More than 10 9%

4. Do you type your own assignments?
   a. Yes 73%
   b. No 27%
   If no, who does? sister, mom, wife, girlfriend, or friend (circle one)

Model 11-8 Commentary

This transmittal letter was written by a management consultant to a client who requested a report.

Discussion

1. Discuss why this transmittal is in the form of a letter rather than a memo.

2. Discuss the purpose of this transmittal letter. Does it merely send the report?

PROFESSIONAL COMMUNICATION ASSOCIATES
P.O. Box 11483
Milwaukee, Wisconsin 53211
(414) 555-2966

May 29, 1989

Mr. Oliver Lawrence
Director of Marketing
Neville Development, Inc.
2555 N. Oakland Avenue
Milwaukee, WI 53211

Dear Mr. Lawrence:

Here is the report of the analysis you authorized on March 6, 1989. The report, "An Analysis of Communication Networks at Neville Development," gives the results of the study conducted on April 6–10, 1989, in your offices.

You'll be interested especially in the description of technical communications problems on pages 8–10. Our study indicates that Neville Development is experiencing particularly good communications among management levels.

It's been a pleasure to work with your staff. After you've read the report, we can meet to discuss the recommendations offered.

Sincerely,

Natalie Chandler
Senior Partner

NC:ss

Model 11-9 Commentary

This portion of a glossary is part of a long technical report. Because some readers are in the company marketing department and not familiar with technical terms, the writer includes a glossary for their information.

Discussion

1. Discuss how the writer identifies both terms and the abbreviations commonly used to represent them.

2. Discuss the visual design of the glossary. How helpful is it to readers who are not familiar with terms used in the report?

GLOSSARY

Accelerometer.................................... An electrical device that measures vibration and converts the signal to electrical output.

Air/Fuel Mixture................................ A ratio of the amount of air mixed with fuel before it is burned in the combustion chamber.

Ammeter.. An electric meter that measures current.

Amplitude.. The maximum rise or fall of a voltage signal from 0 volts.

Battery-Hot.. Circuit fed directly from the starter relay terminal. Voltage is available whenever the battery is charged.

Blown.. A melted fuse filament caused by overload.

Capacitor... A device for holding or storing an electric charge.

DVOM (Digital Volt-Ohmmeter)............................... A meter that measures voltage and resistance and displays it on a liquid crystal display.

ECA (Electronic Control Assembly).. See Processor

Fuse .... A device containing soft metal that melts and breaks the circuit when it is overloaded.

Induced Current................................ The current generated in a conductor as a magnetic field moves across the conductor.

Oscillograph....................................... A device for recording the waveforms of changing currents or voltages.

Chapter 11 Exercises

1. If you have selected a topic for a long report, compile a tentative list of terms that might go into a glossary. Evaluate whether the list is long enough or the terms difficult enough to require a glossary for your readers.

2. Here is a student's informative abstract of the article "Eliminating Gender Bias in Language" in Part II. Read the article and then evaluate this abstract. Write your own informative and descriptive abstracts of the article and submit them to your instructor.

ABSTRACT

There is a lot of concern over bias against women in the English language. This article gives tips on how to avoid using language that might offend women. The writer should be aware of pronouns and words that imply a specific sex.

3. Your supervisor, John Duke, is a research chemist planning to present a paper at a conference. Read the article by Holcombe and Stein in Part II, and write an executive summary of the article. Since Mr. Duke did not ask you to write the executive summary, attach a transmittal memo to the summary that explains to him why you think it will be helpful to him.

4. Your supervisor, Tanya Grant, has to make an oral presentation to the board of directors of a foundation. Ms. Grant is requesting research money to continue a laboratory study. She asks you to find some information on how to make effective oral presentations. Select two articles about oral presentations from Part II, and write an informative abstract of each article. Submit these to Ms. Grant with a transmittal memo, and recommend that she read one of the articles for herself. Tell her why you are recommending the article you selected.

5. Here is a short reference list from a student report. Check the format according to the APA style explained in this chapter, and make any necessary corrections.

REFERENCE LIST

Clark, W. E. (1981). The importance of the Fossil Record, *Science Progress, 6,* 14–21.

Danton, R., and Leigh, J. (1981). Omo research expedition, *Scientific American, 16* No. 3, 20–29.

Leakey, L. S. B. (1959). *World's Oldest Man.* New York, NY: MacMillan.

Mentor, S. B. (1980). The emerging human. *Science,* 122, 500–512.

Mentor, S. B., Brown, J. E., and Anderson, E. A. (1984). Early human migrations from Africa. *Journal of current anthropology. 12,* 116–132, 151.

McGruter, S. L. (1988, 6 January). Cro-Magnon man: disappearance and adaptation. *Time.* pp. 78–79.

Preston, F. (1981). Paleolithic Site in Africa, in T. L. Coles (Ed.), *Out of the Past,* Chicago: Bentsen House, pp. 106–120.

St. John, H. (1987). Our earliest ancestors. New York: Holt, Rinehart and Winston, 3rd ed.

Stratton, D. P. Fire as tool and weapon. Paper presented at meeting of Association of Anthropology, Cleveland, OH, July 1987.

CHAPTER 12

Types of Reports

Understanding Conventional Report Types

Writing a Feasibility Study

- Purpose
- Organization
  - Introduction
  - Comparison of Alternatives
  - Conclusions
  - Recommendations

Writing an Incident Report

- Purpose
- Organization
  - Description of the Incident
  - Causes
  - Recommendations

Writing an Investigative Report

- Purpose
- Organization
  - Introduction
  - Method
  - Results
  - Discussion
  - Conclusion

Writing a Progress Report

- Purpose
- Organization
  - Introduction
  - Work Completed
  - Work Remaining
  - Adjustments/Problems
  - Conclusion

Writing a Trip Report

- Purpose
- Organization
  - Introductory Section
  - Information Sections
  - Conclusions and Recommendations

Writing a Proposal

- Purpose
- Organization
  - Problem
  - Proposed Solution
  - Needed Equipment/Personnel
  - Schedules
  - Budget
  - Evaluation System
  - Expected Benefits
  - Summary/Conclusions

Chapter Summary

Supplemental Readings in Part Two

Models

Exercises

Understanding Conventional Report Types

Over the years, in industry, government, and business, writers have developed conventional organizational patterns for the most frequently written types of reports. Just as tables of contents and glossaries have recognizable formats, so too have such common reports as progress reports and trip reports. Some companies, in fact, have strict guidelines about content, organization, and format for report types that are used frequently. Conventional organizational patterns for specific types of reports also represent what readers have found most useful over the years. For every report, you should analyze reader, purpose, and situation and plan your document to serve your readers' need for useful information. You will also want to be knowledgeable about the conventional structures writers use for common report types.

This chapter explains the purposes and organizational patterns of six of the most common reports. Depending on reader, purpose, situation, and company custom, these reports may be written as informal memos, letters, or formal documents that include all the elements discussed in Chapter 11.

Writing a Feasibility Study

A *feasibility study* provides information to decision makers about the practicality and potential success of several alternative solutions to a problem.

Purpose

Executives often ask for feasibility studies before they consider a proposal for a project because they want a thorough analysis of the situation and all the alternatives. The writer of a feasibility study identifies all reasonable options and prepares a report that evaluates them according to features important to the situation, such as cost, reliability, time constraints, and company or organization goals. Readers expect a feasibility study to provide the information necessary for them to make an informed choice among alternatives. The alternatives may represent choices among products or actions, such as the choice among four types of heating systems, or a choice between one action and doing nothing, such as the decision to merge with another company or not to merge. When you write a feasibility study, provide a full analysis of every alternative, even if one seems clearly more appropriate than the others.

Organization

Model 12-1 presents a feasibility study written by an employee at a day-care center. Because the center is expanding, the current manual recordkeeping system is inadequate. In her report, the writer evaluates two computer systems on the basis of how well they will fit the needs of the center and then recommends one of them. Feasibility studies usually include these sections:

Introduction

The introduction of a feasibility study provides an overview of the situation. Readers may rely heavily on the introduction to orient them to the situation before they read the detailed analyses of the situation and the possible alternatives. Follow these guidelines:

- Describe the situation or problem.
- Establish the need for decision making.
- Identify those who participated in the study or the outside companies that provided information.
- Identify the alternatives the report will consider, and explain why you selected these alternatives, if you did.
- Explain any previous study of the situation or preliminary testing of alternatives.
- Explain any constraints on the study or on the selection of alternatives, such as time, cost, size, or capacity.
- Define terms or concepts essential to the study.
- Identify the key factors by which you evaluated the alternatives.

In Model 12-1, the writer begins with a brief opening summary that states the report purpose and her recommendation for a computer system. The introduction has two subsections in which the writer (1) explains the day-care center's need for a computer system and (2) identifies possible alternatives and the criteria she used in her evaluation.

Comparison of Alternatives

The comparison section focuses equally on presenting information about the alternatives and analyzing that information in terms of advantages and disadvantages for the company. Organize your comparison by topic or by complete subject. For feasibility reports, readers often prefer to read a comparison by key topics because they regard some topics more important than others and they can study the details more easily if they do not have to move back and forth between major sections. Whether comparing by topic or by complete subject, discuss the alternatives in the same order under each topic, or discuss the topics in the same order under each alternative. Follow these guidelines:

- Describe the main features of each alternative.
- Rank the key topics for comparison by using either descending or ascending order of importance.
- Discuss the advantages of each alternative in terms of each key topic.
- Point out the significance of any differences among alternatives.

The writer of the day-care center report in Model 12-1 compares the features of the two computer systems by topic, using subheadings to help readers find specific information. The subheadings represent the standards for evaluating the computers that the writer established in her introduction.

Conclusions

The conclusions section summarizes the most important advantages and disadvantages of each alternative. If you recommend one alternative, do so in the conclusions section or in a separate section if you have several recommendations. State conclusions first, because they are the basis for any recommendations. If you believe some advantages or disadvantages are not important, explain why. Follow these guidelines:

- Separate conclusions adequately so that readers can digest one at a time.
- Explain the relative importance to the company of specific advantages or disadvantages.
- Include conclusions for each key factor presented in the comparison.

Recommendations

The recommendations section, if separate from the conclusions, focuses entirely on the choice of alternative. Your recommendations should follow logically from your conclusions. Any deviations will confuse readers and cast doubt on the thoroughness of your analysis. Follow these guidelines:

- Describe your recommendations fully.
- Provide enough details about implementing the recommendations so that your reader can visualize how they will be an effective solution to the problem.
- Indicate a possible schedule for implementation.

In Model 12-1, the writer combines her conclusions and her recommendation in one section. First, she summarizes the need for a computer system and indicates that either alternative would be acceptable. Because one system is considerably lower in cost, she recommends that one and points out why cost is the determining factor.

Writing an Incident Report

An *incident report* provides information about accidents, equipment breakdowns, or any disruptive occurrence.

Purpose

An incident report is an important record of an event because government agencies, insurance companies, and equipment manufacturers may use the report in legal actions if injury or damage has occurred. In addition, managers need such reports to help them determine how to prevent future accidents or disruptions. An incident report thoroughly describes the event, analyzes the probable causes, and recommends actions that will prevent repetition of such events. If you are responsible for an incident report, gather as many facts about the situation as possible and carefully distinguish between fact and speculation.

Organization

Model 12-2, written by a production supervisor at a manufacturing company, is an incident report about an industrial accident. Although the supervisor was not present during the accident, she is responsible for gathering the facts and writing the company report.

Most large companies have standard forms for reporting incidents during working hours. If there is no standard form, an incident report should include the following sections.

Description of the Incident

The description of the incident includes all available details about events before, during, and immediately after the incident in chronological order. Since all interested parties will read this section of the report and may use it in legal action, be complete, use nonjudgmental language, and make note of any details that are not available. Follow these guidelines:

- State the exact times and dates of each stage of the incident.
- Describe the incident chronologically.

- Name the parties involved.
- State the exact location of the accident or the equipment that malfunctioned. If several locations are relevant, name each at the appropriate place in the chronological description.
- Identify the equipment, by model number, if possible, involved in a breakdown.
- Identify any continuing conditions resulting from the incident, such as the inability to use a piece of equipment.
- Identify anyone who received medical treatment after an accident.
- Name hospitals, ambulance services, and doctors who attended accident victims.
- Name any witnesses to the incident.
- Report witness statements with direct quotations or by paraphrasing statements and citing the source.
- Explain any follow-up actions taken, such as repairs or changes in scheduling.
- Note any details not yet available, such as the full extent of injuries.

The writer in Model 12-2 provides a chronological description of the employee's accident, including date and time. This section also identifies another employee who tried to help the accident victim. For the record, the writer names the ambulance service, hospital, and doctor who attended the injured employee. The writer notes, too, that the employee's recovery cannot be evaluated for several more days.

Causes

This section includes both direct and indirect causes for the incident. Be careful to indicate when you are speculating about causes without absolute proof. Follow these guidelines:

- Identify each separate cause leading to the incident under discussion.
- Analyze separately how each stage in the incident led to the next stage.
- If you are merely speculating about causes, use words such as *appears, probably,* and *seems.*
- Point out the clear relationship between the causes and the effects of the incident.

In analyzing the cause of the accident in Model 12-2, the writer reports the condition of the tar kettle as she found it. Since there were no witnesses to the accident, the supervisor must rely on her own observations as to probable cause. In this case, she concludes that tar buildup prevented the employee from using the safety catch on the kettle cover.

Recommendations

This section offers specific suggestions keyed to the causes of the incident. Focus your recommendations on prevention measures rather than on punishing those connected with the incident. Follow these guidelines:

- Include a recommendation for each major cause of the incident.
- List recommendations if there are more than one or two.
- Describe each recommendation fully.
- Relate each recommendation to the specific cause it is designed to prevent.
- Suggest further investigation if warranted.

The supervisor's report in Model 12-2 recommends two procedures, one to make the equipment safer to use and the other to alert employees to potential danger.

Writing an Investigative Report

An *investigative report* analyzes data and seeks answers to why something happens, how it happens, or what would happen under certain conditions.

Purpose

An investigative report summarizes the relevant data, analyzes the meaning of the data, and assesses the potential impact that the results will have on the company or on specific research questions. The sources of the data for investigative reports can include field studies, surveys, observation, and tests of products, people, opinions, or events both inside and outside the laboratory. Investigations can include a variety of circumstances. An inspector at the site of an airplane crash, a laboratory technician testing blood samples, a chemist comparing paints, a mining engineer checking the effects of various sealants, and a researcher studying the effect of air pollution on children are all conducting investigations, and all will probably write reports on what they find. Remember that readers of an investigative report need to know not only what the data are, but also what the data mean in relation to what has been found earlier and what may occur in the future.

Organization

Model 12-3 is a typical investigative report written by a scientist. The writer investigates a research question or problem, develops a method for testing it, and then analyzes the results to determine how well they answer the original research question. In Model 12-3, the writer investigates the connection between the age at which a student first begins drinking alcohol and the level of consumption and number of drinking problems encountered by the student in college. Because this investigative report is written for a professional journal, the writer follows the journal requirements: an informative abstract and the journal's reference system for citing sources.

Introduction

The introduction of an investigative report describes the problem or research question that is the focus of the report. Since investigative reports may be used as a form of evidence in future decision making, include a detailed description of the situation and explain why the investigation is needed. Readers will be interested in how the situation affects company operations or how the research will answer questions they have. Follow these guidelines:

- Describe specifically the research question or problem that is the focus of the investigation. Avoid vague generalities, such as "to check on operations" or "to gather employee reactions." State the purpose in terms of specific questions the investigation will answer.
- Provide background information on how long the situation has existed, previous studies of the subject by others or by you, or previous decisions concerning it. If a laboratory test is the focus of the report, provide information about previous tests involving the same subject.
- Explain your reasons if you are duplicating earlier investigations to see if results remain consistent.
- Point out any limits of the investigation in terms of time, cost, facilities, or personnel.

The writer in Model 12-3 discusses past research on the question of adolescent drug and alcohol use and cites specific studies and their conclusions. He also explains that these studies have not focused on the precise research question he is investigating. The introduction concludes with a statement of purpose for the present study.

Method

This section should describe in detail how you gathered the information for your investigative report. Include the details your readers will be interested in. For original scientific research, your testing method is a significant factor in achieving meaningful results, and readers will want to know about every step of the research procedure. For field investigations, summarize your method of gathering information, but include enough specific details to show how, where, and under what conditions you collected data. Follow these guidelines where appropriate for your research method:

- Explain how and why you chose the test materials or specific tests.
- Describe any limits in the test materials, such as quantities, textures, types, or sizes, or in test conditions, such as length of time or options.
- List test procedures sequentially so that readers understand how the experiment progressed.
- Identify by category the people interviewed, such as employees, witnesses, and suppliers. If you quote an expert, identify the person by name.
- If appropriate, include demographic information, such as age, sex, race, physical qualities, and so on for the people you used in your study.
- Explain where and when observation took place.
- Identify what you observed and why.
- Mention any special circumstances and what impact they had on results.
- Explain the questions people were asked or the tests they took. If you write the questions yourself, attach the full questionnaire as an appendix. If you use previously validated tests, cite the source of the tests.

The method section in Model 12-3 (1) identifies the tests used in the study, (2) explains the scoring methods, (3) identifies the demographic questions students answered, and (4) explains how students were selected and how the questionnaires were distributed.

Results

This section presents the information gathered during the investigation and interprets the data for readers. Do not rely on the readers to see the same significance in statistics or other information that you do. Explain which facts are most significant for your investigation. Follow these guidelines:

- Group the data into subtopics, covering one aspect at a time.
- Explain in detail the full results from your investigation.
- Cite specific figures, test results, and statistical formulas, or use quotations from interviewees.
- Differentiate between fact and opinion when necessary. Use such words as *appears* and *probably* if the results indicate but do not prove a conclusion.

- If you are reporting statistical results, indicate (1) means or standard deviations, (2) statistical significance, (3) probability, (4) degrees of freedom, and (5) value.
- Point out highlights. Alert readers to important patterns, similarities to or differences from previous research, cause and effect, and expected and unexpected results.

The writer in Model 12-3 provides specific results he collected, identifies the statistical methods he used to analyze the data, and includes two tables illustrating the most important results. He then summarizes the data, explaining that students who began drinking in elementary school and high school have significantly higher scores for consumption and drinking problems than do students who began drinking in college.

Discussion

This section interprets the results and their implications for readers. Explain how the results may affect company decisions or what changes may be needed because of your investigation. Show how the investigation answered your original research questions about the problem being investigated. If your results do not satisfactorily answer the original questions, clearly state this. Follow these guidelines where appropriate for your type of investigation:

- Explain what the overall results mean in relation to the problem or research question.
- Analyze how specific results answer specific questions or how they do not answer questions.
- Discuss the impact of the results on company plans or on research areas.
- Suggest specific topics for further study if needed.
- Recommend actions based on the results. Detailed recommendations, if included, usually appear in a separate recommendations section.

Conclusion

When a research investigation is highly technical, the conclusion summarizes the problem or research question, the overall results, and what they mean in relation to future decisions or research. Nontechnical readers usually rely on the conclusion for an overview of the study and results. Follow these guidelines:

- Summarize the original problem or research question.
- Summarize the major results.

- Identify the most significant research result or feature uncovered in the investigation.
- Explain the report's implications for future planning or research.
- Point out the need for any action.

In Model 12-3, the writer combines a discussion of results and conclusions in one section. He summarizes his results by identifying his main finding—early drinking significantly predicts alcohol-related problems in college—and then he discusses the implications of this finding for planning drug and alcohol education programs. He concludes by suggesting further research on which kinds of intervention would be best for students.

Writing a Progress Report

A *progress report* (also called a *status report*) informs readers about a project that is not yet completed.

Purpose

The number of progress reports for any project usually is established at the outset, but more might be called for as the project continues. Progress reports are often required in construction or research projects so that decision makers can assess costs and the potential for successful completion by established deadlines. Although a progress report may contain recommendations, its main focus is to provide information, and it records the project events for readers who are not involved in day-to-day operations. The progress reports for a particular project make up a series, so keep your organization consistent from report to report to aid readers who are following one particular aspect of the project.

Organization

Model 12-4 is a progress report written by the on-site engineer in charge of a dam repair project. The report is one in a series of progress reports for management at the engineering company headquarters. The writer addresses the report to one vice president, but copies go to four other managers with varying interests in the project. The writer knows, therefore, that she is actually preparing a report that will be used by at least five readers. Progress reports include the following sections.

Introduction

The introduction reminds readers about progress to date. Explain the scope and purpose of the project, and identify it by specific title if there is one. Follow these guidelines:

- State the precise dates covered by this particular report.
- Define important technical terms for nonexpert readers.
- Identify the major stages of the project, if appropriate.
- Summarize the previous progress achieved (after the first report in the series) so that regular readers can recall the situation and new readers can become acquainted with the project.
- Review any changes in the scope of the project since it began.

The writer in Model 12-4 numbers her report and names the project in her subject line to help readers identify where this report fits in the series. In her introduction, she establishes the dates covered in this report and summarizes the dam repairs she discussed in previous reports. She also indicates that the project is close to schedule, but that the expected costs have risen because another subcontractor had to be hired.

Work Completed

This section describes the work completed since the preceding report and can be organized in two ways: You can organize your discussion by tasks and describe the progress of each chronologically, or you can organize the discussion entirely by chronology and describe events according to a succession of dates and times. Choose the organization that best fits what your readers will find useful, and use subheadings to guide them to specific topics. If your readers are interested primarily in certain segments of the project, task-oriented organization is appropriate. If your readers are interested only in the overall progress of the project, strict chronological organization is probably better. Follow these guidelines:

- Describe the tasks that have been completed in the time covered by the report.
- Give the dates relevant to each task.
- Describe any equipment changes.
- Explain special costs or personnel charges involved in the work completed.
- Explain any problems or delays.
- Explain why changes from the original plans were made.
- Indicate whether the schedule dates were met.

The engineer in Model 12-4 organizes her work-completed section according to the repair stages and then lists them in chronological order. She also includes the date on which each event occurred or each stage was completed.

Work Remaining

The section covering work remaining includes both the next immediate steps and those in the future. Place the most emphasis on the tasks that will be covered in your next progress report. Avoid overly optimistic promises. Follow these guidelines:

- Describe the major tasks that will be covered in the next report.
- State the expected dates of completion for each task.
- Mention briefly those tasks which are further in the future.

The work-remaining section in Model 12-4 describes the upcoming stage of the repair work and states the expected completion date.

Adjustments/Problems

This section covers issues that have changed the original plan or time frame of the project since the last progress report. If the project is proceeding on schedule with no changes, this section is not needed. If it is necessary, follow these guidelines:

- Describe major obstacles that have arisen since the last progress report. (Do not discuss minor daily irritations.)
- Explain needed changes in schedules.
- Explain needed changes in the scope of the project or in specific tasks.
- Explain problems in meeting original cost estimates.

The writer in Model 12-4 includes an adjustments section in her report to explain unexpected costs and a short delay because of the deteriorating condition of the dam area.

Conclusion

The conclusion of a progress report summarizes the status of the project and forecasts future progress. If your readers are not experts in the technical aspects of the project, they may rely heavily on the conclusion to provide them with an overall view of the project. Follow these guidelines:

- Report progress on current stages if that is the case.
- Report lack of progress on current stages if that is the case.

- Evaluate the overall progress so far.
- Recommend any needed changes in minor areas of scheduling or planning.
- State whether the project is worth continuing and is still expected to yield results.

The engineer's conclusion in Model 12-4 assures her readers that although she is 3 days behind schedule, she expects no further delays. She also identifies the final stage of the dam repairs and the expected completion date.

Writing a Trip Report

A *trip report* provides a record of a business trip or visit to the field.

Purpose

A trip report is a useful record both for the person who made the trip and for the decision makers who need information about the subjects discussed during the trip. The trip report records all significant information gathered either from meetings or from direct observation.

Organization

Model 12-5 is a trip report written in the opening-summary organization pattern discussed in Chapter 10. The writer has visited another company to learn about a computer system that links production units with design personnel. His report covers both what he observed and what he learned from the company executives he talked to. Trip reports should include the following sections.

Introductory Section

Use the subject line of your trip report to identify the date and location of the trip. Begin your report with an opening summary that explains the purpose of the trip and any major agreements or observations you made. Identify all the major events and the people with whom you talked. Follow these guidelines:

- Describe the purpose of the trip.
- Mention special circumstances connected to the trip's purpose.
- If the trip location is not in the subject line, state the dates and locations you visited.

- State the overall results of the trip or any agreements you made.
- Name the important topics covered in the report.

The opening summary in Model 12-5 reviews the reason for the trip and includes the writer's recommendation that the company seriously consider using the new computer system.

Information Sections

The information sections of your trip report should highlight specific topics with informative headings. Consider which topics are important to your readers, and group the information accordingly. Follow these guidelines:

- State the names and titles of people you consulted for specific information.
- Indicate places and dates of specific meetings or site visits.
- Describe in detail any agreements made and with whom.
- Explain what you observed and your opinion of it.
- Give specific details about equipment, materials, or systems relevant to company interests.

The writer in Model 12-5 divided his information into two sections—one describing how the computer system works and the other describing his talks with company executives.

Conclusions and Recommendations

In your final section, summarize the significant results of your trip, whether positive or negative, and state any recommendations you believe are appropriate. Follow these guidelines:

- Mention the most significant information resulting from the trip.
- State whether the trip was successful or worthwhile.
- Make recommendations based on information gathered during the trip.
- Mention any plans for another trip or further meetings.

In Model 12-5, the writer uses his final section to emphasize his opinion that his company would benefit from using the new computer system. He recommends another visit to the same company by a team of managers also interested in the new computer system.

Writing a Proposal

Proposals suggest new ways to respond to specific company or organization situations, or they suggest specific solutions to identified problems. A proposal may be internal (written by an employee to readers within the company) or external (written from one company to another or from an individual to an organization).

Purpose

Proposals vary a great deal, and some, such as a bid for a highway construction job on a printed form designed by the state transportation department, do not look like reports at all. In addition, conventional format varies according to the type of proposal and the situation. Proposals usually are needed in these circumstances:

1. A writer, either inside or outside a company, suggests changes or new directions for company goals or practices in response to shifts in customer needs, company growth or decline, market developments, or needed organizational improvements.
2. A company solicits business through sales proposals that offer goods or services to potential customers. The sales proposal, sometimes called a *contract bid,* identifies specific goods or services the company will provide at set prices within set time frames.
3. Researchers request funds to pay for scientific studies. The research proposal may be internal (if an organization maintains its own research and development division) or external (if the researcher seeks funding from government agencies or private foundations).

Proposal content and organization usually vary depending on the purpose and reader-imposed requirements. External proposals, in particular, often must follow specific formats devised by the reader. This chapter discusses the conventional structure for a proposal that is written to suggest a solution to a company problem—the most common proposal type.

In general, this type of proposal persuades readers that the suggested plan is practical, efficient, and cost-effective and suits company or research goals. Readers are decision makers who will accept or reject the suggested plan. Therefore, a successful proposal must present adequate information for decision making and stress advantages in the plan as they relate to the established company needs. In a high-cost, complicated situation, a proposal usually has many readers, all of whom are involved with some aspect of the situation.

Proposals are either solicited or nonsolicited. If you have not been asked to submit a proposal for a specific problem, you must consider whether the reader is likely to agree with you that a problem exists. You may need to persuade the reader that the company has a problem requiring a solution before presenting your suggested plan.

Organization

Model 12-6 is an unsolicited proposal written by an employee of a fast-food chain to his supervisor. The writer investigated what he believed was a company weakness in maintaining uniform service and food quality at all outlets and then wrote a proposal to establish companywide quality-control checks. Most proposals that are not restricted by external format requirements use the following organization.

Problem

The introductory section of a proposal focuses on a description of the central problem. Even if you know that the primary reader is aware of the situation, define and describe the extent of the problem for any secondary readers and for the record, and also explain why change is necessary. If the reader is not aware of the problem, you may need to convince him or her that the situation is serious enough to require changes. The first section of the proposal also briefly describes the recommended solution. Include these elements:

- A description of the problem in detail. Do not, for instance, simply mention "inadequate power." Explain in detail what is inadequate about the generator.
- An explanation of how the situation affects company operations or costs. Be specific.
- An explanation of why the problem requires a solution.
- The background of the situation. If a problem is an old one, point out when it began and mention any previous attempts to solve it.
- Deadlines for solving the problem if time is a crucial factor.
- An indication that the purpose of the proposal is to offer recommendations to solve the problem.
- Your sources of data—surveys, tests, interviews, and so on.
- A brief summary of the major proposed solution, for example, to build a new parking deck. Do not attempt to explain details, but alert the reader to the plan you will describe fully in the following sections.
- The types of information covered in the proposal—methods, costs, timetables, and so on.

The writer in Model 12-6 begins by reviewing the problem—inconsistent quality in food and service among company outlets and no standard company quality-control program. Then he explains that the company must resolve this problem if it is to grow, persuasively appealing to the reader's interest in the same goal. The writer also identifies the objectives, and explains how he gathered information on which to base his suggestions.

Proposed Solution

This section should explain your suggestions in detail. The reader must be able to understand and visualize your plan. If your proposal includes several distinct actions or changes, discuss each separately for clarity. Follow these guidelines:

- Describe new procedures or changes in systems sequentially, according to the way they would work if implemented.
- Explain any methods or special techniques that will be used in the suggested plan.
- Identify employees who will be involved in the proposal, either in implementing it or in working with new systems or new equipment.
- Describe changes in equipment by citing specific manufacturers, models, and options.
- Mention research that supports your suggestions.
- Identify other companies that have already used this plan or a similar one successfully.
- Explain the plan details under specific subheadings relating to schedules, new equipment or personnel, costs, and evaluation methods in an order appropriate to your proposal.

The writer in Model 12-6 proposes a new system of quality-control checks by specially trained employees. He uses subheadings in this section to describe his plan relative to needed materials and personnel, scheduling methods, costs, and lines of responsibility.

Needed Equipment/Personnel

Identify any necessary equipment purchases or new personnel required by your proposal. Follow these guidelines:

- Indicate specific pieces of equipment needed by model numbers and brand names.
- Identify new employee positions that will have to be filled. Describe the qualifications people will need in these positions.

- Describe how employee duties will change under the proposed plan and how these shifts will affect current and future employees.

For the fast-food chain in Model 12-6, the writer explains that his plan requires an evaluation form and that the company will have to hire and train new personnel, either as observers or to replace current employees who become observers.

Schedules

This section is especially important if your proposal depends on meeting certain deadlines. In addition, time may be an important element, if, for instance, your company must solve this problem in order to proceed with another scheduled project. Follow these guidelines if they are appropriate to your topic:

- Explain how the proposed plan will be phased in over time or on what date your plan should begin.
- Mention company deadlines for dealing with the problem or outside deadlines, such as those set by the IRS.
- Indicate when all the stages of the project will be completed.

The writer in Model 12-6 suggests that the company could implement his system within two months because personnel training is minimal.

Budget

The budget section of a proposal is highly important to a decision maker. If your plan is costly relative to the expected company benefits, the reader will need to see compelling reasons to accept your proposal despite its high cost. Provide as realistic and complete a budget projection as possible. Follow these guidelines:

- Break down costs by category, such as personnel, equipment, and travel.
- Provide a total cost.
- Mention indirect costs, such as training or overhead.
- Project costs for a typical cycle or time period, if appropriate.

The cost section in Model 12-6 includes the yearly budget for implementing the writer's plan in one region, as well as the yearly budget for establishing his plan in all five company regions. This kind of breakdown is included for the reader's information and to encourage her to consider testing the writer's idea before implementing it companywide.

Evaluation System

You may want to suggest ways to measure progress toward the objectives stated in your proposal and for checking the results of individual parts of the plan. Evaluation methods can include progress reports, outside consultants, testing, statistical analyses, or feedback from employees. Follow these guidelines:

- Describe suggested evaluation systems, such as spot checks, surveys, or tests.
- Provide a timetable for evaluating the plan, including both periodic and final evaluations.
- Suggest by whom these evaluations should be performed and analyzed.
- Assign responsibility for writing progress reports.

The writer in Model 12-6 suggests a simple evaluation system for a trial period. He also suggests which company executives might be appropriate choices to supervise the quality-control program.

Expected Benefits

Sometimes writers highlight the expected benefits of a proposal in a separate section for emphasis. If you have such a section, mention both immediate and long-range benefits. Follow these guidelines:

- Describe one advantage at a time for emphasis.
- Show how each aspect of the problem will be solved by your proposal.
- Illustrate how the recommended solution will produce advantages for the company.
- Cite specific savings in costs or time.

Because the quality-control checks proposed in Model 12-6 will not yield immediate, tangible benefits to the company, the writer emphasizes long-term benefits from increased customer good will.

Summary/Conclusions

The final section of your proposal is one of the most important because readers tend to rely on the final section in most reports and, in a proposal, this section gives you a chance to emphasize the suitability of your plan as a solution to the company problem. Follow these guidelines:

- Summarize the seriousness of the problem.

- Restate your recommendations without the procedural details.
- Remind the reader of the important expected benefits.
- Mention any necessary deadlines.

The writer in Model 12-6 uses his conclusion to summarize the company problem of inconsistency in food quality and service. In addition, he emphasizes the expected benefits of increased sales and improved customer good will if the company accepts his proposal.

CHAPTER SUMMARY

This chapter discusses the conventional content and structure for the six types of reports most commonly written on the job. Remember:

- A feasibility study analyzes the alternatives available in a given situation.
- An incident report records information about accidents or other disruptive events.
- An investigative report explains how a particular question or problem was studied and the results of that study.
- A progress report provides information about an ongoing project.
- A trip report provides a record of events that occurred during a visit to another location.
- A proposal suggests solutions to a particular problem.

SUPPLEMENTAL READINGS IN PART TWO

Buehler, M. F. "Defining Terms in Technical Editing: The Levels of Edit as a Model," *Technical Communication.*

Model 12-1 Commentary

This feasibility study was written by a student in response to Exercise 4 in this chapter. The report evaluates two computer systems as to their appropriateness for a growing day-care center.

The feasibility study includes formal report elements: a transmittal letter, title page, and informative abstract.

Discussion

1. The president of the day-care center is the reader of this report, which she asked for. Discuss the "Background and Problem" section of the report. What kinds of information would the writer probably have included if the reader had not asked for the report and if no previous discussions of the need for computer equipment had taken place?

2. Consider the subheadings in the section "Alternative Computer Systems," and discuss how useful they are to a reader trying to understand the information in order to make a decision about purchasing equipment. In groups, draft new subheadings that might guide the reader to specific types of information more effectively than these do.

3. Discuss the organizational strategies the writer uses. How effective is it to discuss IBM computers ahead of Tandy computers in each section? There are no conclusions in the subsections under "Alternative Computer Systems." Discuss the advantages of including brief statements about why or why not certain features are important for the day-care center.

4. Discuss how effectively the "Conclusions and Recommendations" section presents the case for purchasing a Tandy. Are there any points you would place more emphasis on?

5. After reading the report, consider the informative abstract (63 words). Discuss whether the reader could benefit from some additional information in the abstract. In groups, draft an abstract for this report that provides the reader with more information but does not exceed 100 words.

PEPPERMINT STICK NURSERY SCHOOL
16 West Range Road
River Hills, Wisconsin 53209

April 9, 1989

Ms. Lyla Baldwin
President
Peppermint Stick Nursery School
16 West Range Road
River Hills, WI 53209

Dear Ms. Baldwin:

Here is the feasibility study you requested on computer systems for use at the Peppermint Stick Nursery School. During the last two weeks I researched two computer systems: the IBM Personal System/2 Model 50Z and the Tandy 1000 TX.

Although both computer systems seem capable of meeting the day-care center's office needs, I have concluded that the Tandy 1000 TX would be the best choice because of its lower cost.

If you have any questions, I would be glad to discuss the report further. You may contact me at Extension 1212.

Sincerely,

Janice Knapp
Asst. Director of Services

JK:ss

Enc: Report

Feasibility Study
of Computer Systems for
Peppermint Stick Nursery School

Prepared for

Lyla Baldwin
President

By

Janice Knapp
Asst. Director of Services

April 9, 1989

ABSTRACT

This report presents the results of a feasibility study to evaluate computer systems for the growing needs of Peppermint Stick Nursery School. The two systems evaluated—IBM Personal System/2 Model 50Z and Tandy 1000 TX—were judged on capability, ease of use, maintenance, and cost. Both systems are appropriate for Peppermint Stick, but the Tandy 1000 TX is recommended because of lower cost.

Feasibility Study
of Computer Systems for
Peppermint Stick Nursery School

Introduction

This report assesses two potential computer systems--IBM Personal System/2 Model 50Z and Tandy 1000 TX--for Peppermint Stick Nursery School. After reviewing both, I recommend the Tandy 1000 TX as most appropriate for our day-care center.

Background and Problem

The need for a computer system at Peppermint Stick Nursery School is a result of the increased number of children, new larger quarters, and future anticipated growth. The simple card catalog system for maintaining records and the manual payroll system we use currently are outmoded and inefficient for our operations.

Possible Alternatives

After researching several computer systems manufactured by leading companies, I looked most closely at the IBM and Tandy models. Both systems meet the basic requirements for the day-care center. In deciding which system would best suit the school, I used the following criteria: capabilities of hardware and software, ease of use, service/maintenance availability, and cost. I gathered the information for both computer systems from local dealers and through product information materials.

Alternative Computer Systems

Basic Requirements

Both the IBM model and the Tandy model meet the following basic requirements for the nursery school. The needed hardware consists of the computer with at least 640 kilobytes of memory, a floppy disk drive, a keyboard, a monitor, and a printer. Software requirements include a word processor, an accounting package with a payroll system, and a file management system to keep track of children's records. Our budget for initial investment in a system is $8000 with a yearly maintenance budget of $500.

Capability

The IBM Personal System/2 Model 50Z has extensive hardware capabilities. Main memory starts at 1 megabyte (1024 kilobytes) and can be expanded to 16 megabytes. In addition to a 3.5-in. floppy disk drive, the system contains a 30-megabyte hard disk drive for storing files internally on the computer. A color monitor and near-letter-quality printer are also available at an additional cost.

Three software programs are necessary with the IBM model to serve Peppermint Stick's needs. These are DOS, the IBM Solution-Pac-Business Advisor Financial Accounting, and the IBM Writing Assistant package. DOS is a systems software program that is needed to run other programs on IBM computers. The Solution-Pac provides full-function accounting operations with modules for Accounts Payable, Accounts Receivable, Payroll, and database capabilities. The Writing Assistant package provides word-processing capabilities, including a spell-checker dictionary.

The Tandy 1000 TX also has adequate capabilities. Main memory starts at 640 kilobytes and can be expanded to 2 megabytes (2048 kilobytes). The Tandy 1000 TX has a 3.5-in. floppy disk drive and a 40-megabyte hard disk drive. The system provides a high-resolution color monitor and a near-letter-quality printer.

The Tandy model requires two software programs to meet our needs—the Bedford International Accounting Package and First Choice. The accounting package provides payroll and other financial functions needed for daily business operations. The First Choice software package provides word-processing, spreadsheet, and database capabilities.

Ease of Use

The IBM computer systems are user-friendly, and all hardware and software come with easy-to-read instruction manuals. Peppermint Stick Nursery School's secretary, John Smith, is familiar with the Writing Assistant package and, thus, would not have any problems using the word processor. He does not know the accounting system, however, and would need training. The User's Guide and end-user education supplied by IBM should be sufficient to train him.

The Tandy computer is also user-friendly, and all hardware and software come with instruction manuals. The software itself is very instructional and provides easy-to-read menus and screens. Smith, the secretary, has also worked with the First Choice software package and would need minimal training on it. He would, however, need training on the accounting package, which comes with a 3-hour training session by a qualified Radio Shack employee.

2

Service/Maintenance

IBM offers a 1-year guarantee on all hardware. In that year, repairs are made by the vendor at the vendor's shop. If we purchase a service contract for an annual fee, IBM offers the most extensive service network in the nation. Maintenance specialists are available 24 hours a day, 7 days a week and are usually dispatched within 4 hours of the service call. IBM Customer Engineers are well trained and supplied with the latest diagnostic equipment and parts. All software is supported by a no-charge 800 number for 90 days after delivery.

Installation of the Tandy model includes putting the software packages on the hard drive for easy access. Qualified personnel offer a free, 3-hour training session on the accounting software package. All hardware and software have a 90-day warranty, and a service contract may be purchased for an annual fee.

Both systems offer maintenance assistance, but the IBM service contract is superior to Tandy's.

Cost

The monitor, printer, and all software for the IBM model must be purchased separately. Total cost for the initial investment in IBM is as follows:

Personal System/2 Model 50Z computer	$3300
Color Monitor	600
Pro Printer II	450
DOS version 4.0	$ 150
Business Advisor Financial Accounting	1745
Writing Assistant	195
Annual Service Fee	300
	$6740

The Tandy model is considerably less expensive. All of the necessary hardware and software for the Tandy model is included in the price of the computer:

Tandy 1000 TX	$3752
Annual Service fee	175
	$3927

Conclusions and Recommendations

As Peppermint Stick Nursery School grows in size, so will its information-processing needs. The day-care center needs a computer system that will keep up with this growth and yet be user-friendly and cost-efficient. Both the IBM and Tandy models have adequate capabilities for the day-care center. They are both easy to use, and both systems have sufficient service agreements. Cost for the IBM system, however, is significantly higher than the cost of the Tandy model.

Based on capabilities, ease of use, and service, both systems would be good choices for Peppermint Stick. However, when cost is considered, the Tandy 1000 TX is the better choice, because the Tandy model provides virtually all the features available on the IBM model for about half the price. Although our budget would cover the cost of the IBM model, we are currently expanding our services, and all possible cost savings should be realized. I therefore recommend that we purchase the Tandy 1000 TX computer system.

Model 12-2 Commentary

This incident report, written by a supervisor in a manufacturing company, describes an accident involving personal injury. The writer includes as many details as she can learn from others and also reports what she found when she examined the equipment involved in the injury.

Discussion

Consider the "Recommendations" section of this report and assume that you want to write a memo to the department employees about following these safety procedures. What other kinds of information would you have to include if your were to write an instructional memo to the employees? In groups, draft a memo to all department employees telling them about Metlock's accident and the probable causes and instructing them to follow two safety procedures—cleaning the tar kettles and engaging the safety locks.

TO: Sean Donnelly

FROM: Heather Weber

DATE: September 28, 1990

SUBJECT: Accident Report—Alvin Metlock

Description of the Incident

On September 27, 1990 at 1:20 p.m., Alvin Metlock began checking and cleaning a cold portable tar kettle in the Processing Department, Building #4. While Metlock was leaning over the kettle, the hinged steel cover fell, knocking his hard hat to the ground and his safety glasses into the kettle. Metlock raised the cover again and reached into the kettle to retrieve his glasses. The cover fell a second time and caught his right middle finger between the cover and the rim of the kettle. His finger was nearly amputated. There were no witnesses to the accident. However, Anne Caldwell, working nearby, heard Metlock yell and ran to his aid. Caldwell wrapped the injured finger in a loose bandage and called for medical attention. She reported that Metlock told her he had difficulty setting the cover lock before his accident.

The Waverly Ambulance Service arrived at 1:31 p.m. and took Metlock to Memorial Hospital. Dr. Jake Meyer reattached Metlock's finger. The success of the procedure will not be known for 8 days. Metlock was advised to remain at home for 48 hours and keep his hand in a sling.

Causes

When I examined the tar kettle, I found that the cover latch was badly clogged with hardened tar. There was enough tar clogging the latch to make the hinge unreliable and difficult for the user to be sure the cover was in an upright position. In addition, however, the safety locking device was not engaged, indicating that Metlock did not fully secure the cover before reaching into the kettle.

Recommendations

To prevent similar occurrences in the Processing Department, I suggest the following:

1. Clean the tar kettles with the recommended solvent weekly.
2. Instruct all employees to fully engage the safety locking device before leaving the tar kettle covers open.

Model 12-3 Commentary

This research report is a published investigative study of a specific research question of interest to the readers of a professional journal. Most of the readers also do investigative studies of similar research questions and are interested not only in the results, but also in the method the writer used to achieve results.

Discussion

1. The abstract here is written for other professionals who do research similar to this. Discuss how useful it would be to a nonexpert reader who wanted to know about the study. What adjustments or changes would have to be made? After reading this article, in groups, draft an abstract of the article that would be suitable for nonexpert readers.

2. Discuss the introductory section of the study. What types of information does the writer include? How does this introduction help prepare readers to understand the rest of the report?

3. This published report has no subheadings, probably because the professional journal does not use subheadings. Discuss subheadings you might include in the "Conclusions" section if you wanted to help nonexpert readers find specific information.

Early Onset of Drinking As a Predictor of Alcohol Consumption and Alcohol Related Problems in College*

Gerardo M. Gonzalez, Ph.D.

University of Florida, Gainesville

ABSTRACT

A survey of 4202 students enrolled in the State University System of Florida from 1986 to 1988 showed that 14 percent started drinking in elementary school, 34 percent started in middle school, 45 percent in high school, and 7 percent in college. A General Linear Models analysis of variance procedure showed that the time of first drink significantly predicted

*This research was made possible by a grant from the Florida Department of Health and Rehabilitative Services to the Campus Alcohol and Drug Resource Center, Office for Student Services, University of Florida.

the quantity-frequency of alcohol consumption and the incidence of alcohol-related problems reported by students in college. Students who started to drink while in elementary and middle school reported significantly higher levels of consumption and problems than those who started drinking while in high school or college.

Alcohol and other drug use by children and adolescents has been a source of increasing concern in recent years (1). In addition to the threat that the use of chemical substances presents to the educational and personal development of young people, early onset of alcohol and other drug use is one of the strongest predictors of later dependency problems. In a large-scale epidemiological study, it was found that the likelihood of developing a drug disorder depends heavily on the age at which drug use begins. According to Robins and Przybeck, the single best predictor of substance abuse in early adulthood was onset of use prior to the age of fifteen (2).

Most substance abusers begin with experimental use of alcohol, cigarettes, and marijuana and then progress to heavier and expanded use of other illicit drugs (3). If the onset of these "gateway" drugs could be prevented, or at least postponed until late adolescence, the probability of heavy drug involvement in later years is greatly reduced (4). In a comprehensive review of the literature, Braucht found that between the ages of twelve and seventeen alcohol is the most widely used drug and that problem drinking increases sharply with advancing age during adolescence (5). Moreover, "misusers" of alcohol appear to begin drinking at an earlier age than "users" (6). This finding is particularly troublesome in view of the research which shows that in 1985 over 90 percent of high school seniors reported using alcohol. More than half (56%) of these seniors reported that their first use of alcohol occurred before the 10th grade; and nearly 10 percent of them began drinking as early as the sixth grade (7).

Follow-up studies of graduates who were one to four years past high school showed that patterns of alcohol use among those attending college did not vary substantially from use among nonstudents in the same age group. One exception to this, however, was the frequency of occasional heavy drinking. Among college students, 45 percent drank five or more drinks at a time during the two weeks preceding the survey. By contrast, only 41 percent of the group as a whole and 37 percent of high school seniors drank at this level (7). It is widely recognized that heavy drinking and alcohol-related problems are common among college students (8). However, very few studies have examined how individual and situational factors interrelate to predict college drinking (9). One area in particular where there is a dwarf of research is on the effects of early onset of drinking upon alcohol use in college. Such information might provide some clues as to characteristics of students at risk for alcohol-related problems in college. Furthermore, it might suggest some directions for research on how campus environments interact with personal characteristics of students to produce a high incidence of heavy drinking and alcohol-related problems. Therefore, the purpose of this research was to examine whether age as measured by school level at the time of first drink predicted consumption levels and alcohol-related problems among college students.

METHODS

The instrument used in this study was the Student Drinking Information Scale (SDIS). The SDIS is made up of several subscales, including a section on demographic information and drinking history, a section on Quantity-Frequency (QF) of consumption, and a section on alcohol-related problems commonly reported by college students. The QF scale is a modified version of the original QF scale used by Straus and Bacon (10) to measure consumption of alcohol by college students. In the SDIS version the QF score is obtained by multiplying the number of drinks (12 oz. of beer, 5 oz. of wine, or 1.5 oz. of liquor) students report usually consuming per drinking occasion times the number of occasions a student usually drinks per month. Thus, the QF score is an indication of the average number of drinks a student reports consuming per month. The QF scale has a theoretical score range of 1 to 120.

The incidence of alcohol-related problems is measured through the Problems scale of the SDIS. This scale lists twenty alcohol-related problems commonly reported by college students (e.g., being nauseated from drinking or driving after drinking too much) and asks respondents to indicate how many times—from never to 5 or more—they have experienced these problems in the past twelve months. This method of scoring produces a theoretical range of 0 to 100, and it is a modification of the original Problems scale developed by Engs (11). The reliability and validity for the current version of both the QF and Problems scales were established by the author (12) and have been used extensively with college students.

The demographic information section of the SDIS asks the question: How old were you when you took your first drink? To which students can respond: (1) elementary school age, (2) middle school age, (3) high school age, (4) college age. By pairing the age of first drink with the school level at which it occurred, this method of assessment provides some control for the notoriously high level of difficulty in recalling exact age of first drink.

Beginning in the spring semester of 1986, the SDIS was distributed to a random sample of students at each of the nine institutions in the State University System of Florida. In order to select the sample, a list of all undergraduate courses offered at each institution was obtained from the registrar's office. The list contained the course title, instructor's name, time and date of class meeting, and number of students enrolled. Courses were randomly selected from the list until approximately one percent of the total undergraduate enrollment at each institution was selected. Then the instructor for each course was contacted by a designated research coordinator at each university and permission was obtained to distribute the SDIS during class. In the event that an instructor was unable to accommodate the request for data collection, an alternate course was selected until the desired representation was obtained. The great majority of the instructors contacted agreed to allow the data collection in their courses. The SDIS required about thirty minutes for administration and the same instructions and procedures were followed in each class at each institution. Data were collected at all nine institutions in the State University System in the spring semester of 1986, 1987 and 1988.

RESULTS

There were 1562 collected data in 1986, 1295 in 1987, and 1345 in 1988. For the purpose of this study, the samples collected each year were combined into one overall sample which contained 4202 usable questionnaires. Fifty-seven percent of the overall sample were females and 43 percent males. Seventy-four percent of the students said they were drinkers at the time of the survey. Of these, 14 percent said they had their first drink in elementary school, 34 percent in middle school, 45 percent in high school, and only 7 percent said they had their first drink in college. Using the General Linear Models (GLM) procedure of the SAS statistical package, an analysis of variance was computed to determine if school age at the time of first drink significantly predicted alcohol consumption and alcohol-related problems in college. Table 1 shows the results of the analysis of variance for the QF and the Problems scales.

Table 1 Analyses of Variance for QF and Problems with School Age at the Time of First Drink as a Predictor Variable

	QF			***Problems***		
	DF	***F***	***PR > F***	***DF***	***F***	***PR > F***
School Age	3	57.70	0.0001	3	51.29	0.0001

As shown in Table 1, school age at time of first drink significantly predicted both QF and Problem scores in college. A post hoc Scheffee analysis of means showed that there was no significant difference at the $p < 0.05$ level for either QF or Problem scores between students who started drinking in elementary and students who started drinking in middle school. However, students who started drinking in elementary or middle school had significantly higher QF and Problem scores than those who had their first drink either in high school or in college. Also, students who started drinking in high school had significantly higher QF and Problem scores than those who started drinking in college. The mean QF and Problem scores for each group are shown in Table 2.

CONCLUSIONS

These data showed that early onset of drinking is indeed a significant predictor of alcohol consumption and alcohol-related problems in college. In fact, there appears to be a linear progression whereby the earlier students begin to drink the more they tend to consume and the more alcohol-related problems they experience while in college. This finding has significant implications for alcohol education efforts both in the schools and in college. First, these data support the notion that school-based alcohol and drug education efforts for elementary and middle school children should focus on delaying the onset

Table 2 Mean Scores for QF and Problems Variables According to School Age at the Time of First Drink

School Age	*QF*	*Problems*
Elementary School	29.04	16.83
Middle School	27.48	16.87
High School	18.92	12.31
College	12.33	6.79

of drinking. Saying no to alcohol use at this age may very well mean a reduction in the risk of becoming a heavy drinker and experiencing more alcohol-related problems in young adulthood, a time when the risk for heavy consumption and drinking problems is at its highest (13).

These data also suggest that alcohol education and prevention programs for college students should take into consideration the drinking history of individual students or student groups for whom the programs are intended. It may be that programs and other educational interventions designed to impact the campus environment can have differential effects on students depending on their individual drinking history. If this is so, preventive programs should be specifically designed to meet the needs of carefully targeted student populations. This research suggests that an important item to consider in the design of such programs is the time of onset of drinking among the target population. For example, students could be made aware of how the age at which they began to drink may place them at special risk for drinking problems. Such knowledge may increase the students' perceived susceptibility and serve as a motivating factor for the development of skills necessary to manage that risk. While more research is needed to determine just what kind of interventions are most effective with what kind of student populations, knowledge of how early onset of drinking affects early adulthood drinking patterns brings us closer to the prevention goal of being able to match specific interventions with specific populations.

References

1. S. Griswold-Ezekaye, K. L. Kumpfer, and W. J. Bukoski, *Childhood and Chemical Abuse,* The Haworth Press, New York, 1986.
2. L. N. Robins, and T. R. Przybeck, Age of Onset of Drug Use as a Factor in Drug and Other Disorders, in *Etiology of Drug Abuse: Implications for Prevention,* C. L. Jones and R. J. Battjes (eds.), National Institute on Drug Abuse, Research Monograph 56, pp. 178–192, 1985.
3. D. B. Kandel, *Longitudinal Research on Drug Use: Empirical Findings and Methodological Issues,* John Wiley & Sons, New York, 1978.
4. D. B. Kandel, Epidemiological and Psychosocial Perspectives on Adolescent Drug Use, *Journal of American Academic Clinical Psychiatry, 21,* pp. 328–347, 1982.

5. G. N. Braucht, Problem Drinking Among Adolescents: A Review and Analysis of Psychosocial Research. In *Alcohol and Health Monograph 4,* DHHS Publication No. (ADM) 82-1193, National Institute on Alcohol Abuse and Alcoholism, Rockville, Maryland, 1982.
6. J. V. Rachel, L. L. Guess, R. L. Hubbard, S. A. Maisto, E. R. Cavanaugh, R. Waddel, and C. H. Benrud, Facts for Planning No. 4: Alcohol Misuse by Adolescents, *Alcohol Health and Research World,* pp. 61–68, 1982.
7. L. Johnston, P. M. O'Malley, and J. G. Backman, *Drug Use Among High School Students, College Students, and Other Young Adults: National Trends Through 1985,* DHHS Publication No. (ADM) 86-1450, National Institute on Drug Abuse, Rockville, Maryland, 1986.
8. T. G. Goodale (ed.), *Alcohol and the College Student,* New Directions for Student Services No. 35, Jossey Bass, San Francisco, 1986.
9. R. Saltz and D. Elandt, College Student Drinking Studies 1976–1985, *Contemporary Drug Problems, 13*:1, pp. 117–159, 1986.
10. R. Straus and S. D. Bacon, *Drinking in College,* Yale University Press, New Haven, 1953.
11. R. C. Engs, Drinking Patterns and Drinking Problems of College Students, *Journal of Studies on Alcohol, 38,* pp. 2144–2156, 1977.
12. G. M. Gonzalez, What Do You Mean—Prevention? *Journal of Alcohol and Drug Education, 23*:3, pp. 14–23, 1978.
13. D. B. Kandel and J. A. Logan, Patterns of Drug Use from Adolescence to Young Adulthood: Periods of Risk for Initiation, Continued Use, and Discontinuation, *American Journal of Public Health, 74*:7, pp. 660–666, 1984.

Model 12-4 Commentary

This progress report, written by an on-site engineer, is one of a series of reports on a specific company project—dam repairs. The report will be added to the others in the series and kept in a three-ring binder at the company headquarters for future reference when the company has another contract for a similar job.

The writer addresses the report to a company vice president, but four other readers are listed as receiving copies.

Discussion

1. The writer knows that her readers are familiar with the technical terms she uses. If this report were needed by a nontechnical reader at the company, which terms would the writer have to define?

2. Discuss the information in the introduction and how effectively it supports the writer's assertion that work on the dam is progressing satisfactorily.

3. Discuss how the writer's organization and use of format devices support her report of satisfactory progress. Why did she list items under "Work Completed" but not under "Work Remaining"?

TO: Mark Zerelli
Vice President
Balmer Company

October 16, 1990

FROM: Tracey Atkins
Project Manager

SUBJECT: Progress Report #3--Rockmont Canyon Dam

Introduction

This report covers the progress on the Rockmont Canyon Dam repairs from September 15 to October 15 as reported previously. Repairs to the damaged right and left spillways have been close to the original schedule. Balmer engineers prepared hydraulic analyses and design studies to size and locate the aeration slots. These slots allowed Balmer to relax tolerances normally required for concrete surfaces subjected to high-velocity flows. Phillips, Inc., the general contractor, demolished and removed the damaged structures. To expedite repairs, construction crews worked on both spillways simultaneously. Construction time was further reduced by hiring another demolition company, Rigby, Inc. The project costs rose during the first month when Phillips, Inc., had to build batching facilities for the concrete because the dam site had no facilities.

Work Completed

Since the last progress report, three stages of work have been completed:

1. On September 18, aggregate for the concrete mix was hauled 230 miles from Wadsworth, Oregon. The formwork for the tunnel linings arrived from San Antonio, Texas, on September 20.

2. Phillips developed hoist-controlled work platforms and man-cars to lower workers, equipment, and materials down the spillways. Platforms and man-cars were completed on September 22.

3. Phillips drove two 20-foot-diameter modified-horseshoe-shaped tunnels through the sandstone canyon walls to repair horizontal portions of the tunnel spillways. A roadheader continuous-mining machine with a rotary diamond-studded bit excavated the tunnels in three weeks, half the time standard drill and blast techniques would have taken. The tunnels were completed on October 15.

Mark Zerelli -2- October 16, 1990

Work Remaining

The next stage of the project is to control flowing water from gate leakage. Phillips will caulk the radial gates first. If that is not successful in controlling the flow, Phillips will try French drains and ditches. After tunnels are complete, both spillways will be checked for vibration tolerance, and the aeration slot design will be compared with Balmer's hydraulic model. Full completion of repairs is expected by November 15.

Adjustments

Some adjustments have been made since the last progress report. During construction work on both spillways, over 50 people and 200 pieces of equipment were on the site. Heavily traveled surfaces had to be covered with plywood sheets topped with a blanket of gravel. This procedure added $3500 to the construction costs and delayed work for half a day.

Conclusion

The current work is progressing as expected. The overall project has fallen three days behind estimated timetables, but no further delays should occur. The final stage of the project will require measurements at several areas within both spillways to be sure they can handle future flood increases with a peak inflow of at least 125,000 cfs. Balmer expects to make the final checks by November 25.

TA:ss
c: Robert Barr
Mitchell Lawrence
Mark Bailey
Joseph Novak

Model 12-5 Commentary

This trip report covers a visit made to another company to observe a computer system. The writer addresses the company president, but he also lists three other readers as receiving copies.

Discussion

1. This report includes two information sections—one describing the computer system and the other discussing potential personnel difficulties if the system is adopted. Discuss why the writer included both sections and what kind of reader would be interested in each section. Discuss the headings used in these sections. Why did the writer probably decide to use "Meeting with Gentry Executives" as a heading rather than a heading that reflects the major topic in the section?

2. Notice that in his recommendations the writer suggests sending two other design engineers to observe the computer system. Discuss why the writer probably decided not to send these people copies of his report.

TO: Paige Maitland, President June 15, 1990
FROM: Peter Vochek, Design Engineer
SUBJECT: Trip to Gentry Manufacturing, San Diego, June 12-13

Summary

In order to observe computer-integrated manufacturing in operation, I spent two days at Gentry Manufacturing, studying the use of CIM and talking to Henry Brophy, vice president of production. Mr. Brophy showed the facility to me and explained the system features. Based on my visit to Gentry, I believe the CIM could increase our competitive position in an international market and provide a strategic investment for Barlow-Taylor. I recommend that we explore CIM further.

CIM Operation

Through CIM, machines and computers are linked to design and quality control. When design personnel need to create a new part for a special order, they can display it on a video screen for the customer. When the customer is satisfied, the design personnel transfer that information to the engineers via the data link. The engineers can then analyze the design and make changes. The final specifications for the new part are sent electronically to the machine section, where computers make the changes in the machines to produce the new part. Gentry reduced machining time for multiton engine frame parts from 14 days to 24 hours with CIM.

Meeting with Gentry Executives

My meeting with Henry Brophy included Kimberly Collier and Tyrone Harris, both on the CIM planning team for Gentry. They told me that we need to think of CIM as an investment rather than a technology acquisition. CIM ultimately can affect the entire organization. Departments, such as sales, purchasing, engineering, and production, have to be brought into the initial planning effort. I was also told that many managers will view CIM as a threat to their autonomy. With CIM, all pieces of machinery--production, design, materials handling, computers--must fit into the subsystems that integrate the total CIM system. The emphasis when we present the system to company managers should be on competitive advantage rather than production efficiency.

Paige Maitland -2- June 15, 1990

Conclusions and Recommendations

Because Gentry's operation is similar to ours, I believe we need to explore CIM further. My trip was successful in showing me how effective the system could be with full implementation and full support in a company. I recommend that a team of managers visit Gentry again as well as Darrow Machines, also in San Diego and also with a CIM system. The team should include Janet Harkin and Patricia Samuelson, our senior design engineers. I would be glad to accompany the team on a second visit.

PV:ss

c: Ben Shelley
   Roger Bordon
   Victoria Lord

Model 12-6 Commentary

This proposal was written by a student in response to Exercise 6 in this chapter. The student identified what he believed was a weakness in the company he works for, investigated the situation, and then proposed a new system of quality-control checks.

Discussion

1. The reader of this proposal has not asked for it and may not be aware that any quality-control problem exists. Discuss how effectively the writer makes the case that a problem exists and needs to be solved.

2. Since the writer had no access to actual cost figures, he estimated costs. Discuss the advantages or disadvantages of moving these cost figures to an appendix.

3. Discuss the organization of the "Problem and Objectives" section. Discuss how effective the list of objectives would be at the end of the section.

4. Assume that you are the manager of a Burger Heaven franchise and you read a copy of this proposal. What kinds of information do you think you would want about the suggested procedures?

5. Discuss how effectively the writer presents his proposal and persuades the reader that benefits will be worth the cost. What kinds of information might be needed in the section "Benefits vs. Cost"?

6. In groups, establish standards that you believe are important for any fast-food outlet regarding food quality, cleanliness, and service. Draft a quality-control form that might be used at Burger Heaven if this proposal were approved. Each group member should visit a fast-food outlet before the next class and fill in the sample form according to how well the restaurant meets these standards. Then, in groups, compare results and write a final draft of the form, based on the experiences of the group members.

TO: Kathleen Rescott
Division Manager

November 3, 1989

FROM: Thomas Bowen

SUBJECT: Quality-Control Program for Burger Heaven, Inc.

Problem and Objectives

During the past several months, Burger Heaven has experienced inconsistent quality in food and service at the four franchise outlets in this city. Since franchise operations are committed to following quality guidelines set by Burger Heaven, Inc., this failure to meet those levels reflects negatively on all Burger Heaven operations. Quality variations among the four franchises include fast, friendly service one day and slow, poor service the next; and variations in taste and texture of food products, in heat or crispness of food products, and in facility cleanliness. For Burger Heaven to increase its market size, the franchises must represent consistent high quality in all areas, following official Burger Heaven standards. The most effective way to eliminate lapses in quality standards is through implementing a plan of quality-control checks.

This proposal has two objectives:

1. To establish consistency among all Burger Heaven franchises in food quality and level of service

2. To implement a program of quality-control checks at all franchises on a regular basis

The following plan for a systematic program of quality checks is based on my personal observation of the four area franchises over a four-week period. From September 16 to October 12, I visited each franchise twice at varying times and ordered varying products. Food quality at all four outlets ranged from excellent (crisp fries, hot and juicy meat) to poor (limp fries, cold food, dry meat). Service also ranged from excellent (friendly, within time limits) to poor (sullen, over time limits). On most visits, the cleanliness of the facilities was only fair to poor. Simple, but regular, quality-control checks should eliminate these variations.

Kathleen Rescott -2- November 3, 1989

Suggested Quality-Control Procedures

The proposed quality-control checks involve regularly scheduled visits to each franchise by a trained observer from corporate headquarters. Each franchise should be visited four times a year. Franchises can be visited on the same day, but times of visits should rotate on succeeding visits. The observer will select food products and perform taste tests, also on a rotating basis. In addition, the observer will rate employees for service and the facility for cleanliness. A standard evaluation report will be used and filed at headquarters.

Managers of each franchise will receive the evaluation sheet with comments from headquarters. Expected improvements should be made before the next quality-control check.

Materials

The new quality-control check program will require only a standard Quality-Control Evaluation Report form. This evaluation report will cover food quality, service level, and facility appearance inside and out.

Personnel

For the initial trial, one employee from headquarters could be trained to make the visits to the franchises in a selected area. One additional clerk to process paperwork and prepare schedules will be needed on a part-time basis. If the system is implemented nationwide, one trained observer for each region would be needed, plus two full-time clerks to process paperwork. In addition, another full-time trained observer would be needed at headquarters to communicate with franchise managers who have questions or want to discuss reports. Some rotation of field observers with the headquarters observer would be beneficial to reduce burnout.

Schedule

Because the initial trial requires so little in materials or additional personnel, the new system could be in place by the beginning of next year. The training needed to accurately assess a Burger Heaven franchise is not extensive, since the experience of going to a fast-food outlet is a typical one in most people's lives. Observers will have to be trained in the Burger Heaven standards for each food item. That training probably requires only two weeks.

Kathleen Rescott -3- November 3, 1989

Cost

The initial cost for the trial is minimal. Following are estimated costs for one year's observation of the four franchises in our region:

Observer--4 day's work @ $115/day	$ 460
Travel expenses	160
Part-time clerk 80 hours @ $6.50/hour	520
	$1140

If the program is expanded to cover all Burger Heaven franchises, the first year's cost would involve the following:

Five regional observers @ $35,000/year	$175,000
Travel expenses	70,000
Two full-time clerks @ $22,000/year	44,000
Training time	6,000
Evaluation forms, development and printing	12,000
	$317,000

Travel expenses are estimated on the basis of automobile mileage, food-purchase expense, and lodging at Holiday Inns.

Staff training could be done at corporate headquarters in one week. Anticipated turnover in observers will result in training a new observer each year after the initial six are trained.

For effective evaluation, a consulting company will have to be hired to develop a form suited for our franchises and the kinds of information we wish to gather about food and service.

Kathleen Rescott -4- November 3, 1989

Reporting System

This franchise evaluation system should be reviewed by corporate executives after each day's visit during the initial trial period. When the system is a regular program, the vice president of operations should be responsible for the program. The headquarters observer who handles calls from franchise managers should report to the Director of Consumer Relations.

Benefits vs. Cost

The cost to Burger Heaven for implementing this quality-control program is minimal when compared to the improved customer relations, increased sales volume, and industry reputation that will result. Last year's sales reached $500 million. Through improved quality control and advertising geared to that quality, Burger Heaven should be able to increase sales by $10 million within 18 months. Customer goodwill is probably the most valuable asset a fast-food business has, and Burger Heaven would enjoy the best reputation in a crowded field of fast-food choices.

Conclusion

Burger Heaven right now has a problem with inconsistent quality from franchise to franchise. This variation in food and service hurts business overall, since customers with a bad experience in one outlet may not choose to try another outlet. With the program I have proposed, the high quality-control standards of Burger Heaven can be enforced. We have the potential to increase sales volume and customer goodwill through a relatively simple evaluation system. Trained observers can eliminate the inconsistences that now exist, and Burger Heaven can enjoy the position of being the leader in serving fast-food customers.

Chapter 12 Exercises

1. Assume that you are the safety engineer at Ridgeway Manufacturing Company. Yesterday, a flash fire caused extensive property damage. Sprinklers were effective in getting the blaze under control. Fire brigade action by employees also helped control the blaze. Damage is estimated at $19,000. The fire took place at 11:00 A..M. in Bay 20 of the Machining Unit.

You talk to the people in the unit and discover that the fire started from flammable liquids coming in contact with hot surfaces. The fire department was not called. The fire brigade used a dry chemical to put out the fire. One hose (1½ in.) 250 ft long was used in fighting the fire, and 21 of the automatic sprinklers opened. Loss of production time was 24 hours because two Dynamatic machines producing gear blanks were down. This resulted in failure to produce 4800 gear blanks. While Joe Kelly was in the process of clearing a plugged oil coolant supply line that was disconnected on a Dynamatic machine, the maintenance pipefitter Matt Arnold turned on the oil pump. Unit leader Nico Alvarez tells you that the fire resulted when the pressure created by the pump apparently freed the plugged line causing oil to spray upward hitting the unapproved type light bulb (150 W, 120 V). The light shattered and ignited the oil, flashpoint 332°F. "It all happened in an instant," Nico reports to you. "After that, the fire spread to the Dynamatic machine next to the first one." In addition to the machine damage, the plant roof suffered some minor fire damage. Alvarez tells you that fire- and explosion-proof lighting is needed on the automatic screw machines and throughout the unit. "We work with so much flammable liquid," he says, "we need protection from this kind of accident."

When you write your incident report, you decide to suggest also that supervisors inspect light bulbs daily to be sure they are the approved type and in working order. Because $6000 in downtime is part of the loss, you will have to schedule overtime on three Saturdays with two shifts to catch up on production. Fortunately, no one was hurt. Write an incident report for this situation, and address it to Ellen Hubbler, vice president of operations.

2. This assignment requires a visual inspection of one rest room on your campus. The selection of which room in which building is up to you. Observation or inspection requires that you examine the area under *existing conditions.* Make no attempt to control or manipulate the situation. The purpose of your study is to investigate conditions in one rest room *after 10 A.M.* on a regular school day. Record what *you* see, smell, hear, and so on. You must interview at least one person, either someone using the facility or someone responsible for maintenance, and include that person's comments in your report. You should inspect the following: sinks, toilets/urinals, mirrors, towel and soap containers, hand dryers, light fixtures, doors/entrances, walls/floors. Make notes on the need for repair or replacement. *In addition,* observe the frequency of use during your investigation, and record any need for more equipment, more maintenance, or more supplies. You *will not* make any recommendations in this report, but you can conclude that the room needs improvement in some way or that the room is well maintained. Address your investigation report to Tom Ortega, director of physical facilities. Assume that you are a member of a team of students who have been hired to inspect campus facilities from the student perspective. If your instructor wants you to write a formal report, use the guidelines in Chapter 11.

3. Based on the investigative report you write for Exercise 2, write a short informal proposal for one specific improvement in the campus rest room you observed. Address your proposal to Tom Ortega, director of physical facilities.

4. Study a problem at the company where you work. The problem should be a situation that requires a change in procedures or new equipment. Investigate at least two alternative solutions to this problem. Then write a feasibility study in which you evaluate all the solutions and recommend one of them. Address the report to the appropriate person at your company, and assume that that person asked you to write the report. If your instructor wants you to prepare a formal report, use the guidelines in Chapter 11.

5. Write a progress report for a project you are doing in a science or technology class. Report your progress as of the midpoint of your project. Address the report to the instructor of the science or technology class and say that you will submit another progress report when the project is nearly finished.

6. Write a proposal for a change at your job to your supervisor or the appropriate person. The change you propose may include equipment purchases, reorganizing work duties, a new system, new parking arrangements, and so on. If you do not have a current job, you may use a previous job or propose a change on campus, such as increased food services or reorganization in the bookstore. Address the director of student services. Assume that your supervisor or the director of student services did not ask you for this proposal, so you need to make a persuasive case that a problem exists and needs to be solved. Include at least one graphic aid with your proposal. If your instructor wants you to write a formal proposal, use the guidelines in Chapter 11.

7. You are the senior consultant for recreation facilities at the Tyrolean Forest resort. As consultant, you are responsible for monitoring recreation programs and recommending all purchases connected with recreation. The Tyrolean Forest resort is located in attractive rural land 65 miles from a large city. Thus its guests can easily come for weekend retreats as well as longer stays. As a result, business is booming, and the facilities have expanded rapidly over the past 2 years. The Tyrolean Forest has an Olympic-size swimming pool, racquetball courts, tennis courts, a softball field, indoor bowling alleys, complete workout rooms for men and women (with aerobics instructors), an indoor ice skating rink, a volleyball court, and a golf course. As consultant, you have decided that some new equipment is needed. Select one of the preceding areas (e.g., bowling alley, golf course) and write a feasibility study in which you evaluate *three* different models of equipment suitable for the area. The equipment can be either for individual use by guests (e.g., bowling balls) or for general usability of the area (e.g., tennis court nets). Address your report to Audrey Hardy, vice president, Tyrolean Forest, Inc. You may include a recommendation for one of the models you evaluate, but remember that Hardy will make the decision about what to purchase.

8. Write an investigative report for a research project or experiment you are working on in another class. Address your report to your writing instructor, who is interested in learning about your work in your technical classes. Remember that your reader is not an expert in the technical subject. Include a glossary for your nonexpert reader, using the guidelines in Chapter 11.

9. Prepare an investigative report on a subject that is a major concern in your community, on campus, or in your company. Consider a fairly complex problem that requires you to gather facts from printed sources and interview at least two people for opinions or information. Subjects for this report might be (a) rising crime rates on campus or in the community, (b) the need for day care on campus or in the community, (c) the need for computers, video equipment, or specific machinery in your company, (d) the need to restore an old building with historical significance in your community or on campus, (e) the need for a change in technology at your company, (f) inadequate recreation facilities for teenagers in your community, or (g) any currently important issue. In your investigative report, identify possible solutions to the problem. *Then* study two alternative solutions to the problem, and write a feasibility study in which you identify the various factors that are important in each alternative and recommend one of the alternatives. *Then* write a proposal for the option you have chosen as a solution to the original problem. You may assume that these reports have been solicited and address them to the appropriate person, such as the college provost, city mayor, or president of your company. These three reports should all be formal reports with appropriate front and back matter, as explained in Chapter 11. Your instructor will probably want you to submit the reports one at a time, so keep a copy of each for reference as you work on the next report.

10. Write a progress report to your instructor when you have finished the first two reports required in Exercise 9. Explain what those reports covered and what you expect to accomplish in the third report. Project a timetable for finishing the rough draft and the final draft for the proposal.

11. In connection with your reports in Exercise 2 or 9, write a trip report that covers your meeting and interview with a person you quoted as a source of information. Include your opinion as to whether this person would be a good source of further information on the topic you investigated. Address your trip report to your writing instructor.

CHAPTER 13

Oral Presentations

Understanding Oral Presentations

Purpose

Advantages

Disadvantages

Types of Oral Presentations

Organizing Oral Presentations

Preparing for Oral Presentations

Delivering Oral Presentations

Nerves

Rehearsal

Voice

Professional Image

Gestures

Eye Contact

Notes and Outlines

Visual Equipment

Questions

Joining a Team Presentation

Chapter Summary

Supplemental Readings in Part Two

Exercises

Understanding Oral Presentations

Although most people hate to speak in front of a group, you must be prepared to handle oral presentations gracefully because they are an important, and frequent, element in professional communication on the job. Oral presentations can be internal, i.e., for management or employees, or external, i.e., for clients, prospective customers, or colleagues at a scientific conference. Oral presentations may have different purposes, and they have both advantages and disadvantages.

Purpose

Like written documents, oral presentations often have mixed purposes. A scientist speaking at a conference about a research study is informing listeners as well as persuading them that the results of the study are significant. Oral presentations generally serve these purposes, separately or in combination:

- *To inform.* The speaker presents facts and analyzes data to help listeners understand the information. Such presentations cover the status of a current project, results of research or investigations, company changes, or performance quality of new systems and equipment.
- *To persuade.* The speaker presents information and urges listeners to take a specific action or reach a specific conclusion. Persuasive presentations involve sales proposals to potential customers, internal company proposals, or external grant proposals.
- *To instruct.* The speaker describes how to do a specific task. Presentations that offer instruction include training sessions for groups of employees and demonstrations for procedures or proper handling of equipment.

Advantages

Oral presentations have these advantages over written documents:

- The speaker can explain a procedure and demonstrate it at the same time.
- The speaker controls what is emphasized in the presentation and can keep listeners focused on specific topics.

- A speaker's personality can create enthusiasm in listeners during sales presentations and inspire confidence during informative presentations.
- Speakers can get immediate feedback from listeners and answer questions on the spot.
- Listeners can raise new issues immediately after hearing the presentation.
- If most of the listeners become involved in the presentation—asking questions, nodding agreement—those who may have had negative attitudes may be swept up in the group energy and become less negative.

Disadvantages

Oral presentations are not adequate substitutes for written documents, however, because they have these distinct disadvantages:

- Oral presentations are expensive because listeners may be away from their jobs for a longer time than it would take them to read a written document.
- Listeners, unlike readers, cannot select the topic they are most interested in or proceed through the material at their own pace.
- Listeners can be easily distracted from an oral presentation by outside noises, coughing, uncomfortable seating, and their own wandering thoughts.
- A poor speaker will annoy listeners, who then may reject the information.
- Spoken words vanish as soon as they are said, and listeners remember only a few major points.
- Time limitations require speakers to condense and simplify material, possibly omitting important details.
- Listeners have difficulty following complicated statistical data in oral presentations even with graphic aids.
- Audience size has to be limited for effective oral presentations unless expensive video/TV equipment is involved.

Types of Oral Presentations

An oral presentation on the job is usually in one of these four styles:

Memorized Speech. Memorization is useful primarily for short remarks, such as introducing a main speaker at a conference. Unless it is a dramatic reading, a memorized speech has little audience appeal. A speaker reciting a memorized speech is likely to develop a monotone that will soon have the listeners glassy-eyed with boredom. In addition, in such presentations, if the speaker is interrupted by a question, the entire presentation may disappear from his or her memory.

Written Manuscript. Speakers at large scientific and technical conferences often read from a written manuscript, particularly when reporting a research study or presenting complex technical data. The written manuscript enables the speaker to cover every detail and stay within a set time limit. Unfortunately, written manuscripts, like the memorized speeches, also may encourage speakers to develop a monotone. Usually, too, the speaker who reads from a manuscript has little eye contact or interaction with the audience. Listeners at scientific conferences frequently want copies of the full research report for reference later.

Impromptu Remarks. Impromptu remarks occur when, without warning, a person is asked to explain something at a meeting. Although others in the meeting realize that the person has had no chance to prepare, they still expect to hear specific information presented in an organized manner. You should anticipate being called on at meetings about projects you are involved with in any way. Always come prepared to explain projects or answer questions about specific topics under your authority. Bring notes, current cost figures, and other information that others may ask you for.

Extemporaneous Talk. Most oral presentations on the job are extemporaneous, that is, the speaker plans and rehearses the presentation and follows a written topic or sentence outline when speaking. Because this type of on-the-job oral presentation is the most common, the rest of this chapter focuses on preparing and delivering extemporaneous talks.

Organizing Oral Presentations

As you do with written documents, analyze your audience and their need for information, as explained in Chapter 2. Consider who your listeners are and what they expect to get from your talk. Oral presentations also must be as carefully organized as written documents are. Even though you are deeply involved in a project, do not rely on your memory alone to support your review of the information or your proposal. Instead, prepare an outline, as discussed in Chapter 3, that covers all the points you wish to make. An outline also will be easier to work with than a marked copy of a written report you plan to distribute because you will not have to fumble with sheets of paper.

Use either a topic or a sentence outline—whichever you feel most comfortable with and whichever provides enough information to stir your memory. Remember that topic outlines contain only key words, whereas sentence outlines include more facts:

Topic: Venus—Earth's twin

Sentence: Venus is nearly identical to Earth in size, mass, density, and gravity.

Some people prefer to put an outline on note cards, only one or two topics or sentences per card. Other speakers prefer to use an outline typed on sheets of paper because it is easier to glance ahead to see what is coming next. Oral presentations generally have three sections:

Introduction. Your introduction should establish (1) the purpose of the oral presentation, (2) why it is relevant to listeners, and (3) the major topics that will be covered. Since listeners, unlike readers, cannot look ahead to check on what is coming, they need a preview of main points in the opening so that they can anticipate and listen for the information relevant to their own work.

Data Sections. The main sections of your presentation are those that deal with specific facts, arranged in one of the organizational patterns discussed in Chapter 3. Remember these tips when presenting data orally:

- Include specific examples to reinforce your main points.
- Number items as you talk so that listeners know where problem number 2 ends and problem number 3 begins.
- Refer to visual illustrations and explain the content to be sure people understand what they are looking at.
- Cite authorities or give sources for your information, particularly if people in the audience have contributed information for your presentation.
- Simplify statistics for the presentation, and provide the full tables or formulas in handouts.

Conclusion. Your conclusion should summarize the main points and recommendations to fix them in listeners' minds. In addition, remind listeners about necessary future actions, upcoming deadlines, and other scheduled presentations on the subject. Also ask for questions so that listeners can clarify points immediately.

Figure 13-1 is a sentence outline of an oral presentation at a farm machinery manufacturing company. The speaker here is in the company's forecasting unit attached to the president's office, and a regular duty is to make oral presentations to division managers covering the same information that appears in the written reports to the president.

FORECAST REPORT OF FARM MACHINERY SALES IN 1990s

I. North American farm machinery sales are determined primarily by the farmers' demand for horsepower.
   A. Information in this report updates the previous studies of trends in farm machinery sales.
   B. Variations in yearly sales depend on five factors.
      1. Farmers' incomes control purchases in units and sizes.
      2. Grain export potential affects planted acreage.
      3. Planted acreage affects need for replacement machinery and unit size.
      4. Replacement machinery traditionally is larger than original machinery.
      5. Farm size affects replacement demand, and as average farm size increases, the trend toward larger-horsepower machines continues.
   C. Forecasts for the 1990s are based on trends in horsepower demand and comparisons of dollar sales, unit sales, dollar value, and average horsepower per unit for the segmented market.

II. To project 1990 trends, we must examine the factors that influence demand.
   A. Farmers' ability to purchase machinery is most easily measured by cash receipts.
      1. Farm cash receipts from major crops are related to unit sales of two-wheel- and four-wheel-drive machines.
      2. Farm cash receipts from livestock are related to sales of small, two-wheel-drive machines.
      3. Cash receipt growth in the 1990s should be slow but steady, and sales of larger-horsepower units should advance accordingly.
   B. Grain exports in the 1990s should continue their upward trend and increase demand for large machines.
   C. Because of projected growth in grain exports, crop acreage should increase at least 1% yearly through the 1990s.
   D. Units with more horsepower will replace smaller machines.
   E. Average farm size has been steadily increasing and will continue to, increasing farm machinery sales as labor is replaced.

III. Equipment sales will increase overall each year through the 1990s, with the largest increase in units of 140 horsepower and over.
   (Show pie graphs with unit sales of 1990 versus projected sales for 2000.)
   A. Unit sales of 40 horsepower and under should decline steadily, but dollar sales should remain stable.
   B. Unit sales of 40 to 99 horsepower should continue to decline along with dollar sales as farmers purchase replacement machines.
   C. Unit sales of 100 to 139 horsepower should decline slowly, with dollar sales remaining flat.

Figure 13-1 Sentence Outline for Oral Presentation

D. Unit sales of 140 horsepower and over should continue to grow, reaching 41% of the market by 1994. Dollar sales should increase steadily at 6% yearly.
E. Unit sales of four-wheel-drive machines will continue to increase 9.5% yearly.
(Show table of market shares by manufacturers of 140-horsepower and over machines.)

IV. North American farms will remain competitive and aggressive in world markets.
A. Farm consolidation will boost sales of larger-horsepower units.
B. Average size of machines will increase.
C. Sales of small- and medium-horsepower units will decrease.
D. Overall, annual growth in farm machinery should be 2.5% to 5.5% in the 1990s.
(Distribute sales tables and ask for questions.)

Figure 13-1 cont.

As the outline shows, the introduction (1) identifies the main topic—farm machinery sales projections, (2) indicates the purpose—to update previous reports, (3) identifies the five significant factors in sales, and (4) explains the basis for the forecast. Sections II and III cover the specific data: trends in the five factors and sales trends according to machinery horsepower. The conclusion summarizes the major points in the forecast.

Usually, speakers supplement oral presentations with handouts that highlight specific points and statistics or distribute full written reports on the subject of the oral presentation. Listeners can then check later on items they are most interested in, read sections not fully covered in the presentation, and confirm their memories of details they may be unsure of. Notice that the speaker using the outline in Figure 13-1 has inserted reminders of when to use graphic aids at the beginning and end of Section III, as well as a note about questions at the end. Such notes about when to use graphic aids, distribute handouts, or ask for questions are important reminders, because during a presentation, you may concentrate so much on the information involved that you forget the supplemental materials designed to aid listeners. Many speakers also put extrawide margins in their outlines so that they can jot down more facts in the margins as reminders.

Preparing for Oral Presentations

Because oral presentations usually have time restrictions, concentrate on the factors that your audience is most interested in, and prepare written materials covering other points that you do not have time to present orally. In addition, for oral presentations, you must check on these physical conditions, if possible:

- *Size of the group.* Is the group so large that you need a microphone to be heard in the back, and will those in the back be able to see your graphic aids? Remember that the larger the group, the more remote the listeners feel from the speaker and the less likely they are to interact and provide feedback.

- *Shape of the room.* Can all the listeners see the speaker, or is the room so narrow that those in the back are blocked from seeing either the speaker or the graphic aids? Does the speaker have to stand in front of a large glass wall through which listeners can watch other office activities? An inability to see or an opportunity to watch outside activities will prevent listeners from concentrating on the presentation.

- *Visual equipment.* Does the room contain built-in visual equipment, such as a movie screen or chalkboard, or will these have to be brought in, thus altering the shape of the speaking area? Will additional equipment in the room crowd the speaker or listeners? Are there enough electrical outlets, or will long extension cords be necessary, thereby creating walking obstacles for the speaker or those entering the room?

- *Seating arrangements.* Are the chairs bolted in place? Is there room for space between chairs, or will people be seated elbow-to-elbow in tight rows? If listeners want to take notes, is there room to write? People hate to be crowded into closely packed rows. An uncomfortable audience is an inattentive one.

- *Lighting.* Is there enough light so that listeners can see to take notes? Can the light be controlled to prevent the room from being so bright that people cannot see slides and transparencies clearly?

- *Temperature.* Is the room too hot or too cold? Heat is usually worse for an audience because it makes people sleepy, especially in the afternoon. If the temperature is too cold, people may concentrate on huddling in jackets and sweaters rather than listening to the presentation. A cool room, however, will heat up once people are seated because of body heat. Good air circulation also helps keep a room from being hot, stuffy, and uncomfortable.

You may not be able to control all these factors, but you can make some adjustments to enhance your presentation. You can adjust the temperature control or change the position of the lectern away from a distracting window, for example.

Delivering Oral Presentations

After planning your oral presentation to provide your listeners with the information they need and organizing your talk, you must deliver it. Fairly or unfairly, listeners tend to judge the usefulness of an oral presentation at least partly by the physical delivery of the speaker. Prepare your delivery as carefully as you do your outline.

Nerves

Everyone who speaks in front of a group is somewhat nervous. Fortunately, nerves create energy, and you need energy to deliver your talk. Prepare thoroughly so that you know you are giving your listeners the information they want. Picture yourself performing well beforehand, and remember that no one is there to "catch" you in a mistake. In fact, the audience usually does not know if you reverse the order of transparencies or discuss points out of order. Consider speaking in front of a group not as a dreadful ordeal that you must suffer through, but as an essential part of your job, and develop your ability to make an effective presentation.

Rehearsal

Rehearse your oral presentation aloud so that you know it thoroughly and it fits the allotted time. Rehearsing it will also help you find trouble spots where, for instance, you have not explained in enough detail or your chronology is out of order. If possible, practice in the room you intend to use so that you can check the speaking area, seating arrangements, and lighting.

Voice

If you are nervous, your voice may reveal it more than any mistakes you make. Do not race through your talk on one breath, forcing your audience to listen to an unintelligible monotone, but do not pause after every few words either, creating a stop-and-start style. Avoid a monotone by being interested in your own presentation. Speak clearly, and check the pronunciation of difficult terms or foreign words before your talk. Be sure also to pronounce words separately. For example, say "would have" instead of "wouldof."

Professional Image

When you speak in front of a group, you are in a sense "on stage." Dress conservatively but in the usual office style. Uncomfortable clothes that are too tight or that have unusual, eye-catching decorations are not appropriate for oral presentations. They will distract both you and your audience. Do not hang on the lectern, perch casually on the edge of a table, or sway back and forth on your feet. Be relaxed, but stand straight.

Gestures

Nervous gestures can distract your audience, and artificial "on purpose" gestures look awkward. If you feel clumsy, keep your hands still. Do not fiddle with your note cards, jewelry, tie, hair, or other objects.

Eye Contact

Create the impression that you are speaking to the individuals in the room by establishing eye contact. Although you may not be able to look at each person individually, glance around the room frequently. Do not stare at two or three people to the exclusion of all others, but do not sweep back and forth across the room like a surveillance camera. Eye contact indicates that you are interested in the listener's response, and a smile indicates that you are pleased to be giving this presentation. You should project both attitudes to your listeners.

Notes and Outline

Have your notes organized before arriving in the room, and number each card or page in case you drop them and have to put them in order quickly. If a lectern is available, place your notes on that. In any case, avoid waving your notes around or making them obvious to your listeners. Your listeners want to concentrate on what you are telling them, not on a sheaf of white cards flapping in the air.

Visual Equipment

Practice using any visual equipment before your talk so that you know where the control buttons are and how the machine works. If you know you write slowly on chalkboards and flip charts, put your terms or lists on them before the presentation so that you can simply refer to them at the appropriate time. If you are using videotape machines or cassettes, set the tapes to the exact start spot and know which key word or image signals the place where you want to shut them off. If any equipment breaks during your presentation, simply continue without it. Trying to manage repairs in the middle of a presentation will result in chaos. No matter what kind of visual equipment you use, be sure that all members of the audience can see the visual part of your presentation adequately.

Questions

One of the advantages of oral presentations is that listeners can ask questions immediately. Answer as many questions as your time limitations permit. If you do not know the answer to a question, say so and promise to find the answer later. You may ask if anyone else in the audience knows the answer, but do not put someone on the spot by calling his or her name without warning. If there is a reason you cannot answer a question, such as the information being classified, simply say so. If you are not sure what the questioner means, rephrase the question before trying to answer it by saying, "Are you asking whether . . . ?" In all cases, answer a question completely before going on to another.

Joining a Team Presentation

In some instances, such as sales presentations to potential clients or training sessions, a team of people makes the oral presentation, with each person on the team responsible for a unit that reflects that person's speciality. Although each person must plan and deliver an individual talk, some coordination is needed to be sure the overall presentation goes smoothly. Remember these guidelines:

- One person should be in charge of introducing the overall presentation to provide continuity. That person also may present the conclusion, wrapping up the team effort.
- One person should be in charge of all visual equipment so that, for example, two people do not bring overhead projectors.
- If the question period is left to the end of the team presentation, then one person should be in charge of moderating the questions that the team members answer in order to maintain an orderly atmosphere.
- All members should pay close attention to time limits so that they do not infringe on other team members' allotted time.
- All team members should listen to the other speakers as a courtesy and because they may need to refer to each other's talks in answering questions.

CHAPTER SUMMARY

This chapter discusses making oral presentations on the job. Remember:

- Oral presentations may have these purposes, separately or in combination: informing, persuading, instructing.
- Oral presentations have the advantages of allowing the speaker to create enthusiasm through a dynamic personality as well as allowing listeners to ask questions on the spot.
- Oral presentations have disadvantages in that listeners cannot study materials thoroughly and are easily distracted.
- Oral presentations may be (1) a memorized speech, (2) a written manuscript, (3) impromptu remarks, or (4) an extemporaneous talk.
- Speakers usually use topic or sentence outlines for oral presentations.
- If possible before a presentation, speakers should check the room layout, lighting, temperature, and visual equipment.
- Speakers should (1) control nerves by preparing thoroughly, (2) rehearse aloud, (3) speak clearly, (4) dress professionally, (5) keep hand gestures to a minimum, (6) establish eye contact, (7) number notes and outline sheets, (8) practice with video equipment, and (9) answer audience questions fully.
- For a team presentation, one speaker should handle the general introduction and conclusion.

SUPPLEMENTAL READINGS IN PART TWO

Calabrese, R. "Designing and Delivering Presentations and Workshops," *The Bulletin of the Association for Business Communication.*

Elsea, J. G. "Strategies for Effective Presentations," *Personnel Journal.*

Holcombe, M. W., and Stein, J. K. "How to Deliver Dynamic Presentations: Use Visuals for Impact," *Business Marketing.*

Martel, M. "Combating Speech Anxiety," *Public Relations Journal.*

Chapter 13 Exercises

1. Prepare a 3- or 4-minute oral presentation to give to your class based on one of these assignments from this book:

Exercises 6, 7, 8, or 9 in Chapter 6
Exercises 1 or 5 in Chapter 7
Exercises 5, 8, or 9 in Chapter 8
Any report assignment in Chapters 10 and 12

Prepare a sentence outline of your oral presentation to submit to your instructor.

2. Read two articles in Part II on a particular topic, such as computer documentation, document design, ethics, the job search, or style, and prepare a sentence or topic outline for a 2- or 3-minute oral presentation based on the articles for your fellow students. Do not report on each article separately. Identify the important points in the articles, and organize an oral presentation that groups these ideas by topic. Submit your outline to your instructor. If your instructor wants you to give the presentation to your class, prepare a graphic aid to illustrate a particular point in your talk.

APPENDIX

Guidelines for Grammar, Punctuation, and Mechanics

These guidelines for grammar, punctuation, and mechanical matters, such as using numbers, will help you revise your writing according to generally accepted conventions for correctness. These guidelines cover the most frequent questions writers have when they edit their final drafts. Your instructor may tell you to read about specific topics before revising. In addition, use this Appendix to check your writing before you submit it to your instructor.

Grammar

Dangling Modifiers

Dangling modifiers are verbal phrases, prepositional phrases, or dependent clauses that do not refer to a subject in the sentence in which they occur. These modifiers are most often at the beginning of a sentence, but they may also appear at the end. Correct by rewriting the sentence to include the subject of the modifier.

Incorrect: Realizing the connections between neutron stars, pulsars, and supernovas, the explanation of the birth and death of stars is complete. (The writer does not indicate *who* realized the connections.)

Correct: Realizing the connection between neutron stars, pulsars, and supernovas, many astronomers believe that the explanation of the birth and death of stars is complete.

Incorrect: To obtain a slender blade, a cylindrical flint core was chipped into long slivers with a hammerstone. (The writer does not indicate *who* worked to obtain the blade.)

Correct: To obtain a slender blade, Cro-Magnon man chipped a cylindrical flint core into slivers with a hammerstone.

Incorrect: Born with cerebral palsy, only minimal mobility was possible. (The writer does not indicate *who* was born with cerebral palsy.)

Correct: Born with cerebral palsy, the child had only minimal mobility.

Incorrect: Their sensitivity to prostaglandins remained substantially lowered, hours after leaving a smoke-filled room. (The writer does not indicate *who* or *what* left the room.)

Correct: Their sensitivity to prostaglandins remained substantially lowered, hours after the nonsmokers left a smoke-filled room.

Misplaced Modifiers

Misplaced modifiers are words, phrases, or clauses that do not refer logically to the nearest word in the sentence in which they appear. Correct by rewriting the sentence to place the modifier next to the word to which it refers.

Incorrect: The financial analysts presented statistics to their clients that showed net margins were twice the industry average. (The phrase about *net margins* does not modify *clients.*)

Correct: The financial analysts presented their clients with statistics that showed net margins were twice the industry average.

Squinting Modifiers

Squinting modifiers are words or phrases that could logically refer to either a preceding or a following word in the sentence in which they appear. Correct by rewriting the sentence so that the modifier refers to only one word in the sentence.

Incorrect: Physicians who use a nuclear magnetic resonance machine frequently can identify stroke damage in older patients easily. (The word *frequently* could refer to *use* or *identify.*)

Correct: Physicians who frequently use a nuclear magnetic resonance machine can identify stroke damage in older patients easily.

Correct: Physicians who use a nuclear magnetic resonance machine can identify stroke damage in older patients frequently and easily.

Parallel Construction

Elements that are equal in a sentence should be expressed in the same grammatical form.

Incorrect: With professional care, bulimia can be treated and is controllable. (The words *treated* and *controllable* are not parallel.)

Correct: With professional care, bulimia can be treated and controlled.

Incorrect: The existence of two types of Neanderthal tool kits indicates that one group was engaged in scraping hides, while the other group carved wood. (The verbal phrases after the word *group* are not parallel.)

Correct: The existence of two types of Neanderthal tool kits indicates that one group scraped hides, while the other group carved wood.

Incorrect: The social worker counseled the family about preparing meals, cleaning the house, and gave advice about childcare. (The list of actions should be in parallel phrases.)

Correct: The social worker counseled the family about preparing meals, cleaning the house, and caring for children.

Elements linked by correlative conjunctions (*either . . . or, neither . . . nor, not only . . . but also*) also should be in parallel structure.

Incorrect: Such symptoms as rocking and staring vacantly are seen not only in monkeys that are deprived of their children but also when something frightens mentally disturbed children. (The phrases following *not only* and *but also* are not parallel.)

Correct: Such symptoms as rocking and staring vacantly are seen not only in monkeys that are deprived of their children but also in mentally disturbed children who are frightened.

Pronoun Agreement

A pronoun must agree in number, person, and gender with the noun or pronoun to which it refers.

Incorrect: Each firefighter must record their use of equipment. (The noun *firefighter* is singular, and the pronoun *their* is plural.)

Correct: Each firefighter must record his or her use of equipment.

Correct: All firefighters must record their use of equipment.

Incorrect: Everyone needs carbohydrates in their diet. (*Everyone* is singular, and *their* is plural.)

Correct: Everyone needs carbohydrates in the diet.

Correct: Everyone needs to eat carbohydrates.

Correct: Everyone needs carbohydrates in his or her diet.

Incorrect: The sales department explained their training methods to a group of college students. (*Sales department* is singular and requires an *it.* Correct by clarifying *who* did the explaining.)

Correct: The sales manager explained the department training methods to a group of college students.

Correct: The sales manager explained her training methods to a group of college students.

Incorrect: The International Microbiologists Association met at their traditional site. (*Association* is singular and requires an *it.*)

Correct: The International Microbiologists Association met at its traditional site.

Pronoun Reference

A pronoun must clearly refer to only one antecedent.

Incorrect: Biologists have shown that all living organisms depend on two kinds of molecules—amino acids and nucleotides. They are the building blocks of life. (*They* could refer to either *amino acids, nucleotides,* or both.)

Correct: Biologists have shown that all living organisms depend on two kinds of molecules—amino acids and nucleotides. Both are the building blocks of life.

Incorrect: Present plans call for a new parking deck, a new entry area, and an addition to the parking garage. It will add $810,000 to the cost. (*It* could refer to any of the new items.)

Correct: Present plans call for a new parking deck, a new entry area, and an addition to the parking garage. The parking deck will add $810,000 to the cost.

Reflexive Pronouns

A *reflexive pronoun* (ending in *-self* or *-selves*) must refer to the subject of the sentence when the subject also receives the action in the sentence. (Example: She cut herself when the camera lens cracked.) A reflexive pronoun cannot serve instead of *I* or *me* as a subject or an object in a sentence.

Incorrect: The governor presented the science award to Dr. Yasmin Rashid and myself. (*Myself* cannot serve as the indirect object.)

Correct: The governor presented the science award to Dr. Yasmin Rashid and me.

Incorrect: Mark Burnwood, Christina Hayward, and myself conducted the experiment. (*Myself* cannot function as the subject of the sentence.)

Correct: Mark Burnwood, Christina Hayward, and I conducted the experiment.

Reflexive pronouns are also used to make the antecedent more emphatic:

The patient himself asked for another medication.
Venus itself is covered by a heavy layer of carbon dioxide.

Sentence Faults

Comma Splices

A *comma splice* results when the writer incorrectly joins two independent clauses with a comma. Correct by (1) placing a semicolon between the clauses, (2) adding a coordinating conjunction after the comma, (3) rewriting the sentence, or (4) creating two sentences.

Incorrect: The zinc coating on galvanized steel gums up a welding gun's electrode, resistance welding, therefore, is not ideal for the steel increasingly used in autos today.

Correct: The zinc coating on galvanized steel gums up a welding gun's electrode; resistance welding, therefore, is not ideal for the steel increasingly used in autos today.

Correct: The zinc coating on galvanized steel gums up a welding gun's electrode, so resistance welding, therefore, is not ideal for the steel increasingly used in autos today.

Correct: Because the zinc coating on galvanized steel gums up a welding gun's electrode, resistance welding is not ideal for the steel increasingly used in autos today.

Correct: The zinc coating on galvanized steel gums up a welding gun's electrode. Resistance welding, therefore, is not ideal for the steel increasingly used in autos today.

Fragments

A *sentence fragment* is an incomplete sentence because it lacks a subject or a verb or both. Correct by writing a full sentence or by adding the fragment to another sentence.

Incorrect: As the universe expands and the galaxies fly further apart, the force of gravity is decreasing everywhere. According to one imaginative theory. (The phrase beginning with *According* is a fragment.)

Correct: According to one imaginative theory, as the universe expands and the galaxies fly further apart, the force of gravity is decreasing everywhere.

Incorrect: Apes are afraid of water when they cannot see the stream bottom. Which prevents them from entering the water and crossing a stream more than 1 ft deep. (The phrase beginning with *which* is a fragment.)

Correct: Apes are afraid of water when they cannot see the stream bottom. This fear prevents them from entering the water and crossing a stream more than 1 ft deep.

Run-on Sentences

A *run-on sentence* occurs when a writer links two or more sentences together without punctuation between them. Correct by placing a semicolon between the sentences or by writing two separate sentences.

Incorrect: The galaxy is flattened by its rotating motion into the shape of a disk most of the stars in the galaxy are in this disk.

Correct: The galaxy is flattened by its rotating motion into the shape of a disk; most of the stars in the galaxy are in this disk.

Correct: The galaxy is flattened by its rotating motion into the shape of a disk. Most of the stars in the galaxy are in this disk.

Subject/Verb Agreement

The verb in a sentence must agree with its subject in person and number. Correct by rewriting.

Incorrect: The report of a joint team of Canadian and American geologists suggest that some dormant volcanos on the West Coast may be reawakening. (The subject is *report,* which requires a singular verb.)

Correct: The report of a joint team of Canadian and American geologists suggests that some dormant volcanos on the West Coast may be reawakening.

Incorrect: The high number of experiments that failed were disappointing. (The subject is *number,* which requires a singular verb.)

Correct: The high number of experiments that failed was disappointing.

Incorrect: Sixteen inches are the deepest we can drill. (The subject is a single unit and requires a singular verb.)

Correct: Sixteen inches is the deepest we can drill.

When compound subjects are joined by *and,* the verb is plural.

Computer-security techniques and plans for evading hackers require large expenditures.

When one of the compound subjects is plural and one is singular, the verb agrees with the nearest subject.

Neither the split casings nor the cracked layer of plywood was to blame.

When a compound subject is preceded by *each* or *every,* the verb is singular.

Every case aide and social worker is scheduled for a training session.

When a compound subject is considered a single unit or person, the verb is singular.

The vice president and guiding force of the company is meeting with the Nuclear Regulatory Commission.

Punctuation

Apostrophe

An apostrophe shows possession or marks the omission of letters in a word or in dates.

The engineer's analysis of the city's water system shows a high pollution danger from raw sewage.

The extent of the outbreak of measles wasn't known until all area hospitals' medical records were coordinated with those of private physicians.

Chemist Charles Frakes' experiments were funded by the National Institute of Health throughout the '70s.

Do not confuse *its* (possessive) with *it's* (contraction of *it is*).

The animal pricked up its ears before feeding.

Professors Jones, Higgins, and Carlton are examining the ancient terrain at Casper Mountain. It's located on a geologic fault, and the professors hope to study its changes over time.

Colon

The colon introduces explanations or lists. An independent clause must precede the colon.

The Olympic gymnastic team used three brands of equipment: Acme, Dakota, and Shelby.

Do not place a colon directly after a verb.

Incorrect: The environmentalists reported finding: decreased oxygen, pollution-tolerant sludgeworms, and high bacteria levels in the lake. (The colon does not have an independent clause preceding it.)

Correct: The environmentalists reported finding decreased oxygen, pollution-tolerant sludgeworms, and high bacteria levels in the lake.

Comma

To Link. The comma links independent clauses joined by a coordinating conjunction (*and, but, or, for, so, yet,* and *nor*).

The foreman stopped the production line, but the damage was done.

Rheumatic fever is controllable once it has been diagnosed, yet continued treatment is often needed for years.

To Enclose. The comma encloses parenthetical information, simple definitions, or interrupting expressions in a sentence.

Gloria Anderson, our best technician, submitted the winning suggestion.

The research team, of course, will return to the site after the monsoon season.

The condition stems from the body's inability to produce enough hemoglobin, a component of red blood cells, to carry oxygen to all body tissues.

If the passage is essential to the sentence (restrictive), do not enclose it in commas.

Acme shipped the toys with red tags to the African relief center. (The phrase *with red tags* is essential for identifying which toys were shipped.)

To Separate. The comma separates introductory phrases or clauses from the rest of the sentence and also separates items in a list.

> Accurate to within 30 seconds a year, the electronic quartz watch is also water resistant.
>
> Because so many patients experienced side effects, the FDA refused to approve the drug.
>
> To calculate the three points, add the static head, friction losses, velocity head, and minor losses. (Retain the final comma in a list for clarity.)

The comma also separates elements in addresses and dates.

> The shipment went to Hong Kong on May 29, 1989, but the transfer forms went to Marjorie Howard, District Manager, Transworld Exports, 350 Michigan Avenue, Chicago, Illinois 60616.

Dash

The dash sets off words or phrases that interrupt a sentence or that indicate sharp emphasis. The dash also encloses simple definitions. The dash is a more emphatic punctuation mark than parentheses or commas.

> The foundation has only one problem—no funds.
>
> The majority of those surveyed—those who were willing to fill out the questionnaire—named television parents in nuclear families as ideal mother and father images.

Exclamation Point

The exclamation point indicates strong emotion or sharp commands. It is not often used in technical writing, except for warnings or cautions.

> Insert the cable into the double outlet.
>
> Warning! Do not allow water to touch the cable.

Hyphen

The hyphen divides a word at the end of a typed line and also forms compound words. If in doubt about where to divide a word or whether to use a hyphen in a compound word, consult a dictionary.

Scientists at the state-sponsored research laboratory are using cross-pollination to produce hybrid grains.

Modifiers of two or three words require hyphens when they precede a noun.

Tycho Brahe, sixteenth-century astronomer, measured the night sky with giant quadrants.

The microwave-antenna system records the differing temperatures coming from the Earth and the cooler sky.

Parentheses

Parentheses enclose (1) nonessential information in a sentence or (2) letters and figures that enumerate items in a list.

The striped lobsters (along with solid blue varieties) have been bred from rare red and blue lobsters that occur in nature.

Early television content analyses (from 1950 to 1980) reported that (1) men appeared more often than women, (2) men were more active problem solvers than women, (3) men were older than women, and (4) men were more violent than women.

Parentheses also enclose simple definitions.

The patient suffered from hereditary trichromatic deuteranomaly (red-green color blindness).

Quotation Marks

Quotation marks enclose (1) direct quotations and (2) titles of journal articles, book chapters, reports, songs, poems, and individual episodes of radio and television programs.

"Planets are similar to giant petri dishes," stated astrophysicist Miranda Caliban at the Gatewood Research Institute.

"Combat Readiness: Naval Air vs. Air Force" (journal article)

"Science and Politics of Human Differences" (book chapter)

"Feasibility Study of Intrastate Expansion" (report)

"Kariba: The Lake That Made a Dent" (television episode)

Place commas and periods inside quotations marks. Place colons and semicolons outside quotation marks. See also "Italics" under "Mechanics" in this Appendix.

Question Mark

The question mark follows direct questions but not indirect questions.

> The colonel asked, "Where are the orders for troop deployment?"
>
> The colonel asked whether the orders for troop deployment had arrived.

Place a question mark inside quotation marks if the quotation is a question, as shown above. If the quotation is not a question but is included in a question, place the question mark outside the quotation marks.

> Why did the chemist state "The research is unnecessary"?

Semicolon

The semicolon links two independent clauses without a coordinating conjunction or other punctuation.

> The defense lawyer suggested the arsenic could have originated in the embalming fluid; Bateman dismissed this theory.
>
> All the cabinets are wood; however, the countertops are of artificial marble.

The semicolon also separates items in a series if the items contain internal commas.

> The most desirable experts for the federal environmental task force are Sophia Timmons, Chief Biologist, Ohio Water Commission; Clinton Buchanan, Professor of Microbiology, Central Missouri State University; and Hugo Wysoki, Director of the California Wildlife Research Institute, San Jose.

Slash

The slash separates choices and parts of dates and numbers.

> For information about the property tax deduction, call 555-6323/7697. (The reader can choose between two telephone numbers.)
>
> The crew began digging on 9/28/89 and increased the drill pressure ¾ psi every hour.

Mechanics

Acronyms and Initialisms

An *acronym* is an abbreviation formed from the first letters of the words in a name or phrase. The acronym is written in all capitals with no periods and is pronounced as a word.

> The crew gathered in the NASA conference room and used the WATS line to call the engineering team.

Some acronyms eventually become words in themselves. For example, *radar* was once an acronym for *radio detecting and ranging*. Now the word *radar* is a noun meaning a system for sensing the presence of objects.

Initialisms are also abbreviations formed from the first letters of the words in a name or phrase; here each letter is pronounced separately. Some initialisms are written in all capital letters; others are written in capital and lowercase letters. Some are written with periods; some are not. If you are uncertain about the conventional form for an initialism, consult a dictionary.

> The technician increased the heat by 4 Btu.
>
> The C.P.A. took the tax returns to the IRS office.

When using acronyms or initialisms, spell out the full term the first time it appears in the text and place the acronym or initialism in parentheses immediately after to be sure your readers understand it. In subsequent references, use the acronym or initialism alone.

> The Federal Communication Commission (FCC) has authorized several thousand licenses for low-power television (LPTV) stations.

In a few cases, the acronym or initialism may be so well known by your readers that you can use it without stating the full term first.

> The experiment took place at 10:15 A.M. with the application of DDT at specified intervals.

Plurals of acronyms and initialisms are formed by adding an *s* without an apostrophe, such as YMCAs, MBAs, COLAs, and EKGs.

Brackets

Brackets enclose words that are inserted into quotations by writers or editors. The inserted words are intended to add information to the quoted material.

"Thank you for the splendid photographic service [to the Navy, 1976–1982] and the very comprehensive collection donated to the library," said Admiral Allen Sumner as he dedicated the new library wing.

Brackets also enclose a phrase or word inserted into a quotation to substitute for a longer, more complicated phrase or clause.

"The surgical procedure," explained Dr. Moreno, "allows [the kidneys] to cleanse the blood and maintain a healthy chemical balance."

Capital Letters

Capital letters mark the first word of a sentence and the first word of a quotation. A full sentence after a colon and full sentences in a numbered list may also begin with capital letters if the writer wishes to emphasize them.

Many of the photographs were close-ups.

Pilots are concerned about mountain wave turbulence and want a major change: Higher altitude clearance is essential.

Capital letters also mark proper names and initials of people and objects.

Dr. Jonathan L. Hazzard

Ford Tempo

USS Fairfax

Capital letters also mark nationalities, religions, and linguistic groups.

British

Lutheran

American Indian

Afro-American

Capital letters also mark (1) place names, (2) geographical and astronomical areas, (3) organizations, (4) events, (5) historical times, (6) software, and (7) some calendar designations.

San Diego, California (place)

Great Plains (geographic area)

League of Women Voters (organization)

Robinson Helicopter Company (company)

World Series (event)
Paleocene Epoch (historical time)
WordPerfect (software)
Tuesday, July 12 (calendar designations)

Capital letters also mark brand names but not generic names.

Tylenol (acetaminophen)

Capital letters mark the first, last, and main words in the titles of (1) books, (2) articles, (3) films, (4) television and radio programs, (5) music, and (6) art objects. Do not capitalize short conjunctions, prepositions, or articles (*a, an, the*) unless they begin the title or follow a colon. *Note:* If you are preparing a reference list, see Chapter 11 for the correct format.

The Panda's Thumb (book)
"Information Base: On Line with CompuServe" (article)
Gunfight at Smokey Ridge (film)
"The Search for the Dinosaurs" (television program)
Concerto in D Major, Opus 77, for Violin and Orchestra (music)
The Third of May (art)

Companies and institutions often use capital letters in specific circumstances not covered in grammar handbooks. If in doubt, check the company or organization style guide.

Ellipsis

An *ellipsis* (three spaced periods) indicates an omission of one or more words in a quotation. When the omission is at the end of a sentence or includes an intervening sentence, the ellipsis follows the final period.

General Donald Mills stated, "It is appalling that those remarks . . . by someone without technical, military, or intelligence credentials should be published."

At the fund-raising banquet, Chief of Staff Audrey Spaulding commented, "All of the volunteers—more than 600—made this evening very special. . . ."

Italics

Italics set off or emphasize specific words or phrases. If italic type is not available, underline the words or phrases. Italics set off titles of (1) books, (2) periodicals, (3) plays, films, television and radio programs, (4) long musical works, (5) complete art objects, and (6) ships, planes, trains, and aircraft. The Bible and its books are not italicized.

The Cell in Development and Inheritance (book)
Chicago Tribune (newspaper)
Scientific American (journal)
Inherit the Wind (play)
Blade Runner (film)
60 Minutes (television program)
Carmen (long musical work)
The Thinker (art object)
City of New Orleans (train)
Apollo III (spacecraft)

Italics also set off words and phrases discussed as words and words and phrases in foreign languages or Latin. Scientific terms for plants and animals also are italicized.

The term *mycobacterium* refers to any one of several rod-shaped aerobic bacteria.

These old fossils may be closer to the African *Australopithecus* than to *Homo.*

See also "Quotation Marks" under "Punctuation" in this Appendix.

Measurements

Measurements of physical quantities, sizes, and temperatures are expressed in figures. Use abbreviations for measurements only when also using figures and when your readers are certain to understand them. If the abbreviations are not common, identify them by spelling out the term on first use and placing the abbreviation in parentheses immediately after. Do not use periods with measurement abbreviations unless the abbreviation might be confused with a full word. For example, *inch* abbreviated needs a period to avoid confusion with the word *in.* Use a hyphen when a measurement functions as a compound adjective.

Workers used 14-in. pipes at the site.

The designer wanted a 4-oz decanter on the sideboard.

Ship 13 lb of feed. (Do not add *s* to make abbreviations plural.)

Ship several pounds of feed. (Do not use abbreviations if no numbers are involved.)

The pressure rose to 12 pounds per square inch (psi).

If the measurement involves two numbers, spell out the first or the shorter word.

The supervisor called for three 12-in. rods.

Here are some common abbreviations for measurements:

C	centigrade	kw	kilowatt
cc	cubic centimeter	l	liter
cm	centimeter	lb	pound
cps	cycles per second	m	meter
cu ft	cubic foot	mg	milligram
dm	decimeter	ml	milliliter
F	Fahrenheit	mm	millimeter
fl oz	fluid ounce	mph	miles per hour
fpm	feet per minute	oz	ounce
ft	foot	psf	pounds per square foot
g	gram	psi	pounds per square inch
gal	gallon	qt	quart
gpm	gallons per minute	rpm	revolutions per minute
hp	horsepower	rps	revolutions per second
in.	inch	sq	square
kg	kilogram	t	ton
km	kilometer	yd	yard

Numbers

Use figures for all numbers over ten. If a document contains many numbers, use figures for all amounts, even those under ten, so that readers do not overlook them.

The testing procedure required 27 test tubes, 2 covers, and 16 sets of gloves.

Do not begin a sentence with a figure. Rewrite to place the number later in the sentence.

Incorrect: 42 tests were scheduled.
Correct: We scheduled 42 tests.

Write very large numbers in figures and words.

2.3 billion people
$16.5 million

Use figures for references to (1) money, (2) temperatures, (3) decimals and fractions, (4) percentages, (5) addresses, and (6) book parts.

$4230	$.75
72°F	0°C
3.67	3/4
12.4%	80%

2555 N. 12th Street
Chapter 7, page 178

Use figures for times of day. Because most people are never certain about P.M. and A.M. in connection with 12 o'clock, indicate noon or midnight in parentheses.

4:15 P.M.
12:00 (noon)
12:00 (midnight)

Add an *s* without an apostrophe to make numbers plural.

4s 1980s 33s

Symbols and Equations

Symbols should appear in parenthetical statistical information but not in the text. Use symbols with figures, but write out the word if you are not using figures.

Incorrect: The *M* was 2.64.
Correct: The mean was 26.

Incorrect: The sample population (Number = 450) was selected from children in the second grade.

Correct: The sample population ($N = 450$) was selected from children in the second grade.

Incorrect: We found a high % of error.

Correct: We found a high percentage of error.

Do not try to create symbols on a typewriter or computer by combining overlapping characters. Write symbols not on the keyboard by hand in ink.

$$^{2}\ (N = 916) = 142.64 \qquad p < .001$$

Place simple or short equations in the text by including them in a sentence.

The equation for the required rectangle is $y = a + 2b - x$.

To display (set off on a separate line) equations, start them on a new line after the text and double space twice above and below the equation. Displayed equations must be numbered for reference. The reference number appears in parentheses flush with the right margin. In the text, use the reference number.

$$F\ (9{,}740) = 2.06 \qquad p < .05 \tag{1}$$

If the equation is too long for one line, break it before an operational sign. Place a space between the elements in an equation as if they were words.

$$3v - 5b + 66x$$

$$+\ 2x - y = 11.5$$

Very difficult equations that require handwritten symbols should always be displayed for clarity. Highly technical symbols and long equations in documents intended for multiple readers are best placed in appendixes rather than in the main body of the document.

Part Two

Technical Writing: Advice from the Workplace

How to Avoid Costly Proofreading Errors

Carolyn Boccella Bagin and Jo Van Doren

Proofreading errors cost businesses money. This article tells you how to produce error-free copy. Try finding all the mistakes in the sample copy in the article, and check your results against the corrected version.

ARE PROOFREADING ERRORS COSTING YOU MONEY? Some organizations have sad stories to tell about the price of overlooked mistakes.

- One insurance firm reported that an employee mailed a check for $2,200 as a settlement for a dental claim. Payment of only $22.00 had been authorized.
- One executive wasted $3 million by not catching a hyphen error when proofreading a business letter. In originally dictating the letter to his secretary, the executive said, "We want 1,000-foot-long radium bars. Send three in cases." The order was typed, "We want 1,000 foot-long radium bars."
- A magazine accidentally ran a cake recipe in which "3/4 cup" was printed as "1/4 cup." Irate readers sent complaint letters and cancelled their subscriptions.

Your company's image could be marred by unfortunate mistakes that find their way into the documents you produce. Developing good proofreading techniques and systems can save time, money, and embarrassment.

HOW CAN YOU PRODUCE ERROR-FREE COPY?

- Never proofread your own copy by yourself. You'll tire of looking at the document in its different stages of development and you'll miss new errors. (If you must proof your own copy, make a line screen for yourself or roll the paper back in the typewriter so that you view only one line at a time. This will reduce your tendency to skim your material.)
- Read everything in the copy straight through from beginning to end: titles, subtitles, sentences, punctuation, capitalizations, indented items, and page numbers.
- Read your copy backward to catch spelling errors. Reading sentences out of sequence lets you concentrate on individual words.
- Consider having proofreaders initial the copy they check. You might find that your documents will have fewer errors.
- If you have a helper to proof numbers that are in columns, read the figures aloud to your partner, and have your partner mark the corrections and changes on the copy being proofread.

- If time allows, put your material aside for a short break. Proofreading can quickly turn into reading if your document is long. After a break, reread the last few lines to refresh your memory.
- Read the pages of a document out of order. Changing the sequence will help you to review each page as a unit.
- List the errors you spot over a month's time. You may find patterns that will catch your attention when you proofread your next document.
- If you can, alter your routine. Don't proofread at the same time every day. Varying your schedule will help you approach your task with a keener eye.
- Not everyone knows and uses traditional proofreading marks. But a simple marking system should be legible and understandable to you and to anyone else working on the copy.

Where Do Errors Usually Hide?

- Mistakes tend to cluster. If you find one typo, look carefully for another one nearby.
- Inspect the beginning of pages, paragraphs, and sections. Some people tend to skim these crucial spots.
- Beware of changes in typeface—especially in headings or titles. If you change to all uppercase letters, italics, boldface, or underlined copy, read those sections again.
- Make sure your titles, subtitles, and page numbers match those in the table of contents.
- Read sequential material carefully. Look for duplications in page numbers or in lettered items in lists or outlines.
- Double-check references such as, "see the chart below." Several drafts later, the chart may be pages away from its original place.
- Examine numbers and totals. Recheck all calculations and look for misplaced commas and decimal points.
- Scrutinize features that come in sets, such as brackets, parentheses, quotation marks, and dashes.

Try Your Hand (and Eye) at This Test

Mark the mistakes and check your corrections with our marked copy on page 462.

It is improtant to look for certain item when proofing a report , letter, or othr document. Aside from spelling errors, the prooffer should check for deviations in format consitent use of punctuation, consistent use of capitol letters, undefined acronyms and correctpage numbers listed in the the Table of contents.

After checking a typed draft againts the original manuscript one should also read the draft for aukward phrasing, syntactical errors, and subject/

verb agreement and grammatical mistakes. paralell structures should be used im listings headed by bullets or numbers: ie, if one item starts with the phrase ''to understand the others should start with to plus a verb.

The final step in proofing involves review of the overall appearance on the document. Are the characrters all printed clearly Are all the pages there? Are the pages free of stray marks ? Is the graphics done? Are bullets filled in? All of the above items effect the appearance of the the document and determine whether the document has the desired effect on the reader.

How Did You Do?

Check your markup against ours. (We only show corrections in the copy and not the margin notes that most proofreaders typically use.)

It is improtant to look for certain item when proofing a report; letter, or othr document. Aside from spelling errors, the proofer should check for deviations in format consitent use of punctuation, consistent use of capitol letters, undefined acronyms and correctpage numbers listed in the the Table of contents.

After checking a typed draft againts the original manuscript one should also read the draft for aukward phrasing, syntactical errors, and subject/ verb agreement and grammatical mistakes. paralell structures should be used im listings headed by bullets or numbers: ie, if one item starts with the phrase ''to understand the others should start with to plus a verb.

The final step in proofing involves review of the overall appearance on the document. Are the characrters all printed clearly Are all the pages there? Are the pages free of stray marks ? Is the graphics done? Are bullets filled in? All of the above items effect the appearance of the the document and determine whether the document has the desired effect on the reader.

Writing Visually: Design Considerations in Technical Publications

Philippa J. Benson

Readers find documents most usable when the visual design guides them to the information they need. This article provides guidelines for using visual devices, such as typeface, headings, and graphics, to produce a document your readers can use easily.

WRITERS AND GRAPHIC DESIGNERS are working more closely together as each learns more about how the other approaches the task of planning and structuring a document. The field of document design has developed from this synthesis of the processes of writing and design: writers and designers have come to agree that the most effective documents are those that use both words and design to reveal and reinforce the structure of information in a text.

The synthesis of writing and design is supported not only by the developing bond between writers and designers, but also by the conclusions of applied research. Research in cognitive psychology, instructional design, reading, and graphic design indicates that documents are most usable when the information in them is apparent both visually and syntactically. In particular, the results of usability tests of documents are providing concrete evidence that readers find documents most usable when the visual aspects of a document support the hierarchy of the information in a text.[1–5]

The synthesis of writing and design has also been evolving because computers increasingly put control over typography and page format in the writer's domain. Although many text-formatting programs limit the design decisions that we can make, they do allow us to simultaneously write and design information.

The increasing use of computers has also led to the development of a new concern for document designers: online information. Designers of online documentation should be concerned with many of the same issues as page designers —that is, the legibility and accessibility of information.

UNDERSTANDING CONSTRAINTS ON DESIGNING A DOCUMENT

How effectively writers and designers can work together depends not only on how well they can communicate but also on the practical constraints of their working environment. Before you begin planning a document, find out about the limits of your schedule and budget, and the flexibility of the method you'll be using to produce the end product (printer, typesetter, printing process). Your schedule, budget, and method of printing may limit what you can do with the design of your document. Constraints may affect these aspects of your book:

- Page count, page size
- Packaging (binding, cover, separators, etc.)
- Typeface, type size
- Range of available fonts
- Use of color and graphics
- Number of heading levels available
- Kinds of graphics you can use (tables, flowcharts, illustrations)

If you find that the constraints depart from the principles of good document design, you should consider trying to change them. For example, you may be told you must use a type size of smaller than 8 or 9 points or set a sizable portion of text in uppercase. The research cited in this article provides evidence you can use to convince your managers to change the constraints that would make your document less usable. Once you know your constraints, you can choose design options that are compatible with them.

Designing for Your Audience(s) and Their Tasks

A primary step in planning a document is understanding the audience for the document and their purpose in reading it. For example, if you are writing a book about how to service a software product on a computer system, you may have several audiences. Some readers may simply want to know the procedures they need to follow to service the product, while other readers may want to know how these procedures affect the computer system as a whole.

Research indicates that document designers should consider how readers will use the document. In a document with multiple audiences, each with different purposes and tasks in mind, research suggests that you should design the document so that members of each audience can quickly find the specific information they need.[6–8]

Depending on the reading situation, readers come to a technical document with assumptions and expectations about its structure, in terms of both its language and its design. For example, readers of a tutorial manual may expect to see step-by-step procedural instructions, while readers of a reference manual may expect to see lots of definition lists. The tutorial or reference manual you design, however, may be very different from what the reader initially expects. The design options you use can affect the assumptions and expectations readers have when they begin using a book and, in doing so, can affect the way the readers understand and use it. To design a document well, you need to imagine what linguistic and visual organizers will help readers understand how the text is structured.[9,10]

Here are some guidelines you can use to help readers find, understand, use, and remember the information in a document:

Guideline: Provide Readers with Road Maps Through the Document

Think about how you can design the tables of contents, glossaries, indexes, appendixes, and other materials that support the body of text so that readers can quickly find the information they need. One type of road map is a separator between each section of a document, marked with icons, followed by text explaining the information that will be covered in that section.

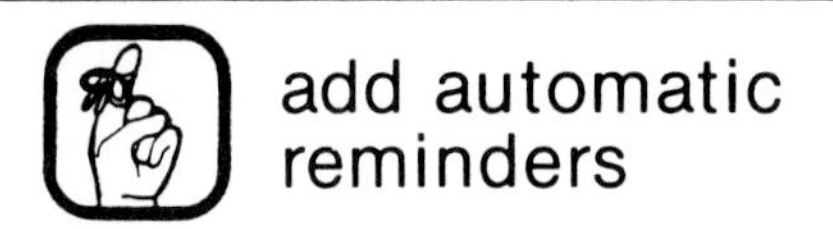

One type of roadmap is a section heading with an icon.

Another type of road map is a table that directs readers to specific information. Here is an example of a table road map:

If you want to	*Go to*
Install product x	Section A
Use product x	Section B
Service product x	Section C

Guideline: Make Your Document Aesthetically Pleasing

Many readers are motivated to use a document that is aesthetically pleasing. Documents should be attractively bound and packaged, have well-balanced page layouts, contain some graphics and illustrations to break up the text, and be legible.

A good and a poor page layout.

Guideline: Be Consistent in the Visual Formats and Prose Styles You Use

Readers can use a document more effectively when it contains consistent visual patterns, such as use of boldface or italic type, and a consistent linguistic style, such as use of the active voice or direct address. Readers use these patterns and styles to help them structure, comprehend, and remember the relationships between different levels and types of information in a text.

Designing the Visual Structure of Your Document

You can think of the structure of a document on several levels: as a whole, as a set of interrelated parts (chapters/sections), as a series of pages, and as text. Because the audience, purpose, and information for each document are unique, the relationships between these levels are different for each document you produce. Your challenge as a document designer is to choose design options that will support the document on all levels.

Research on design options suggests guidelines on how to graphically reinforce the structure of a text and, therefore, how to improve its usability. These guidelines cover

- Serif vs. Sans-Serif Typeface
- Lowercase vs. Uppercase Type
- Type Weight and Font
- Type Size, Line Length, and Leading
- Justification of Lines, Size of Margins, Use of White Space
- Headings
- Graphs, Tables, Flowcharts, and Diagrams

Serif vs. Sans-Serif Typeface. Guideline: Choose between a serif and sans-serif typeface according to the visual tone of the document you want. Note: Use a type size 10 points or larger, 2 or more points of leading between lines, and a moderate line length.

Serif typefaces have small strokes added to the edges of each letter. Serifs reinforce the horizontal flow of letters and make each letter distinct. Letters in sans-serif typefaces are less distinct from each other not only because they lack the extra, ornamental strokes of serif type, but also because the widths of the lines that constitute each sans-serif letter are often uniform. Sans-serif typefaces are generally thought to have a cleaner, more modern look than serif type.

Some studies indicate that readers can read text set in serif typefaces more quickly than text in sans-serif type. Readers also seem to prefer text set in serif type.

However, research also shows that the legibility of serif and sans-serif typefaces is approximately the same if the text follows general standards for type size and leading (10-point or larger type size with 2–4 points of leading). Because the legibility of serif and sans-serif print is so similar when properly designed, many researchers and typographers suggest that document designers choose a

Choose a serif or sans-serif typeface according to the visual texture you want your document to have.

Choose a serif or sans-serif typeface according to the visual texture you want your document to have.

Serif and sans-serif typefaces.

serif or sans-serif typeface according to what visual texture they want the text to have.[11-13]

Lowercase vs. Uppercase Type. Guideline: Avoid using use all uppercase type. When you want to emphasize a word or portion of text, use boldface type or italics.

Research supports the conclusion that in printed text words set in lowercase are faster to read than words set in uppercase.[14-17] Text set in lowercase is easier to read than text in uppercase because

1. Lowercase letters take up less space and, therefore, allow readers to take in more words as they scan a line of text.
2. Each lowercase word has a distinct outline that aids recognition and recall of the word.

However, some research does indicate that in some situations, such as when readers are scanning text or reading short informational statements, words set in uppercase are easier for readers to locate and read.[18,19] Therefore, you can put words in uppercase, like "WARNING!" when it is appropriate.

LOWERCASE WORDS TAKE UP LESS SPACE THAN UPPERCASE WORDS AND ALSO HAVE A DISTINCT OUTLINE.

Lowercase words take up less space than uppercase words and also have a distinct outline.

Text set in all uppercase and mixed upper- and lowercase.

Type Weight and Font. Guidelines: Use a medium type weight. Use boldface type to emphasize words or short portions of text.

Research suggests that when a text has a logical structure, typographic distinctions such as changes in type weight or typeface may help readers understand the structure. Research specifically indicates that readers notice changes in type weight (heavy, medium, light) more readily than they notice changes in typeface and that readers find very light or very heavy type tiring and difficult to read.[20]

Readers may find large blocks of text set in very heavy or very light type difficult and tiring to read.

Readers may find large blocks of text set in very heavy or very light type difficult and tiring to read.

Text set in a very heavy type and a light type.

Type Size, Line Length, and Leading. Guidelines: Use 9- to 11-point type for text. Use between 2 and 4 points of leading between lines of text. Use a moderate line length, approximately twice the length of the alphabet of the typeface you use.

Type size, line length, and leading all affect the legibility of print. The actual point size of a typeface does not necessarily reflect the perceived size of a letter. Words set in the same type size may appear larger or smaller because of variations in the height of the lowercase letters.

☐ And then there was light

☐ And then there was light

☐ **And then there was light**

Same typesize with different typestyles.

However, research supports a conclusion that type sizes larger than 9 points are easier to read than smaller type sizes—when they are set with an appropriate amount of leading, usually 2 to 4 points.[21,22]

Although researchers have developed some rules of thumb for considering combinations of line length and leading, these rules are not hard and fast.[14,16,21,23] For example, if a text is set in a typeface that appears small, reducing leading between lines may well have no effect on the legibility of the text.

When you consider options of type size, line length, and leading, remember:

- Sans-serif typefaces may be easier to read with slightly more leading.
- Long lines may be easier to read with more leading between them.
- As type size increases, the effect of leading on legibility decreases.

Justification of Lines, Size of Margins, Use of White Space. Guidelines: Use ragged-right margins (unjustified text). Use white space in margins and between sections.

Unjustified text is less costly to produce and easier to correct than justified text. Also, many readers prefer unjustified text. However, many publications still set text with both left and right justified margins.

Unjustified text is less costly to produce and easier to correct than justified text. Also, many readers prefer unjustified text. However, many publications still set text with both left and right justified margins.

Unjustified and justified text.

Many researchers and typographers advocate the use of ragged-right margins both because research indicates that justified text is more difficult for poor readers to read and comprehend, and because many readers prefer text with ragged-right margins.[24–26] Unjustified text is also less costly to produce and easier to correct than justified text.

Headings. Guidelines: Use informative headings. Make all headings consistent and parallel in structure.

Research suggests that readers understand and remember text preceded by titles and/or headings better than text with no headings or titles. Poor readers are significantly aided by headings that are full statements or questions.[27]

Clear, active, and specific headings can also alter how readers comprehend a text. Headings can affect both what sections of text readers choose to read and how readers remember the organization of the text.[28–32]

For example, which of the following headings do you find easier to understand and remember?

Subpart B: Applications and Licensee
- 83.20 Authorization required.
- 83.22 General citizenship requirements.
- 83.24 Eligibility for license.

or

2—How to get a license
- 2.1 Do I need a license?
- 2.2 How do I apply for a license?
- 2.3 May I operate my car while my applications are being processed?

Graphs, Tables, Flowcharts, and Diagrams. Guideline: When you are considering using graphics to convey quantitative or procedural information, you should consider

- the reader
- the amount and kinds of information the graphics must display
- how the reader will use the graphics
- what graphic forms you can use (tables, graphs, charts)

If you decide it is appropriate to use graphics, keep the following guidelines in mind:

1. Reinforce all graphics with supporting text.
2. Use typographic cues (changes in type weight and style) to distinguish between different types of information, such as headings or different kinds of data.
3. Use 8- to 12-point type and adequate spacing between printed items.
4. Include informative labels with all types of graphic devices.

5. Use only those lines, grid patterns, and other ink markings necessary to make the information clear.
6. Whenever possible, arrange items vertically so they can be easily scanned.

Because readers' skills in interpreting graphics vary so widely, and because graphic displays can be designed and used in so many different ways, a tremendous amount of research has been done to explore the most effective ways to present quantitative and procedural information.[33–40]

This research strongly suggests that many readers can understand and use complex information more quickly and accurately when it is presented in well-designed tables, graphs, flowcharts, or diagrams rather than solely in text. However, these graphic devices are even more effective when they are reinforced with explanatory text.[41–44]

Some research suggests that bar charts are easier to understand than line, pie, or surface charts; or text.[36,45]

(A)

Range (in miles)		***Payload (in lbs)***
878	100	710
630	200	598
459	300	471
388	400	326

(B)

Range (in miles)	***Payload (in lbs)***	***Payload (in lbs)***	***Range (in miles)***
100	878	100	710
200	630	200	598
300	459	300	471
400	388	400	326

Table B is easier to use than Table A because the columns are explicitly labeled.

WHAT DOES THE RESEARCH SUGGEST TO THE PRACTITIONER?

Although the relationship between writers and designers is often a difficult one to develop, research in document design (and the practical experience of designers and writers) supports the conclusion that to produce good technical documentation, designers and writers must work together from the beginning of the document design process. As you begin to plan a document, you should not only

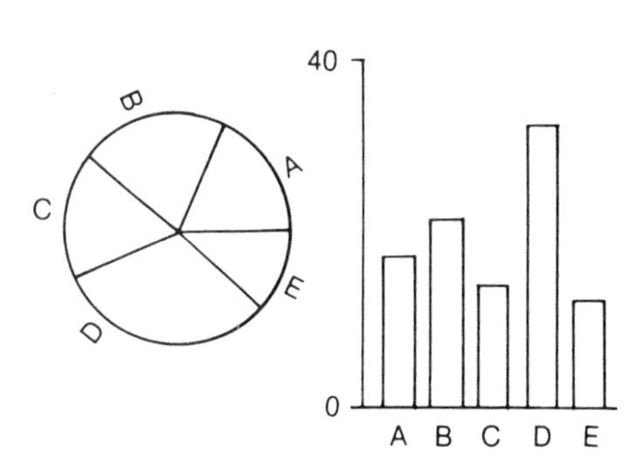

This bar chart reveals the difference in proportions more clearly than the pie chart.

consider who your audience is and how they will use the document, but also how you can use the visual presentation of the text to reinforce the structure of the information.

To develop a sound working relationship, writers and designers should make conscious efforts to learn about how the other approaches planning and preparing a document so they can communicate effectively.

REFERENCES

1. J. Hartley, *Designing Instructional Text* (New York: Nichols Publishing Company, 1978).

2. D.H. Jonassen, ed., *The Technology of Text* (Englewood Cliffs, NJ: Educational Technology Publications, 1982).

3. P. Wright, "Usability: The Criterion for Designing Written Information," *Processing Visible Language, Vol. 2,* ed. Koler, Wrolstad, and Bouman (New York: Plenum Press, 1979), pp. 182–205.

4. L.T. Frase, "Reading Performances and Document Design," Bell Laboratories, paper given at Society for Applied Learning Technologies, Washington, DC, June, 1976.

5. R. Waller, "Typographic Access Structures for Educational Texts," *Processing Visible Language, Vol. 1,* ed. Koler, Wrolstad, and Bouman (New York: Plenum Press, 1979), pp. 175–187.

6. E.E. Miller, *Designing Printed Instructional Materials: Content and Format* (Alexandria, VA: Human Resources Research Organization, 1975).

7. W. Diehl and L. Mikulecky, "Making Written Information Fit Workers' Purposes," *IEEE Transactions on Professional Communication* PC-24 (1981).

8. J. Redish, R. Battison, and E. Gold, "Making Information Accessible to Readers," *Research in Non-Academic Settings,* ed. Odell and Goswami (Guilfor Press, in press).

9. P. Carpenter and M. Just, "Reading Comprehension as Eyes See It," *Cognitive Processes in Comprehension,* ed. Just and Carpenter (Hillsdale, NJ: Lawrence Erlbaum Assoc., 1977), pp. 109–139.

10. A. Marcus, "Icon Design Requires Clarity, Consistency," *Computer Graphics Today*, November 1984.

11. R. McLean, *The Thames and Hudson Manual of Typography* (London: Thames and Hudson, Ltd., 1980), pp. 42–48.

12. D. Robinson, M. Abbamonte, and S. Evans, "Why Serifs Are Important," *Visible Language* 4 (1971).

13. R.F. Rehe, *Typography: How to Make It Most Legible* (Carmel, IN: Design Research International, 1981), p. 32.

14. A. Seigel, *Designing Readable Documents: State of the Art* (New York: Seigel and Gale, 1978).

15. M.A. Tinker and D.G. Paterson, "Influence of Type Form on Speed of Reading," *Journal of Applied Psychology* 7 (1928).

16. J. Hartley, "Space and Structure in Instructional Text," paper given at the NATO Conference of the Visual Presentation of Information, Het Vennenbos, Netherlands, 1978.

17. E.C. Poulton and C.H. Brown, "Rate of Comprehension of Existing Teleprinter Output and Possible Alternatives," *Journal of Applied Psychology* 52 (1968).

18. E.C. Poulton, "Searching for Newspaper Headlines Printed in Capitals or Lower-case Letters," *Journal of Applied Psychology* 51 (1967).

19. A.G. Vartabedian, "The Effects of Letter Size, Case, and Generation Method in CRT Display Search Time," *Human Factors* 13 (1971).

20. H. Spencer, L. Reynolds, and B. Coe, *A Comparison of the Effectiveness of Selected Typographic Variations* (Readability of Print Research Unit, Royal College of Art, 1973).

21. M.A. Tinker, *Legibility of Print* (Ames, Iowa: Iowa State University Press, 1963), pp. 88–107.

22. D. Felker, *Document Design: A Review of the Relevant Research* (Washington, DC: American Institutes for Research, 1980), pp. 104–106.

23. M. Gray, "Questionnaire Typography and Production," *Applied Ergonomics* 6 (1975).

24. M. Gregory and E.C. Poulton, "Even versus Uneven Right-Hand Margins and the Rate of Comprehension in Reading," *Ergonomics* 13 (1970).

25. J. Hartley and P. Burnhill, "Experiments with Unjustified Text," *Visible Language* 3 (1971).

26. T.E. Pinelli, et al, "Preferences on Technical Report Format: Results of Survey," *Proceedings*, 31st International Technical Communication Conference (Society for Technical Communication, 1984).

27. J. Hartley, P. Morris, and M. Trueman, "Headings in Text," *Remedial Education* 16 (1981).

28. E. Kozminsky, "Altering Comprehension: The Effect of Biasing Titles on Text Comprehension," *Memory and Cognition* 5 (1977).

29. J. Hartley, J. Kenely, G. Owen, and M. Trueman, "The Effect of Headings on Children's Recall from Prose Text," *British Journal of Educational Psychology* 50 (1980).

30. H. Swartz, L. Flower, and J. Hayes, "How Headings in Documents Can Mislead Readers" (Pittsburgh, PA: Document Design Project, Carnegie-Mellon University, 1980).

31. P. Wright, "Presenting Technical Information: A Survey of Research Findings," *Instructional Science* 6 (1977), pp. 96–100.

32. L.T. Frase and F. Silbiger, "Some Adaptive Consequences of Searching for Information in a Text," *American Educational Research Journal* 7 (1970).

33. W.S. Cleveland and R. McGill, "Graphical Perception: Theory, Experimentation, and Application to the Development of Graphical Methods," *Journal of the American Statistician* 79 (1984).

34. W.S. Cleveland, "Graphical Methods for Data Presentation: Full Scale Breaks, Dot Charts, and Multibased Logging," *Journal of the American Statistician* 38 (1984).

35. A.S.C. Ehrenberg, "What We Can and Can't Get from Graphs, and Why," paper presented at American Statistical Association Meeting, Detroit, 1981.

36. M. MacDonald-Ross, "Graphics in Texts," in *Review of Research in Education,* Volume 5, ed. Shulman (Itasca, IL: F.E. Peacock Publications, 1978).

37. P. Wright and K. Fox, "Presenting Information in Tables," *Applied Ergonomics,* September 1970.

38. P. Wright and K. Fox, "Explicit and Implicit Tabulation Formats," *Ergonomics* 15 (1972).

39. P. Wright, "Tables in Text: The Subskills Needed for Reading Formatted Information" (Cambridge, UK: MRC Applied Psychology Unit).

40. E.R. Tufte, *The Visual Display of Quantitative Information* (Chesire, CT: Graphics Press, 1983).

41. P. Wright and F. Reid, "Written Information: Some Alternatives to Prose for Expressing the Outcomes of Complex Contingencies," *Journal of Applied Psychology* 57 (1973).

42. P. Wright, "Writing To Be Understood: Why Use Sentences?" *Applied Ergonomics* 2 (1971).

43. R. Kammann, "The Comprehensibility of Printed Instructions and the Flowchart Alternative," *Human Factors* 17 (1975).

44. D. Felker, *Guidelines for Document Designers* (Washington, DC: American Institutes for Research, 1981).

45. D.D. Feliciano, R.D. Powers, and B.E. Keare, "The Presentation of Statistical Information," *Audio-Visual Communication Review* 11 (1963).

How to Get Users to Follow Procedures

Elizabeth Berry

Procedures teach readers how to perform a specific task. This article explains nine formats for presenting procedures effectively. Use the "Procedure Format Guide" in the article to help you decide which format would be most appropriate for your readers and purpose.

EVERYONE FOLLOWS MANY PROCEDURES DAILY. Initially, procedures are learning formats which teach us how to successfully complete a given task. When performed over a period of time, some procedures become unconscious habits.

For instance, we all follow a specific procedure for tying our shoe laces. We no longer consciously think about the procedure: "To tie your shoe laces, hold the left lace in your left hand, hold the right lace in your right hand, cross the right lace over the left. . . ." But, as very young children, we did consciously follow a similar procedure before we could master the task of tying our shoe laces.

In business, some simple linear procedures do become habit with users, but most do not. More often, business procedures involve complex operational tasks which have to be spelled out to the user in written forms. How do you get users to follow these written forms? Users will follow procedures if certain criteria are met. The three basics for getting users to follow written procedures are:

- The procedure must use a format that is appropriate for the procedure.
- The procedure must not assume knowledge that the user does not have.
- The procedure must be fair to the user.

FORMATS APPROPRIATE FOR THE PROCEDURE

Formats for procedures should vary according to the nature of the procedure. Many procedures fail because the writer uses an inappropriate format. The following presents nine basic procedure formats, when they are to be used, and gives format examples.

1. Narrative Formats

Narrative formats use sentences and paragraphs to separate steps of the procedure. Sometimes steps are also separated by commas within sentences. Many narrative formats for procedures are ineffective. This format encourages writers to include a lot of unnecessary description and padding. Narrative formats are also more demanding of the user—the individual steps of the procedure do not stand out clearly. Narrative formats should be used:

- if all the steps of the procedure do not exceed 125 words (about 1/2 a typewritten page).
- if the procedure is a simple linear pattern.
- if the procedure has only a few steps.

An example of a narrative format used to show employees the procedure for punching time cards:

> Time cards must be punched four times daily. They must be punched when you start the work day, when you begin your lunch period, before you return to work after lunch and when you have completed the day.

2. Step-By-Step Listing Formats

Step-by-step listing formats are lists of actions which are performed sequentially. The format begins with the numeral 1 for the first action of the procedure and numbers each action sequentially through the final action of the procedure. These formats should be used:

- if the procedure is linear.
- if the procedure has more than ten steps.
- if the procedure must be contained in the limited space of a label placed on a machine.
- if the procedure is performed by one person.

An example of a step-by-step list showing the procedure for installing a Carter's typewriter ribbon:

1. Move load lever to load position.
2. Wrap leader over guide posts and ribbon guides.
3. Place ribbon cartridge between spring clips.
4. Press down on both sides.
5. Insert leader through ribbon guides.
6. Turn knurled knob in direction of arrow until inked portion of ribbon is past right ribbon guide.
7. Adjust load level to type position.

3. Step-By-Step Listing and Visual Aid Formats

This format is a variation of the step-by-step listing format. Line drawings or photography are added to clarify the directions of the procedure steps. These formats are best:

- if the procedure outlines a complicated assembly task.
- if the procedure outlines linear operations for technical or complex equipment.

4. Playscript Formats

Playscript is a variation of the step-by-step list. This format was developed by Leslie H. Matthies in the late '50's. Playscript is more explicitly defined in his book, *The Playscript Procedure: A New Tool of Administration.* The Playscript format fits best:

- if the procedure is linear.
- if the procedure requires the involvement of several users performing different tasks.

For example:

SUBJECT: 3600 Operating Procedure

Responsibility	Action
I/O Control	1. Pick up diskette and tape at 10:00 A.M.
	2. Return box to set-up clerk.
Set-up Clerk	3. Sign-in returned diskette tape into log book.
	4. Remove label from tape.
	5. Return tape to tape librarian.
	6. File diskette and transmittal form.

5. Question List Formats

This format is a list of questions covering the procedure. The users answer the questions either yes or no. Then, on the basis of their answer, they are given instructions to go on to another step. In order for question lists to be effective, the users must be able to make certain accurate observations. For instance, an error will occur if the user mistakenly answers a question NO when, in fact, the YES situation exists. The user will go to the NO response direction when the YES response situation is the one that really is to be followed. This format is used:

- if certain decisions must be made before the user can determine what the next step will be.
- if the procedure involves only one user.

For example:

SETTING UP A TYPEWRITER FOR TYPING TASKS

1. Is the typewriter plugged in?
   YES—go to 2
   NO—plug it in—go to 2
2. Is the ON button depressed?
   YES—go to 3
   NO—depress ON button—go to 3

3. Do you plan to type a letter?
   YES—set margins at 15 & 77—go to 8
   NO—go to 4
4. Do you plan to type a report?
   YES—set margins at 12 & 72—go to 9
   NO—go to 5
5. . . .

6. Flowchart Formats

As a general rule, flowcharts are not good formats to use for presenting procedures. Even though any process which can be performed can also be flowcharted, the primary purpose of a flowchart is to show *what* is to be done, to graphically illustrate the sequence and flow of an activity. The primary purpose of a procedure is to show *how* an activity is done. This format should be used:

- if the procedure is nonlinear, but only involves a few decisions.
- if the users are familiar with flowcharts.
- if the flowcharted procedure can be contained on one, 8 1/2 x 11″ sheet of paper (no fold-outs for procedures!).
- if the flowchart can be limited to a few basic symbols.

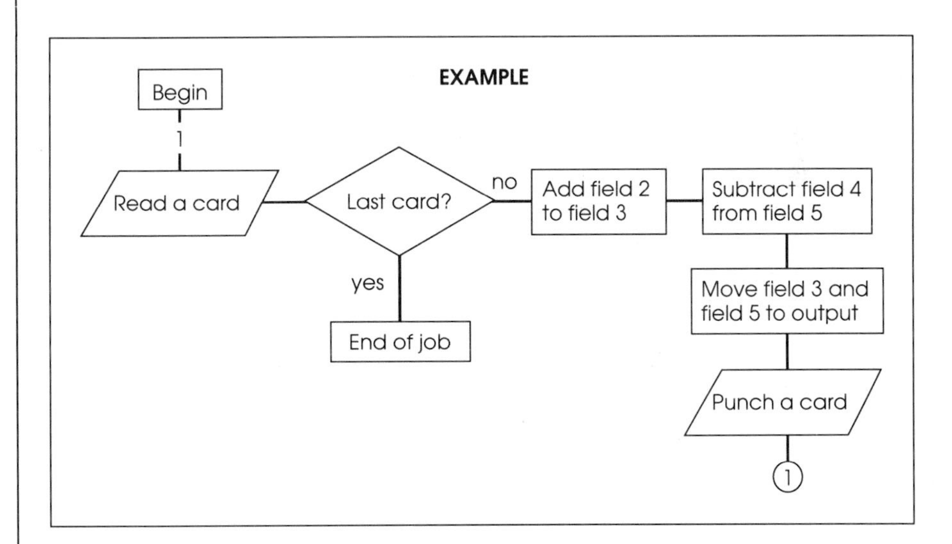

7. Narrative Flowchart Formats

Narrative flowcharts are variations of the flowchart format. Like flowcharts, they too show the sequence and flow of complex operations. They differ from flowcharts in that they do not use a complicated diagramming schema. The various steps of the procedure are presented in narrative blocks. This format works best:

- if the procedure is nonlinear.
- if the procedure involves at least one decision, but usually not more than three.
- if the user has a nontechnical background.

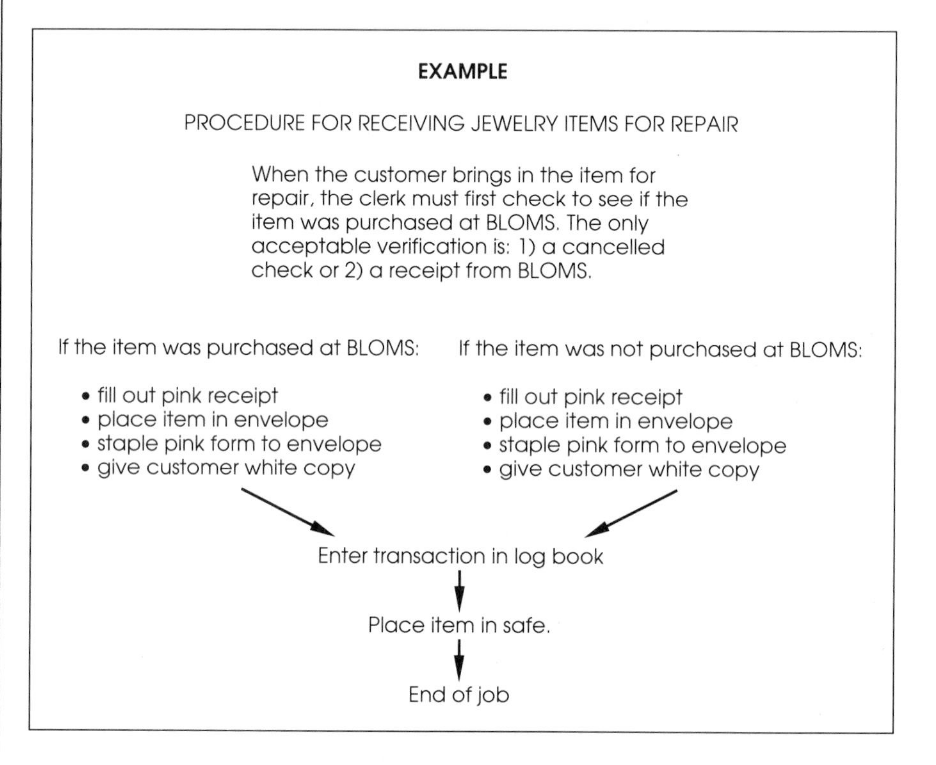

Decision Logic Table Formats

Decision logic tables present the procedure in a tabular format. The design of these tables vary, but they all are divided into two basic parts: the conditions and the actions. The use of decision logic table formats are best:

- if the procedure is nonlinear.
- if the procedure involves many "if" conditions followed by many "then" actions.
- if the procedure is needed to make "on the spot" decisions.

Example: Procedure Format Guide

Procedure Description / **User Requirements**	**Linear, few steps, less than 125 words**	**Linear, more than 10 steps**	**Linear, outlining assembly or operation of equipment**	**Linear with decisions determining**	**Nonlinear with a few decisions**	**Nonlinear (showing outcomes of decisions)**	**Nonlinear with many "if" conditions and "then" actions**
One nontechnical user required to complete procedure	NARRATIVE	STEP-BY-STEP LIST	STEP-BY-STEP LIST AND VISUAL AID	QUESTION LIST	DECISION TREE	DECISION TREE; NARRATIVE FLOWCHART	DECISION LOGIC TABLE
More than one nontechnical user required to complete procedure	PLAY SCRIPT	PLAY SCRIPT	PLAY SCRIPT with VISUAL AID	NARRATIVE FLOWCHART or DECISION TREE	FLOWCHART; DECISION TREE	DECISION TREE; NARRATIVE FLOWCHART	DECISION LOGIC TABLE
One technical user required to complete procedure	NARRATIVE	STEP-BY-STEP LIST	STEP-BY-STEP LIST AND VISUAL AID	QUESTION LIST	FLOWCHART; DECISION TREE	DECISION TREE; NARRATIVE FLOWCHART	DECISION LOGIC TABLE
More than one technical user required to complete procedure	PLAY SCRIPT	PLAY SCRIPT	PLAY SCRIPT with VISUAL AID	NARRATIVE FLOWCHART or DECISION TREE	FLOWCHART; DECISION TREE	DECISION TREE; NARRATIVE FLOWCHART	DECISION LOGIC TABLE

Decision Tree Formats

Decision trees are procedure formats which combine aspects of the flowchart with the decision logic table. The main difference between decision trees and flowcharts is that decision trees are limited to few symbols and the decision branching is more obvious. Use decision tree formats:

- if the procedure is nonlinear.
- if the procedure involves only a few decisions.
- if it is important to show outcomes of the decisions.

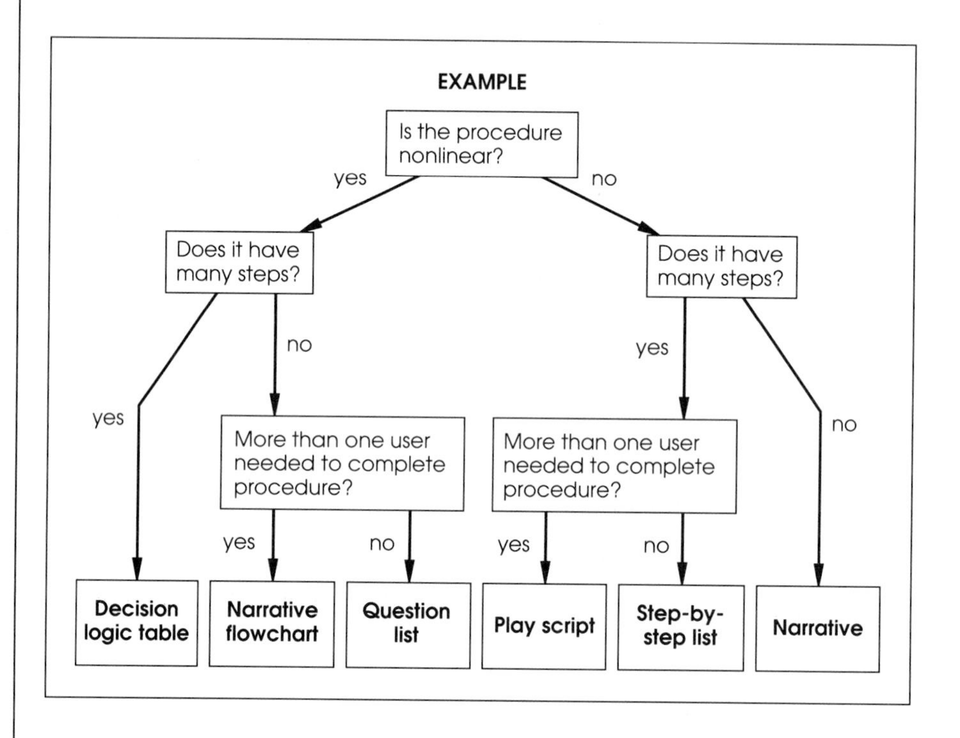

Procedures Should Not Assume Knowledge

This rule seems like a simple one to follow, but often it is a rule that is overlooked by the procedure writer because most often procedures are written by people who have (1) either been following the procedure for a number of years, or (2) they are the people who designed the equipment or system. Their familiarity with the task is so "second nature" to them that they often leave out steps that may be critical for a novice user to know.

This problem can be avoided by:

1. The procedure writer can sit down with the completed procedure and follow each step through to the completion of the task.

or (and this is the best test)

2. The procedure writer can submit the procedure to a user test before distributing the procedure company-wide.

PROCEDURES MUST BE FAIR TO THE USER

Some procedures fail because of their viewpoint. In order to be followed, procedures should be written with the user's viewpoint and convenience in mind. Many procedures are written with slanted viewpoints—viewpoints that obviously are only for the benefit of a particular department. Procedures written with a slanted viewpoint—a viewpoint that is obviously for the benefit of someone other than the user—will only arouse user resistance and sabotage of that procedure.

Procedures that cross department lines should always distribute the work tasks as evenly as possible. Here are three questions to test the fairness of a procedure:

1. Does the procedure have many more steps for one department than for another?
2. Are these steps, because of their nature, steps which can only be done by this department? If so, leave the procedure as it is. The users will understand and know that the tasks are unique functions of their department.
3. Can some of the steps be handled just as well by another department? If so, redistribute them more equally among the other departments.

SUMMARY

To insure that users will follow procedures, three basic criteria must be met. First, the procedure must use a format that is appropriate. This article has presented nine basic procedure formats for writers to follow. Second, the procedure must not assume knowledge that the user does not have. Failure to follow this guideline often results in the elimination of essential steps of the procedure. Third, the procedure must be fair to the user. Being fair to the users translates: consider their viewpoints and distribute work tasks as equally as possible.

Greater Concern for Ethics and the "Bigger Backyard"

S. J. Blank

Corporations face an increasingly broad range of ethical responsibilities to customers, employees, and the community. This article discusses the ethical decisions several large companies have made. In general, companies find that ethical behavior is good business.

BUSINESSES LARGE AND SMALL have traditionally played an upstanding role in their communities. Rotary clubs have been holding luncheons for eons now, and the number of softball teams sponsored by Main Street concerns over the years is downright startling. Even the great patron Exxon endorses high-school concerts-on-the-green in the small New Jersey town that houses its headquarters.

This type of business conscience, however, belongs to a simpler time. Our increasing global awareness is expanding our backyards to include the likes of Soweto and Bhopal. Because of this, corporate ethics and philanthropic considerations have become, and will continue to be, more complex issues. "Corporate America," according to Dr. Clarence Walton, professor of ethics and the professions at The American College, "has entered a new era of soul-searching. No longer is it a matter of dealing fairly and ethically with customers and employees. Corporate ethics have evolved into a much broader range of responsibilities.

"Because of recent legislation, court decisions, higher public expectations, and new problems such as hostile takeovers and corporate raiders, questions over when a deal is a deal, erratic interpretations of tort law, and foreign competition, there is an intensified management interest in ethics."

CORPORATE CREDOS

Because of this, Dr. Walton feels that companies should devise a written corporate philosophy, a standard of ethics that is taught to employees. "This is necessary for two reasons. First, the very act of putting important company values in writing is a useful exercise for management. Second, the written document can serve as a guide to all employees and the public at large."

Managers, says Dr. Walton, should also be guided by two commandments when it comes to ethical behavior. "One, take most seriously the nature of your fiduciary obligations to customers, employees, stockholders, competitors, and the general public. Two, engage in periodic examinations of the corporate conscience to see if your company has lived up to expectations."

The experience of Johnson & Johnson comes immediately to mind when one discusses corporate ethics and sticking by corporate credos. "Anyone knowing

Johnson & Johnson's corporate philosophy could have predicted that the company would have risked a $120-million loss to pull Tylenol capsules from the shelves, rather than risk harming their customers," attests Walton.

Johnson & Johnson's credo, established by Robert Wood Johnson, son of the company's founder, asserts that a high-quality product, commitment to the customer, equal opportunity, safe working conditions, corporate philanthropy, and community responsibility are the company's highest values. Profit-making is considered "our final" obligation. "When we operate according to these principles," the credo holds, "the stockholders should realize a fair return."

There is one problem with having a corporate credo such as this—and that is the danger of *not living* by it. James O'Toole feels that "most companies, through their credos, are artificially trying to create a corporate culture that doesn't exist." As is the case with Johnson & Johnson, culture comes from the behavior of top management, and from the philosophy of the founders of the organization. "It is not a reflection of what people say, but of what they *do*."

Levi Strauss and Company provides another good example of ethical and philanthropic policies that work because they are firmly based upon the corporate philosophy. Commitment to the customer and to quality once guided Levi to reject an exclusive right to a new product—a blend of denim and lycra spandex—because there were insufficient quality standards for the fabric at the time. In addition, Levi's philanthropic activities center around its Community Involvement Teams—groups of workers empowered with the control of funds provided by the company for distribution to worthy causes in their own communities.

ETHICAL PAYOFFS

Many businesspeople share the sentiment that "ethical or philanthropic behavior is all very nice, but the bottom line *must* be any company's overriding concern." However, as Dr. Walton points out, "studies have indicated that companies which adhere to a set of ethical standards are healthier in the long run." Amy Domini and Peter Kinder, coauthors of *Ethical Investing*, also report results from the Franklin Research and Development Corporation to the effect that special, ethical accounts run by President Joan Bavaria were up over 50 percent from mid-1982 through mid-1983, while the Dow was up 44 percent over the same period. U.S. Trust Company of Boston's "socially sensitive" accounts also beat the Dow, 1981–'83, by a small but significant margin.

But businesses do not, of course, act ethically solely for the return on investment. American companies with high ethical standards hold them for no other reason than because it is unquestionably the right thing to do. In this light, profit is "the means, not the end, of corporate activity," as James O'Toole has said.

On the other hand, the recent troubles of ethical/philanthropic visionaries such as Control Data or Atlantic Richfield, whose corporate giving far surpasses that of organizations with comparable resources, have caused some to discredit an "overzealous commitment to do good." Yet, according to O'Toole, Control Data's or Atlantic Richfield's programs had "nothing to do with why they got into trouble . . . nothing to do with the fact that they were thinking in the long term, or that they were trying to provide job security to their employees. . . ."

INCREASED GIVING

According to a recent AMA Council report, corporate philanthropy has grown from a $350-million business in 1950 to some $3.5 billion today. It is still growing, as companies' awareness of their role in and responsibility to society grows.

General Electric Foundations president Paul M. Ostergard, in an annual report essay titled "Facing Reality . . . with a Heart," notes that "If America is to sustain its standard of living, so long the envy of the world, we Americans will have to make even more extensive changes in our way of life. We will have to improve the way we educate our young, care for our health, protect our natural environment, nurture our arts; the way we assist people and communities whose jobs have been lost to competition, and provide for those least able to provide for themselves. GE's contributions (employee, company, and foundation) to charities, education, and grants totaled $54 million for 1985.

Similarly, Johnson & Johnson's efforts to "face reality with a heart" have committed it to help New Brunswick, the somewhat downtrodden New Jersey city in which it historically has been based. Previous Johnson & Johnson chairman Dick Sellars has said, "Our origins are in New Brunswick. Our responsibility is not to run away from the problems of a decaying inner city, but to help try and rejuvenate that city."

President David Clare more than backed this up in a comment from David Freudberg's new book, *The Corporate Conscience:* "Take the community situation —responsibility to the community at large, but also to the community in which we live and work. To what extent are we participating in that community activity? To what extent are we individually, and/or as a company, contributing to the health of the community?"

Atlantic Richfield, too, has a reputation to be "determined to do well by the people who work there and by the people who live in its plant communities," according to the authors of *The 100 Best Companies to Work for in America.* It has broken with other oil companies on several political and social issues; has developed an inordinately large philanthropic budget for its relative size; and has helped to rescue publications such as *Harper's* and London's *The Observer* from losing control or folding altogether.

As Thornton Bradshaw, chairman of RCA and former president of Atlantic Richfield, has said, "[what] should be the fulfillment of everyone . . . is to merge your own set of ethics and values into what you do during the day. If you have to draw a curtain down when you go to work in the morning, and spend eight hours or so doing something that you don't believe in, then you're in trouble."

The Art of the Interview

Eivind Boe

A job interview is a sales presentation of your best features. This article explains how to prepare for a job interview by anticipating questions, practicing answers, and finding information about the company. Try preparing the answers to some of these standard questions before your next job interview.

THE AD WAS FOUND; the résumé was sent; the eagerly awaited phone call was received: You're to go for an interview. Unless you're unquestionably the best or absolutely the worst insofar as qualifications go, your performance at the interview is what makes you or breaks you as far as the job is concerned. There is no getting around this.

As with everything else in life, there is no way to ensure success at an interview. You can buy the fanciest clothes, but they will not buy you the job. The firm handshake is not an automatic in; nor is any connection or pull you might have. But aside from your qualifications and maturity, some fundamental concerns will greatly increase the percentages in your favor.

For one thing, study up on your résumé, your reasons for wanting a job at that particular company, and the points in your past that you want to talk about. The best way to be ready for an interview is to have something to say. Having nothing to say at your interview is about as self-defeating as can be.

Perhaps the greatest temptation for young people at an interview is to act, to try to adjust their personality for each interviewer. Aside from creating impossible problems, you're best off being you. Any company worth working for isn't looking for only exuberant people; nor do they want meek submission, nor frenzied aggressiveness, nor Laputan eggheadism. They want mature, capable human beings. *That* is the only category you have to fit into. The trick is to be the best *you* you can be.

Manners are those things mother taught you that you dispensed with in college. They are really codified courtesy, and courtesy is one of those deeper qualities that make people want to have you around. Things like standing when the interviewer stands and forgoing smoking if the interviewer doesn't offer it—and doing these things with an easy grace—are what interviewers notice. Remember it is the interviewer's prerogative to be irascible, abrupt, or prying: Maintain your courteous demeanor—even if you must demand courteous treatment, which sometimes happens.

SELLING YOURSELF

Remember your manners when you make that all important "sales call." For even if you never, EVER, wanted to go into sales, when you interview, you have something to sell: yourself. To sell effectively you need three things: a good presentation, a unique product, and an unshakable belief in that product.

A good presentation consists of simply this: as many reasons to hire you, and as few reasons not to hire you, as possible. The interviewer is mainly concerned with if you *can* do the job, if you *will* do the job, and if you're going to *get along* with the rest of the company. Inspire the interviewer with confidence in you; handle every question with aplomb. If the interviewer points out that you haven't done such and such, which is an integral part of the job, you can say, "This is true, but I've done this and this, which are quite similar. In the past I've always been able to adapt this way to new situations." Consider every question the interviewer puts to you an opportunity to show how good you really are.

A product has to be distinct from other products, and for it to sell, it must be distinctly better. If you feel like everyone applying for a position has exactly what you have, say, a B.S.E.E. and 4 years of experience, don't worry: No one else has the unique background you have. Your work experience inevitably has had special projects, unusual situations, trying circumstances. Your education includes special courses and activities (even those unrelated to engineering). Everything else you've ever done defines your unique character, and can show some positive attribute in you: Your participation in team sports shows a willingness to work with others; your active involvement in town politics shows your willingness to get involved. An interest in trout fishing evinces patience, even strategy; an interest in music manifests a desire to organize, a penchant for system. Every positive aspect about you can be turned to account.

Believing in the product is paramount. Why apply for a job if you don't think you can do it well? You have already passed many tests in life: What is it about this job that, through perhaps some effort, you couldn't master? You do, of course, already know a lot; you can, of course, learn almost anything you care to. Nothing really stops you from succeeding—at anything.

Try bringing some of these attitudes with you to an interview. They are at least as true as any negative attitudes you may have, and are far more likely to get you the job. Yet the problem may not be one of lack of confidence, but nervousness due to pressure.

If you're nervous before an interview, that's *good.* The stampeding herd of butterflies in your stomach, the apple in your throat, and the 2-lb weight tied to your tongue indicate that your mind feels doubt and is telling your endocrine system to pump out loads of adrenaline. Naturally it won't do to let this unfocused energy mess up your interview. The trick is to channel the energy and make it work for you. Learn how to turn it to positive action, rather than negative reaction.

Awareness of interviewers' standard stock of questions is another way to beat nervousness and build confidence. Being prepared for these questions allows you to free your mind to concentrate on more challenging issues. Here are the questions and some suggestions on how to handle them.

1. *Tell me about yourself.* Actually not a question, of course. In answering this, stick to your life at work: This tack reinforces your interest in the job.
2. *What is it about the job that interests you most?* If you launch into extolling the benefits you'll accrue if you get the job, you will have lost. Make the

interviewer believe that you are interested in performing the work, using your knowledge and energy to help the company.

3. *Where do you see yourself five years from now?* A classic question. Probably best handled by something like: "Doing more of the same, only better." Perhaps you can think of something more imaginative. Just leave the impression that you know what you want.
4. *What do you consider your major strengths/weaknesses?* You know your strengths, so tell them without sounding boastful. As for your weaknesses, make them sound like strengths: Sometimes you're just too much of a perfectionist, or you get impatient when coworkers don't work as hard as you do.
5. *Why do you want to leave your present job?* Avoid saying that you hate your boss, or that you don't get along with your coworkers. Such remarks leave a bad taste in the interviewer's mouth. Wanting more money or a higher position, and being able to do more at a higher level than your present company can reasonably give you, are good answers here.

Rest assured that these aren't the only questions you'll hear. Nor are the suggested answers the only ones you can give. But be prepared for these specific questions, and you'll have started off the interview right.

WARMING UP

How does one practice interviewing? Try this simple idea: Get yourself a tape recorder, tape yourself in an interview situation, and listen to yourself. Either ask some of the standard interview questions yourself (in a different voice if you choose) or have a friend ask the questions, and then answer them.

There are certain things to listen for in the recording. Are you answering the interviewer's questions directly, or do you sound too much like you're dodging? Are there some bad habits there that you can correct, such as an overuse of hems, ums, or y'knows? Do you have a ready answer for every question? Are you perhaps talking too fast?

Granted, this is not a realistic situation. After all, you already know the questions, whereas in an interview you don't. But in most interviews there are stock questions, which are really opportunities for you to glow. Using a tape recorder can allow you to practice the answers to those questions.

KNOW THY PROSPECTIVE COMPANY

Of inestimable value when you go into an interview is knowledge about that particular company. The more you know, the more you appear to be the perfect, ready-made employee. You'll inspire your interviewer with respect for you, and by having something to say, will simultaneously remove a great deal of the burden of the interview from his or her shoulders, a fact that will dispose the interviewer to liking you.

The next question is where to find this knowledge. First of all, a company library or even the local library is a good start. Simply ask the reference librarian to help you. Also, check out the Dun & Bradstreet reports, and the *Reader's Guide to*

Periodical Literature for any articles that may have been written on the company. For the latter source, don't give up if you don't find anything under the first heading you look at; try several headings, as many as you can think of. At least look under the company name, electronics, electrical engineering, microwaves, and radio frequency. The second source to tap is the company's annual report. Get one of these by contacting the company. A third source of information is through friends and contacts: Grill them!

CALL THE COMPANY

Do not despair if next to no information can be located. If this is the case, try calling the company itself, explain that you have an interview with them, and that you'd like to find out more about the company. Many executives and personnel officers will often be pleased to send you brochures, pamphlets, and other information. They will, in addition, remember your name come interview time.

It's a good idea to keep in mind that this information search takes a little time. When preparing for the interview, allow for the time it takes for mail to travel, and remember that libraries are open only during certain hours.

Having gotten this information from your various sources, carefully examine it to determine company trends, policies, and concerns. Look for the company's strengths and weaknesses. Set the knowledge against the background of your general knowledge of the field. Express an interest in these concerns during the interview, and to a certain extent, identify with these concerns. Think of possible solutions to the company's problems—or better yet, how *you* could help the company improve, meet goals, achieve important successes in the market.

These are the weapons you gird yourself with for confronting the question, "Why are you interested in *our* company?" This is the way to show the interviewer, "This guy's no slouch—he's done his homework."

Japanese-American Communication: Mysteries, Enigmas, and Possibilities

Joel P. Bowman and Tsugihiro Okuda

As international business becomes more important, businesspeople have to learn about cultural differences in order to communicate effectively. This article explains some of the customs that differ between the Japanese and American cultures, such as proper greetings and eye contact.

WILL THE JAPANESE AND AMERICANS ever really understand one another? Sometimes it does seem as though "East is East and West is West and never the twain shall meet." In addition to overcoming the barriers caused by different languages, Japanese and American cultures are very different. This difference distorts the perceptions each has of the other and alters the standard rules of behavior for each, which causes even greater confusion than the differences in languages.

For a variety of reasons, it is imperative that individual Japanese and Americans learn to communicate with each other, and members of The Association for Business Communication and the Japanese Business English Association can play a pivotal role in this learning process. Because the authors recently have had the opportunity to work together at Western Michigan University and to correspond on a regular basis, we wanted to describe our experiences as we struggled to understand one another.

As a teacher of business English and English literature at Oita University, Professor Okuda was fluent in English before arriving in the United States, so we did not have to spend much time discussing the literal meanings of our words. Nevertheless, we did spend a great deal of time discussing semantics and checking dictionaries to see why each of us had a particular idea of what a word *should* mean. If the English language *per se* did not present a significant communication problem, cultural differences caused misunderstandings from first meeting through our last exchange of letters.

We thought that it might be interesting and useful for readers of *The Bulletin* of The ABC if we explained some of the differences between Americans and Japanese and showed how these differences influence perception and communication.

THE BOW AND THE HANDSHAKE

The nonverbal rituals of bowing and shaking hands are perhaps symbolic of the differences between the two cultures. Just as most Japanese have studied English and speak it at least fairly well, the Japanese have learned to shake hands, though they find it virtually impossible to shake hands without bowing at the same time.

Americans, however, not only think that everyone should speak English, but also believe that bowing is at best a strange custom. Many Americans refuse to bow, believing it to be an unnatural act. Most who do try, do it badly.

When a Japanese and an American meet for the first time, their first misunderstanding is likely to result from failure of each to understand and properly execute the rituals of bowing and handshaking. What often happens is the Japanese is bowing while the American is extending his or her hand, and then the Japanese is extending his or her hand while the American is working at returning the bow. Both end up feeling awkward.

POLITENESS

The bow is an expression of Japanese politeness in relationships. The bow shows respect for the other person. The handshake is the American expression of the nature of relationships. The bow says, "I am willing to respect you," or "I acknowledge you with respect," whereas the handshake says, "I will meet you half way." One of the greatest hindrances to communication between Japanese and Americans is their differing perceptions about what is polite. To the average American, the average Japanese seems excessively polite—so polite as to be inscrutable. To the average Japanese, the average American seems impolite—so impolite as to be almost rude.

When asked a direct question, for example, a Japanese will often give an indirect answer if he believes that an honest, direct answer may offend. This tendency can drive Americans crazy. Discovering whether a Japanese would prefer coffee or tea following dinner, for example, can require twenty minutes of skillful questioning. Learning how to discover whether a Japanese guest is comfortable or whether he needs something may require a year's training in diplomacy. The Japanese are careful to avoid giving offense and causing inconvenience. A Japanese would probably starve to death before telling an American host that he can't stand hamburgers and French fries.

Americans, on the other hand, will frequently offer an honest, direct opinion when none is called for because they are so convinced that their opinions are right. An average American would not be shy about telling a Japanese host that the vegetables looked like sea weed and that the fish should have been cooked longer, never for a minute considering the possibility that the Japanese think that eating sea weed and raw fish is a perfectly normal thing to do. The American view often seems to be, if it doesn't happen in My Home Town, it either doesn't happen or (because Americans often take a moral position on cultural matters) it *shouldn't* happen.

Compounding this difficulty are the two cultures' differing attitudes toward eye contact. For Americans, eye contact ranks with Godliness and cleanliness as hallmarks of the trustworthy individual. Eye contact is not nearly so important for the Japanese. Too much eye contact, in fact, would be considered impolite. Historically, Japanese people used to show their respect for prominent people, such as Emperors, shoguns, and regional governors, by *not* looking them in the eye. The tradition of showing respect by avoiding eye contact remains.

For these reasons, a Japanese and an American may be suspicious of each other's motives. The Japanese may think that the American does not respect him or her, whereas the American may think that the Japanese is sneaky and untrustworthy.

THOUGHTFULNESS

The Japanese patterns of conversation also differ from those of Americans. The Japanese are much less afraid of silence than Americans are. In Japan, it's not uncommon for a bus to run for 20 minutes with none of the passengers talking. Americans, on the other hand, seem to have a compulsion to fill all silences with some kind of noise. Who but an American would desire music piped in to elevators to avoid even a few moments of silence? Japanese visitors to America are often shocked to hear an American crack a joke among a group of strangers to avoid the threat of silence during a single-floor elevator ascent. The Japanese are inclined to keep silent unless they have something significant to say.

Americans are frequently surprised to learn that a Japanese person will actually think about what someone has said before responding. Americans, eager to make a point, are much more likely to interrupt before the other person has had a chance to finish. But even when they aren't so eager as to interrupt, Americans have a pathological dread of silent periods in conversation. In a conversation between a Japanese and an American, it is possible for the American to do all the talking—guessing at what the Japanese is thinking, offering his or her own thoughts, and then filling in for the Japanese once again when an immediate response is not forthcoming.

If the matter is inconsequential, this pattern of conversation can be amusing. When Japanese and American business representatives are negotiating major contracts, however, the pattern can be extremely detrimental to the American negotiators. Assuming that the silent reflection on the part of the Japanese signals a rejection of the initial offer, Americans give away the store, making concessions left and right, before the Japanese have had a chance to reply. The Japanese end up with a much better deal than they expected, and the Americans go away thinking of the Japanese as "tough" negotiators.

FORMALITY

The average Japanese, especially those from a traditional background, are much more formal in their relationships than Americans. The Japanese, for example, call each other by their last (family) names and always use a courtesy or professional title (Mr., Mrs., Dr., Professor). Americans, on the other hand, prefer to use first (given) names, which seem more intimate. Japanese and Americans who work together frequently waste a great deal of energy trying to figure out what to call each other.

To the Japanese, using the professional or courtesy title and the last name seems perfectly natural. Americans, however, can't feel fully comfortable until they are on a first-name basis with their colleagues. In working with Japanese, Americans find not only that the Japanese strongly resist being called by their first names, but also that Americans can't pronounce most Japanese first names anyway.

Americans may attribute the Japanese reluctance to be on a "first-name basis" to unfriendliness, while the Japanese may attribute the American's desire to use a form of address other than the title and last name as a sign of lack of respect.

Social Relationships

Americans tend to be more casual in their social relationships than do the Japanese. In the United States, professors and students may even call each other by their first names, whereas in Japan only parents have the right to refer to their sons and daughters by their first names in either talking or writing.

American relationships are perhaps best illustrated by the variety and informality of greetings: "Hi," "Howdy," "How'r ya' doing'," "What's happenin'," "How's everything," and so on. American professors have even been known to greet their students by saying, "Happy Monday," or "Happy Friday." Departures in America are also more casual than they are in Japan. Americans might say, "See you later," "Have a good weekend," or even "Have a good one."

The Japanese take greetings and departures much more seriously. Unlike Americans, who are inclined to greet practically anyone within sight, the Japanese do not greet strangers. Greetings are exchanged between friends and acquaintances in Japan. The circumstances in which greetings are exchanged are formal, and the phraseology is limited: "Good morning," "Good afternoon," "Good evening," or "Good night." There are few deviations from the standard expressions. The Japanese do not use each other's names as part of the greeting, but they may include some remarks about the weather.

Although the Japanese may be casual when leaving their friends, saying simply, "So long" or "Goodbye now," they are formal when departing from someone they don't know well. A Japanese might say, "I'm afraid I must excuse myself. Please give my regards to your wife (son or daughter)." Even when they know each other well, however, Japanese would not be so casual as to encourage a departing friend to "Enjoy the weather," especially when the weather is not especially enjoyable.

Even professional and social events are handled differently in Japan and the United States. In the U.S., a typical professional meeting begins with what is loosely called a "social hour," followed by dinner and a program, including a presentation by a guest speaker. The atmosphere is more formal in Japan than in the U.S. When socializing for business purposes, people bow and do not shake hands, and they are acknowledged in descending order of importance. During the process, people wait quietly for the proceedings to finish, and there would be no guest speakers. Should a presentation be required, that would be held in a different location or on a different occasion. When the Japanese have presentations, they have presentations only; when they socialize, they socialize.

Specificness and Exaggeration

Japanese people and Americans have different conceptions of when to be specific and when to exaggerate. Consider, for example, the Japanese who was told that the next town was "a stone's throw down the road," only to discover after walking five or six miles that this "stone's throw" happened to be 30 or 40 miles. Or think about the Japanese who was cautioned about the heat in parts of the U.S., where the temperatures "rise into the hundreds Fahrenheit."

American professors might tell their students that they have been "teaching for 100 years," even though they don't look nearly that old, or that a student has made "millions of mistakes," even though a Japanese observer can spot only ten. What is a Japanese to believe if told that there are 10,000 lakes in Minnesota? On the other hand, where a Japanese would say, "for about five minutes," when he or she wanted to indicate a relatively short time, an American might say, "3 minutes and 37 seconds," meaning an equally vague short time.

LANGUAGE

Very few Americans learn Japanese, seeming to believe that everyone can understand English when it is spoken loudly and slowly. The Japanese have, however, worked hard to learn English. Even so, many English metaphors are confusing to the Japanese. A university might be referred to as a "suit case" college if the students go home on the weekends, but think of how a Japanese might interpret that expression.

When a new American businessperson is "hanging out a shingle," a Japanese might wonder whether holding the nails would be of any help. Or if a real estate firm "puts some buildings on the block," the Japanese might wonder where they were before. Consider the television performer who said that he needed a *bomb* to wake up or the news commentator who had a *beef* with one football player but considered another a *recruiting plum,* and imagine what a Japanese might think.

Some of these confusing words and expressions can be clarified by consulting a dictionary. A Japanese who needs to understand *Pollyannaish* or a *Punch & Judy* act can look it up, but how is a Japanese ever to understand *party store*? A Japanese would logically conclude that a *party store* is either a store where one parties or where one may purchase parties. It would be hard for a Japanese to determine why somebody would want to rob one.

HOLIDAYS

Even holidays can cause problems. Because of the small area of Japan—about the size of Montana—the homogeneity of its population and the centralized system of government, the Japanese are characterized by "uniformity" and "regularity." Even the holidays are predictable in Japan, whereas in America, every person seems to observe a separate holiday system.

In Japan, schools, companies, and government offices all close on the 12 national holidays. Those holidays take place on a regular basis, and everything closes on a regular basis. It's difficult for a Japanese to understand why the post office is closed on the third Monday of February when classes are conducted at the local university. This would never happen in Japan. Both the post office and the university are state run, and both are closed on the same days to maintain the *synchronicity* of the social system.

The Japanese are born into and accustomed to a school system that observes a six-day week. The American admonition to "Enjoy the weekend," does not make much sense to the Japanese, who must plan weekends and holidays to take full advantage of the time available. An American might begin a three-week tour of the country by simply pointing his or her car in a direction and taking off, enjoying the unstructured time available. A Japanese, on the other hand, would plan the trip down to the last detail, ensuring that he or she would make the most effective use of precious vacation time.

POSSIBILITIES

Any who has read *Shogun* or seen the miniseries on TV knows that Japanese and American histories are very different. Japanese tradition is based on classification, rank, order, and harmony. American tradition is based on declassification, equality, exploration, and adventure. Both cultures, of course, contain elements of the opposite tendencies, but the Japanese place a much higher premium on the planned and the orderly than do Americans, who prefer spontaneous and unfettered energy to the restrictions of an imposed order. It is not by accident that the Japanese value the discipline of *haiku* poetry while Americans value free verse.

In spite of these differences in culture and language, the Japanese and Americans have made amazing progress in learning how to communicate with each other over the past 40 years. Because the cultures are so different, they have much that they can learn from each other, and the best way for this learning to take place is through individual representatives from the cultures who are willing to learn to communicate with and to understand one another. Although cultural differences can constitute communications barriers, problems will be minor and by no means insurmountable so long as good will is maintained along with a steady flow of communication, and as long as both parties have the attitude, "If you come 40 percent down the road, I'll travel the other 60."

People of good will can manage to reach agreements and to develop friendships in spite of difficulties encountered in the communication process. The willingness to accept differences, to suspend cultural prejudices, and to persevere in the face of misunderstandings can overcome even great differences in perceptions and expectations.

The Japanese need to understand that American casualness is a sign of friendliness and openness rather than a lack of respect, and Americans need to develop an appreciation for the Japanese conceptions of order and harmony. Both need to learn that difficulties in communicating are no one's *fault* but are the natural result of cultural and linguistic differences. Both also need to learn that these differences make communication between the Japanese and Americans richer and more interesting than it would be without them.

Is Ethics Good Business?

Abby Brown

Corporations are paying more attention to ethics issues. This article presents an interview with the founding director of the Center for Business Ethics and the executive director of the Ethics Resource Center. Both believe that companies are using ethical considerations to identify potentially risky projects. Read the two boxed ethical dilemmas in the article and consider how you would respond.

Personnel Administrator interviewed Dr. W. Michael Hoffman and Gary Edwards, two leaders in a growing movement to institutionalize ethics in business organizations. Hoffman is professor and chairman of the philosophy department at Bentley College in Waltham, Mass., and is the founding director of the Center for Business Ethics, a nonpartisan forum for the exchange of ideas in business ethics. Edwards is executive director of the Washington, D.C.–based, nonprofit Ethics Resource Center, which tries to spread ethics programs in the business community.

PA: *Gentlemen, in recent years, Americans have witnessed a rash of very questionable corporate activities. Among them were checkkiting scams, defense contract fraud, cover ups of health risks, and unfair takeover tactics. Some would also mention the lack of adequate safety standards leading to the tragedy in Bhopal. Do these disclosures indicate that, generally speaking, corporations are just not interested in ethics—that, ultimately, corporate greed simply takes priority over moral responsibility?*

HOFFMAN: I expect the majority of the public would answer yes. At least that's what recent polls show. Certainly, these opinions aren't without foundation. Without discounting the seriousness of these unethical activities, I think it's fair to say that most businesspeople don't want to see themselves or be seen by others in this way. Nor are people in business inherently less ethical than people in other professions.

EDWARDS: I agree. Surveys show that corporations are paying more attention to the institutionalization of ethics within their organizations. This is happening, in part, because they've become aware of the enormous costs of unethical activity—in fines and penalties, in increased governmental regulations, and in damage to their public image. But they also believe that ethical behavior is good business. It's unfair and naive for anyone to believe that most people working in the corporate world are not themselves concerned, even outraged, about unethical practices occurring in their profession.

PA: *Why then are we seeing so many outbreaks of corporate wrongdoing?*

HOFFMAN: We're seeing them, I think, not because businesspeople are less ethical than others, but because business gives so little thought to developing a moral corporate culture within which individuals can act ethically. Causes of unethical actions are quite often systemic and not simply the result of rotten apples in the corporate barrel. Ethical people can be brought down by serving in a bad organization, just as people with questionable ethical integrity can be uplifted, or at least held in check, by serving in a good one.

EDWARDS: I think it's also important to understand, that much of the unethical conduct we see going on in organizations results from a misreading by managers of the culture of the organization. Management by objective, focuses on outcomes: meeting the quarterly profits in performance objectives or exceeding them on the way up the corporate ladder. Somehow the system isn't tuned to the corners people have to cut, but only the ends they achieve. The message to middle managers—the ones often tangled in unethical acts—has been that what matters is achieving those objectives, and that somehow top management doesn't care how. Usually, that's not the case; it's simply that the system isn't tuned to the nuances of its policies.

PA: *What would be an example of such a "mistuning"?*

EDWARDS: The problem can be illustrated by an occurrence about five years ago at General Motors. Plant managers were found to have installed riostatic devices in their offices in order to speed up the assembly line. This was in clear violation of their labor agreement and of anybody's sense of fair play. The managers involved in the incident conceded they were wrong but said they felt the real sanctions ought to come down on their bosses, the people who had set performance objectives that were unrealistic. What the managers read from those performance objectives was: "Do what it takes." Their creative response to this reading wound up embarrassing them and holding the company up to public ridicule, if not criminal liability. All of this was obviously unforeseen by the company. Partly as a result of incidents like this, companies are beginning to look inward at how good people get compromised, and at the responsibility of top management to avoid putting people into no-win situations—into environments where employees feel they have to compromise their integrity to get the job done.

PA: *Before we probe further into the kinds of activities this corporate introspection is leading, I'd like to ask a question that is probably on the minds of many of our readers. Why should there be any focus on the morality of institutions at all? Doesn't this depersonalize the issue, obscuring the primary role of individuals? After all, individuals made the kinds of decisions that got General Dynamics and E.F. Hutton into trouble. You can't have institutional integrity without focusing on individual integrity—and isn't that the traditional domain of home and church?*

HOFFMAN: There's some truth to what you say, but it nevertheless overlooks the essential dynamics and reciprocity between individuals and organizations. Individuals don't operate in a vacuum. They gain meaning, direction and purpose by belonging to and acting out of organizations, out of social cultures that are formed around common goals, shared beliefs and collective duties. Corporations, like other social organizations, can and do influence individual decisions and actions; they have character that can exercise good or bad influences depending on goals, policies, structures, strategies and other characteristics that formalize relations among the individuals who make up a corporation. Therefore, when 60 to 70 percent of the managers of two major corporations feel pressure to sacrifice their own personal ethical integrity for corporate goals, as *Business Week* has reported, it's necessary and appropriate that we direct our attention to issues of corporate integrity.

FRATERNITY VERSUS SENIORITY

The A.O. Smith Corporation supplies frames and other parts for General Motors cars and trucks. Federal Labor Union 19806 represents about 4000 workers in A.O. Smith's Milwaukee automotive products division. When sales of cars and trucks declined in the fall of 1966, the A.O. Smith Corporation management decided that a manpower cut was necessary, and informed the union offices of their decision. The labor contract stipulated that when a 40-hour work week could not be provided for all union employees, the change had to be approved by the union.

A.O. Smith's management offered the union two alternatives: (1) that 200 employees be laid off for about six to eight weeks; or (2) that all production facilities be shut down, and none of the 4000 employees work on Friday, December 23, and Friday, December 30. Since Christmas Eve is a paid holiday but New Year's Eve is not, the workers would receive five-days' pay in the week before Christmas and four-days' pay the following week.

At the request of the union officials, management agreed to give the employees a choice between the two-day closing for 4000 and a six-to-eight week layoff for 200. If the latter were chosen, the layoff would start on December 19. Union leaders recommended the two-day closing to their members.

How do you predict the balloting will go?

The balloting was held at the headquarters of the union on Monday, December 12. The vote was 1220 to 954 in favor of laying off the 200.

PA: *How are companies doing this? Through the old corporate code of conduct?*

EDWARDS: No. This doesn't mean that corporate codes of conduct—so many of which were drafted in the post-Watergate era of the '70s—are being abandoned. We still see lists of do's and don'ts corresponding to clearly illegal or unethical actions such as bribery, price-fixing, conflicts of interest, improper use of company funds, improper accounting practices and the acceptance of gifts. But we're also seeing a shift to codes consisting largely of general statements putting forth the corporate goals and responsibilities, a kind of credo expressing the company's philosophy and values. The better codes consist of both. Rules of conduct without a credo lack meaning; credos without rules of conduct lack specific content.

HOFFMAN: But codes of ethics shouldn't be written to imply that whatever is not strictly prohibited is thereby allowed. There's no way that all ethical or unethical conduct can be exhaustively listed and mandated through a code, nor ought it to be. Business ethics, like all areas of ethics, has its grey areas that require individual discretion. A good corporate code of ethics should include certain managerial and employee guidelines for ethical decision making. They might include the principles and factors that one ought to think about before arriving at a decision. They might include sources both inside and outside the corporation through which advice could be offered. They could even include cases based on history that might clarify a future ethical dilemma. Whatever these guidelines include, they should make people aware that there may very well be some difficult ethical judgments—for which they're accountable—that will have to be made from the spirit of the code, rather than from its letter. This will place a greater sense of personal ethical responsibility on corporate employees and send a clear message that corporate integrity depends on individual integrity.

PA: *How many organizations are incorporating ethical values and concerns into their operations?*

HOFFMAN: The Center for Business Ethics made this a lead question in a lengthy questionnaire to the 1984 *Fortune* 500 industrial and 500 service companies. Of the 279 responding companies, almost 80 percent indicated that they were taking steps to incorporate ethics. Furthermore, the goal of being a socially responsible corporation was listed more often than any other as the primary reason for building ethics into the organization, by far overshadowing the goal of simply complying with state and federal guidelines. These findings corroborate what I felt I was learning piecemeal as director of the Center, namely that more and more of America's major corporations are trying in various ways to institutionalize ethics.

However, the survey also shows that they need to go much further before they'll be successful.

PA: *How so?*

HOFFMAN: Writing a code of ethics is an important step toward building an ethical corporation, but it's just that—a first step. To be effective, it has to be backed up by other kinds of support structures throughout the organization to ensure its adequate communication, oversight, enforcement, adjudication and review. We just didn't see this kind of support system in most of the companies we surveyed. Although 93 percent of the responding companies taking ethical steps have written codes of ethics in place—representing almost a 40 percent rise over a study for the Conference Board 20 years ago—only 18 percent have ethics committees, only 8 percent have an ethics ombudsman and only three have judiciary boards. It's difficult to understand how codes can be overseen and enforced adequately without a committee or ombudsman assigned to that task or how alleged violations of codes can be adjudicated effectively and fairly without a board committee for that purpose.

Furthermore, communication of the codes is inadequate. The survey showed that almost all the companies communicate them to their employees through printed materials, but only 40 percent do so through advice from a superior, only 34 percent through an entrance interview, and only 21 percent through workshops or seminars. Only 11 percent post them in the workplace. So we still have a way to go in these areas.

PA: *What role do you see for human resource managers in this effort to communicate and institutionalize corporate ethics?*

HOFFMAN: A large one. Increasingly, CEOs are involving their top corporate human resource managers in the planning, implementation and oversight of corporate ethics programs.

One area of major importance is ethics training and development. Human resource managers have a vital role to play in ensuring that the ethics programs are well conceived and fully integrated into the rest of the program, and that they include all levels of the corporation, not just upper management. The Center's survey revealed that only 35 percent of those companies engaged in ethics training involved hourly workers. In fact, ethics workshops should mix different levels of the corporation, from hourly workers to executive officers to members of the board to promote better understanding and communication among all members of the corporation relating to the ethical commitments of the organization. This, in turn, would build a stronger and more unified ethical corporate culture.

EDWARDS: On a more fundamental level, there's another reason why human resource managers are beginning to play a large role in this corporate ethics effort. The heart of the concern here is really not corruption but people. It's mainly good people who find themselves compromised, who find themselves in impossible situations. These are people companies are now trying to protect—and that's a human resource focus.

We're seeing, too, a shift in the focal point for company ethics from corporate counsel, with its written codes of conduct and policy documentation, to ethics as a management tool to help organizations identify ethics risks they face in their day-to-day business. Companies still need documents that articulate policies and procedures pertaining to proper conduct—how to handle proprietary information, for example. And they still need to state the major areas of business law that apply to them, such as antitrust and bribery. But organizations today increasingly are moving beyond codes of conduct dealing with crimes to dealing with ethics as a management tool to articulate corporate values and to manage risks of unethical conduct. This shift in focus—which puts considerable emphasis on training and development—has put human resource managers in the driver's seat; they're the people who are expected to make it happen.

PA: *Mr. Hoffman?*

HOFFMAN: I would agree that human resource managers play a vital role in building ethical corporations. Certainly there's an obvious role for them in ethics training. But I'd like to see a larger role for them in the administration of corporate ethics programs. They need to give more teeth to programs: improve codes and ensure that organizations have ethics committees, judiciary boards and a strong ombudsman.

Let's face it, it's impossible totally to prevent unscrupulous people from committing wrongful acts in any organization, but by working toward the development of the ethical corporation, human resource managers can lessen such acts and strengthen individual integrity.

THE DOUBLE EXPENSE ACCOUNT

Heinrich Picaro is a senior in the School of Business at Kruger College. Although he has had many job offers, he continues to go through interviews arranged by the college placement service. He reasons that the interview experience will be valuable and may even turn up a better offer. In fact, Heinrich has discovered a way to make money from job interviews.

On one occasion, two firms invited him to New York for a tour of the home office. He managed to schedule both firms on the same day, and then billed each of them for his full travel expenses. In this way he was able to pocket about $100. When a friend objected that this was dishonest, Heinrich replied that each firm had told him to submit an expense account, so he was not taking something he had no right to. One firm had not even asked for bills, which he interpreted as meaning they really intended to make him a gift of the money.

If you were personnel manager of one of these two firms and discovered what Heinrich had done, would you hire him? Or would you hire another student who had the same talents, grades, and personality as Heinrich, but who had split the total expense for his trip between your company and the other company?

Thirty-nine Chicago businessmen read of the action of Heinrich Picaro. Four said they would hire Heinrich, 28 said they would hire the other student, 7 said they saw no reason for preferring one student over the other.

Defining Terms in Technical Editing: The Levels of Edit as a Model

Mary Fran Buehler

The term editing *has been used to mean everything from reorganizing a document to correcting punctuation. This article discusses nine types of editing, grouped into five distinct levels of revision. These levels include* substantive editing, *which deals with the content of the report, and* integrity editing, *which checks to see that all the necessary parts of the document are in the final draft. Consider how you would apply these levels of editing to your own writing.*

We struggle all our days with misunderstandings, and no apology is required for any study which can prevent or remove them. —I. A. Richards[1]

THE GOAL OF THE TECHNICAL EDITOR is clear communication. But when we—as technical editors—try to communicate clearly about technical editing, we find that we have not yet agreed on standard meanings for our own terms. We may not even know whether we agree or not, because we may not know the meaning that someone else has attached to a given term. How can we know what we are talking and writing about if we do not have commonly understood meanings for the terms we use?

True, our field seems especially plagued with slippery terms. Take the word *copy,* for example. Or *style,* or *format,* or *composition.* (I remember telling an author that we would put his manuscript into composition. "Oh, no," he said. "It's already composed. I finished writing it last night.") But probably the slipperiest term of all is *editing*—which can mean anything from rearranging commas in a sentence to reorganizing a book-length treatise.

Clearly, publications people have recognized this difficulty; they have tried to solve it with such terms as *copy editing, manuscript editing, mechanical editing,* and so on. But these terms do not offer a general solution unless there is a common understanding of exactly what they mean. And, as we shall see, that has not been achieved.

TERMINOLOGY AND MISUNDERSTANDING

Take the term *copy editing.* The word *copy* seems a particularly unfortunate choice to restrict the meaning of *editing,* because all aspects of editing in which the editor works with a manuscript (or, more likely, a typescript) involve editing copy. But what other term was available? The range in meanings attached to the term *copy editing* can be illustrated with John B. Bennett's *Editing for Engineers*[2] and the venerable *Manual of Style* of the University of Chicago Press.[3]

Bennett divides the editorial process into three functions: styling, copy editing, and revising. Styling, he says, is the simplest and most mechanical, making sure that a manuscript will follow a prescribed format and style, and is "often the responsibility of an experienced secretary." Of copy editing, he says:

> *Copy editing* is more demanding but can also be carried out by an experienced secretary. Copy editing ensures that such conventions as grammar, spelling, and punctuation will be observed. Let us say, for example, that the organization asks its writers to avoid split infinitives. Determining that there are no split infinitives in a manuscript or correcting those that occur is copy editing. At a more demanding level copy editing ensures, for example, that the sequence of tenses is proper and that typographical errors have been found and corrected.[4]

We may ask why finding typographical errors (ordinarily a proofreading operation) is more demanding than ensuring correct grammar and punctuation, but we have a rather clear picture of the level of effort implied here by the term *copy editing.*

By contrast, the *Manual of Style* describes copy editing as follows:

> Copy editing . . . is the editor's most important and most time-consuming task. It requires close attention to every detail in a manuscript, a thorough knowledge of what to look for and of the style to be followed, and the ability to make quick, logical, and defensible decisions.[5]

The *Manual of Style* then goes on to detail the requirements of copy editing in slightly more than 11 pages, including instructions for editing text, footnotes, bibliographies, tables, illustrations, and front matter, as well as preparing indexes.

Still another aspect of copy editing is revealed in the Foreword of the *Prentice-Hall Author's Guide:*

> The duty of the "copy editor" is to go over your manuscript for errors or shortcomings in grammar, spelling, punctuation, sentence structure, organizational detail, inconsistency, redundancy, and like matters. The production editor assigns this work to a copy editor and corresponds with the author about the edited manuscript. The copy editor has no direct contact with the author and usually remains anonymous. We may often employ a professional free-lance copy editor—a practice common in the industry.[6]

But not all editors who do "copy editing" are anonymous. Editors often work closely with other publications people, attend meetings, establish schedules, and deal in a variety of ways with authors. These functions may make up an enormously important part of an editor's working life. As Robert R. Rathbone tells us,

> Good editors are bound to spend a harassed existence. Their job requires more than an ability to recognize and to repair poorly constructed manuscripts. To do a good job, an editor must also be a salesman, a teacher, a diplomat, a psychologist, and at times even a chaplain.[7]

This description—which is only too true, as any experienced editor knows—carries another important implication: that editors work on other people's manuscripts. Editing, in other words, involves at least two people, the editor and the author. And, if the editor is doing a good job, the image of the reader is always hovering in the picture, too. This rhetorical situation, with the editor squarely in the middle between the author and the potential audience, has given rise to the useful metaphor of the editor as bridge-builder and bridge-tester, as envisioned by Lola M. Zook in her early, classic paper, "Training the Editor: Skills Are Not Enough":

> [The editor] must learn that he is now, basically, a bridge between two people—an author and a reader—and that he has responsibilities to both of them. These two people want to bring about a transfer of ideas or information from the first individual to the second, and the editor's place is to help them succeed.[8]

In this usage, the term *editing* becomes a subclass of the more general term *revision,* which refers to changes that are made in one's own material, as well as in other people's. I have generally tried to use the terms *revision* and *editing* in this sense—restricting *editing* to work with other people, and using *revision* as the more general term.[9] But this distinction is far from universal. To quote Rathbone again,

> All of us, whether we realize it or not, are editors. Perhaps not *good* editors, but editors nevertheless. Every time we read over something we have written we become editors; every time we change a word or add a phrase we are editing.[10]

And Bennett, as we have noted above, considers *revision* a subclass of *editing,* instead of the other way around.

So the questions remain: not only "What is copy editing?" but "What is any kind of editing?" and even "What is editing?"

The questions are not academic. Editing is not a theoretical exercise, though it may be based on theory. Editing is behavior, a series of actions. The actions an editor takes, in working on a manuscript or in communicating with other people, are specific, and they have specific results, for good or ill.

Moreover, the editor's behavior can be considered not only as a series of actions but as a series of responses to variables—variables that include the subject matter, the condition of the manuscript, the knowledge level of the audience, the purpose of the message, constraints of time and money, and so on. These variables are usually built in—they must be lived with, coped with, responded to.

How do definitions fit into this maze of editorial behavior? *The role of definitions is to stabilize responses to a term.*[11] As we have seen, the term *copy editing* can represent a variety of responses the editor may be expected to make. Unless *copy editing* (or any other term) is carefully spelled out by means of specific actions (i.e., unless it is carefully defined), the editor is unable to select the correct responses to the term with any confidence of success.

If, then, the editor's conception of the editorial function is also a variable—because there is no agreed-upon definition of just what the editor is supposed to do—the entire effort can become a mass of uncertainties. And chaos, which often lurks just out of sight in any sufficiently muddled situation, is invited to move in a little closer.

Because the notion of stabilizing responses may sound mechanical—while good editing, obviously, is not—a word of clarification is in order. Definitions set forth the *structure* of editing, *what* is to be done. This fact in no way disparages the *art* of editing, *how* it is to be done.

Indeed, sound structure would seem to make art more likely. The less time and energy the editor requires to determine what to do, the more can be spent on the art of doing it.

TERMINOLOGY AND THE LEVELS OF EDIT

I do not believe it is possible, or even desirable, to legislate standard terminology in technical communication. Still, in light of our obvious lack of commonly understood terminology—and in the interest of agreeing on some broad definitions—I offer in this article the concepts and terms worked out at the Jet Propulsion Laboratory (JPL) and reported in *The Levels of Edit.*[12] The levels-of-edit concept grew out of a study of technical editing that Robert Van Buren and I conducted at the Laboratory in the mid-1970s. We analyzed the editorial function into nine types of edit, which we combined into five levels. The types of edit we identified are Coordination, Policy, Integrity, Screening, Copy Clarification, Format, Mechanical Style, Language, and Substantive. Each type of edit covers a different domain of editorial activity and is set forth as a list of specific instructions to the editor.

In more than 5 years' experience with the levels-of-edit concept at JPL, we have found that the concept and the terminology work: the types of edit cover the field of what we do, without any obvious losses or overlaps. In fact, we have stabilized our definitions to the extent that we collect publications costs according to the level of edit performed on each document (that is, the aggregate of the types of edit), and we use the computer printouts of these historical costs in estimating new work.

And while we at JPL do not presume that the terminology of the levels-of-edit concept is the best that could be devised, we know that other people in technical publications organizations have found the concept useful, too. In addition, the levels-of-edit concept is being discussed in workshops on editing[13] and has been incorporated into a recently published textbook.[14] Since more acceptance seems to be building for the levels-of-edit terminology as time goes on, the types of edit may provide a foundation for some agreement on editorial definitions.

Although the levels-of-edit concept was formally developed at the Jet Propulsion Laboratory, it does not represent a unique achievement in technical communication; others have devised or suggested different levels of editorial effort to fit different needs. For example, Dr. Frank R. Smith, Corporate Manager of Technical Information at McDonnell Douglas Corporation, had derived a stratified scheme of editorial definitions several years before the levels-of-edit concept was formulated at JPL.[15] And Lola Zook, in the 1967 paper cited above, pointed out that

> *the editor needs to work, consciously, at many different levels.* The writer—though he will do well to keep in mind the audience he wishes to reach—is primarily concerned with getting onto paper the things that are in his mind. It is the editor who must make sure that the things the author wants to say

are said in a way that will reach the reader—the particular readers of that particular item.[16]

Our effort at JPL was directed not toward defining terminology as such, but toward specifically identifying the domains of editorial activity and establishing the boundaries between them. In the process, we found we had also established the boundaries among the terms we used to identify the domains (or types of edit)—we had established a discrete, nonoverlapping terminology. A more complete discussion can be found in the article "Controlled Flexibility in Technical Editing: The Levels-of-edit Concept at JPL,"[17] or in *The Levels of Edit* itself. The types and their terminology are discussed below.

Coordination Edit. The only type of edit that does not deal with copy or manuscripts, the Coordination Edit deals with people, budgets, schedules, paperwork, meetings, planning, liaison, monitoring, and other aspects of the publications process, as well as the explanation and interpretation of publications procedures.

By identifying the Coordination Edit as a discrete activity, we reveal the heavy administrative and communicative responsibilities inherent in producing publications, as well as the editor's need for skills in communicating effectively with *people*—an element that can be easily obscured by the somewhat common image of an editor as someone who spends the day bent quietly over a manuscript.

The free-lance copy editor who remains anonymous, as mentioned above, would presumably be concerned with fewer elements of the Coordination Edit, although schedules and budgets would seem to be universal. But for any editor who is not anonymous in relation to authors and others, the Coordination Edit is a crucially important activity—perhaps, in the last analysis, the most important activity of all.

Policy Edit. The Policy Edit enforces those elements of institutional policy that are more important than simple "house rules" on format or mechanical style. Elements of Policy Edit can almost be defined by saying "These are the things we had better do or we will be in trouble." Because the authority for the Policy Edit comes from the individual institution or its parent organization, each set of Policy Edit rules may be different. At JPL, for example, one Policy rule dictates that distribution lists of (except in some cases) disclaimer statements may not be included in external technical reports. In other companies or institutions, exactly the opposite rules may apply—the point being that, in either case, the Policy Edit conforms to the requirements laid down by management.

Integrity Edit. The Integrity Edit, primarily a matching function, makes sure, for example, that when the text of a report says "See Figure 1," there really is a Figure 1. The same matching activity is applied to tables, references, footnotes, appendixes, and the sections and subsections of the report. The Integrity Edit also ensures that there are no incorrectly identified sequences; that is, no two elements have the same alphanumeric designation and there are no gaps in the sequences (e.g., if there are 15 figures, Figure 13 is not missing) and no two figure captions or table titles are identical.

What the Integrity Edit does not do is to determine that the "Figure 1" mentioned in the text is *actually* the graph or diagram identified as "Figure 1." This is an important distinction; the Integrity Edit, as a simple matching activity, does not delve into the technical meaning of the report, which is the province of Substantive Edit. To determine whether the "Figure 1" included in the report is actually the "Figure 1" cited in the text, the editor may be required to have an extremely sophisticated knowledge of the scientific discipline being reported, particularly if a series of very similar graphs is grouped together, as can easily happen. In such a case, the author of the report may be the only person who can say with certainty that "Figure 1" is correctly cited.

Screening Edit. The Screening Edit represents the minimum editorial standard that is acceptable in a JPL external report. Primarily, the Screening Edit is a minimal language edit, although it also requires the correction of certain other unacceptable items (e.g., hand-drawn graphs).

The Screening Edit allows the Laboratory to save time, money, and effort by accepting author-prepared text and artwork for publication (assuming it is sufficiently well executed) and by giving this input a minimal screening to take out any errors that fall below the level of acceptability.

The elements of the Screening Edit are specified: no misspelled words, no subject-verb disagreement, no incomplete sentences, no incomprehensible statements. (By *incomprehensible,* we mean sentences in which words have obviously been left out or garbled.) The Screening Edit could have been called a Minimal Edit—which is what it is—but that terminology could have been confusing, because *minimal* is also used generically: authors can ask for "just a minimal amount of editing," without knowing what, exactly, they will get in return. The Screening Edit lets both authors and editors know what to expect, thus stabilizing responses to the term on both sides.

Because the elements of the Screening Edit are institutionally imposed, it is possible that each publications organization could set up different standards for its Screening Edit, depending on how the institution defined minimal acceptable quality.

Copy Clarification Edit. The Policy, Integrity, and Screening Edits can all be performed on publications that the author will prepare. When the manuscript must be more fully edited and composed, however, additional types of edit come into play. Copy Clarification and Format Edits are mandatory in order to convert manuscripts into publications. The Copy Clarification Edit provides clear, unambiguous instructions to the compositor or illustrator. Handwritten Greek letters are identified; subscript and superscript positions are clearly marked in mathematics; acceptable equation breaks are indicated; the tops of photographs are marked; and so on. The Copy Clarification Edit may seem mechanical, but it is not: it requires the editor to look at every aspect of the report from the point of view of other people (compositors, illustrators) who may not be familiar with the subject matter and who must never have to guess what a symbol or an instruction may mean.

Format Edit and Mechanical Style Edit. Because so much confusion exists in technical communication between the terms *format* and *style,* the Format Edit and the Mechanical Style Edit are discussed together here, in order to emphasize the difference between the two. It will probably never be possible to dispel the style-format confusion completely, but enough progress has been made under the levels-of-edit concept to establish a working understanding of the differences.

The term *style* has the further disadvantage of many different, sometimes overlapping, shades of meaning: *style* can refer to literary production, "house style" as set forth in style guides, and typographic style, among others. To help stabilize our meaning, we chose the term Mechanical Style—no other use of the term *style* is made.

As types of edit, Format and Mechanical Style both refer to graphic and typographic treatment, but in different ways. The Format Edit is a macro approach concerned with the physical display of publications elements; it considers heading systems, type fonts, column widths, image areas, and similar matters. The essential questions answered by the Format Edit are "How does it look on the page?" and "Where is it placed in the book?" Some considerations in the Format Edit may be aesthetic and subjective—the choice of type font or column arrangement, for example—and some format decisions may be limited mechanically by what is available or by traditional usage. But whenever the Format Edit designates similarities and differences in content, it operates on *blocks of meaning* (e.g., sections, headings, paragraphs, listings, footnotes, appendixes) rather than on individual units of meaning (e.g., words or numbers).

The Format Edit, for example, may specify that all first-level heads are set in all-capitals, with second-level heads in caps and lowercase. Or the abstract of a report may be given a specified location and set in a different column width from that of the report body. These format instructions will apply regardless of the individual words in the headings or the text. But the format typography will signal to the reader the hierarchical organization of the material, the relation of subsections to larger sections, and so on, providing a typographical road map by which readers can quickly orient themselves in the terrain of an unfamiliar report.

The Mechanical Style Edit, by contrast, is a micro approach typographically and is applied to *individual units of meaning:* it answers the question "Are similar units of meaning treated consistently in graphic or typographic terms?"

For example, the Mechanical Style Edit makes sure that a publication does not randomly refer to "Reference 1," "reference 1," and "Ref. 1." Because the reference number citations represent similar units of meaning, they are treated consistently. A unit of meaning, in this sense, can be a single letter (W for *watts*, rather than *w*), a numeral (4 rather than *four*), a word (*nonparallel* vs. *non-parallel*), or a group of words (*solid state device* vs. *solid-state device*).

As they are defined at JPL, the Format Edit and the Mechanical Style Edit can easily be performed separately, and, in fact, the addition of the Mechanical Style Edit (along with the Language Edit) raises the editorial effort from Level 3 to Level 2. Perhaps the most crucial difference between Format Edit and Mechanical Style Edit is that the Format Edit *must* be performed whenever material is to be composed or illustrated. The format instructions must be given to the compositor or the artist. The Mechanical Style Edit, however, is optional; it represents a desirable level of publications quality—typographic consistency on a micro scale.

Language Edit. The Language Edit is a full treatment of grammar, punctuation, usage, parallelism, and similar components of what is often considered to be an editor's major—or even only—function. The Language Edit incorporates the Screening Edit but goes a great deal further. Because usage is variable and because editors—being only human—have different degrees of expertise in different aspects of language, the Language Edit is based on specific references: *Webster's Third*[18] for spelling, the *U.S. Government Printing Office Style Guide*[19] for punctuation, and Follett's *Modern American Usage*[20] for usage. Without specific references, we may find that some editors, for example, are meticulous about Fowler-style distinctions between *that* and *which,* while other editors may never have read Fowler[21] or may consider Fowler to be old-fashioned. Or some editors may be carefully concerned about dangling participles, while others adopt a permissive approach. Standardized references, like standardized definitions of terms, tend to stabilize these variables.

Substantive Edit. The Substantive Edit deals with the content of the report, although not as review for technical accuracy. (If an organization specifies that the editor does review publications for technical accuracy, that review would be a component of the Substantive Edit.) As defined at JPL, the Substantive Edit emphasizes appropriate organization, ensuring that the material is grouped and subdivided in a rational manner, that apparent contradictions or inconsistencies are resolved, and that the presentation is complete and coherent. These requirements are set forth in a comprehensive list of actions concerning such elements as figures and tables, as well as the publication as a whole.

Interestingly, the University of Chicago Press *Manual of Style* identifies *substantive editing* as one of two elements making up the editorial function—the other being *mechanical editing,* which seems to combine the Mechanical Style Edit with elements of the Screening Edit. The *Manual* defines substantive editing to include "rewriting, reorganizing, or suggesting other ways to present material," and goes on to say that "The editor will know by instinct and learn from experience when and how much of this kind of editing to do on a particular manuscript." The *Manual* further states that "Since every manuscript is unique in the amount and kind of substantive editing desirable, no rules can be devised for the editor to follow."[22]

Analysis, Synthesis, and Definition

The terms of the levels-of-edit concept can be used straight—on a one-to-one basis—to denote the specific editorial domains. But the terms can also be used both analytically and synthetically to define other terms. Analytically, the terms can be used to break down fuzzy and ambiguous words into understandable, discrete units. Synthetically, the terms can be used to build up terminology that can be specified in detail. An example of each process is given below.

Analysis. Consider the word *style*: a recent article, "Four Kinds of Style," in the compendium *Scientific Information Transfer: The Editor's Role,* stated that "Style may be the most common piece of jargon used by writers and editors; it may also be the most confusing,"[23] and asserted that the word *style* has four distinct meanings as used by authors and editors:

1. Editorial style (generally speaking, what is included in a Mechanical Style Edit)
2. Typographic style (included in a Format Edit)
3. Literary style (described as the author's distinctive mode of writing)
4. Usage style (defined as dealing with "grammar, syntax, idiom and shades of meaning," which would fall within a Language Edit).

The levels-of-edit terminology, as demonstrated above, can be used to differentiate these four types of "style," giving specific definitions to the three types that are of primary concern to the editor, and leaving the remaining type to be called Literary Style—the province of the author—which seems a useful and relatively unambiguous term.

Synthesis. The terms can also be used to build up a definition (as we use them to build up levels of edit). If we wished to use the term *copy editing,* we could easily define the term by saying that copy editing includes, for example, the Integrity, Copy Clarification, and Format Edits. Or we could add Mechanical Style and Language Edits to *copy editing.* The various ways in which the term *copy editing* is used today suggest that even these possibilities may not exhaust the range of meanings currently attached to the words—but we could pin down the meanings if we wished to.

The Need for Definition. And so we return to the question "What is editing?" As elements of the levels-of-edit concept, the types of edit specify what a technical editor does. Because these types indicate distinct domains of activity, they can be constituted with different specifics by different organizations and still carry a commonly understood meaning—Format Edit would still refer to blocks of meaning, for example, and Mechanical Style Edit would still operate on units of meaning. Language Edit and Substantive Edit could still play their individual roles.

Whether we in technical communication use the terminology of the types and levels of edit or not, we must find some commonly understood terms with which we can communicate, if we are to prevent or remove our misunderstandings—unless we are content to struggle with misunderstandings for all our days.

ACKNOWLEDGMENTS

I am grateful to Professor Walter R. Fisher, Department of Communication Arts and Sciences, and Professor Edward Finegan, Department of Linguistics, both of the University of Southern California, for their valued comments on this article. To Robert Van Buren, my colleague and the senior author of *The Levels of Edit,* go my thanks for his always perceptive criticism, especially on the distinctions between Format and Mechanical Style Edits.

The research described in this article was carried out in part by the Jet Propulsion Laboratory, California Institute of Technology, under contract with the National Aeronautics and Space Administration.

REFERENCES

1. I. A. Richards, *The Philosophy of Rhetoric* (New York: Oxford University Press, 1936, reprinted 1976), p. 3.

2. John B. Bennett, *Editing for Engineers* (New York: Wiley-Interscience, John Wiley & Sons, 1970).

3. *A Manual of Style,* 12th edition (Chicago: The University of Chicago Press, 1969).

4. Bennett, p. 2.

5. *A Manual of Style,* p. 40.

6. *Prentice-Hall Author's Guide* (Englewood Cliffs, New Jersey: Prentice-Hall, Inc., 1975), p. x.

7. Robert R. Rathbone, *Communicating Technical Information* (Reading, Massachusetts: Addison-Wesley Publishing Company, 1967), p. 88.

8. Lola M. Zook, "Training the Editor: Skills Are Not Enough," in *Technical Editing: Principles and Practices,* Anthology No. 4 (Society for Technical Communication, 1975), p. 13. (Originally presented at the 14th International Technical Communication Conference, Chicago, Illinois, May 1967.)

9. For example, in "Creative Revision: From Rough Draft to Published Paper," *IEEE Transactions on Professional Communication,* December 1976, pp. 26–32.

10. Rathbone, p. 84.

11. W. K. Wimsatt, Jr. and Monroe Beardsley, "The Affective Fallacy," in *The Verbal Icon* (University of Kentucky Press, 1967), p. 22.

12. Robert Van Buren and Mary Fran Buehler, *The Levels of Edit,* 2nd edition, Publication 80-1 (Pasadena, California: Jet Propulsion Laboratory, 1980), (available from the Government Printing Office).

13. For example, the 28th Technical Writers' Institute, Rensselaer Polytechnic Institute, Troy, New York, June 1980; and the Publications Editing Workshop, University of Illinois, Urbana, Illinois, February 1981.

14. Carolyn J. Mullins, *The Complete Writing Guide* (Englewood Cliffs, New Jersey: Prentice-Hall, Inc., 1980), pp. 217–235.

15. Frank R. Smith, McDonnell Douglas Corporation, St. Louis, Missouri, personal communication.

16. Zook, p. 14.

17. Mary Fran Buehler, "Controlled Flexibility in Technical Editing: The Levels-of-edit Concept at JPL," *Technical Communication,* Vol. 24, no. 1 (First Quarter 1977), pp. 1–4.

18. *Webster's Third New International Dictionary of the English Language, Unabridged* (Springfield, Massachusetts: G. & C. Merriam Company, 1965).

19. *U.S. Government Printing Office Style Manual,* Revised Edition (Washington, D.C.: Government Printing Office, 1973).

20. Follett, W., *Modern American Usage* (New York: Hill & Wang, Inc., 1966).

21. Fowler, H. W., *A Dictionary of Modern English Usage,* Second Edition (New York: Oxford University Press, 1965).

22. *A Manual of Style,* p. 40.

23. Wendell Cochran and Mary Hill, "Four Kinds of Style," in *Scientific Information Transfer: The Editor's Role,* Proceedings of the First International Conference of Scientific Editors, April 24–29, 1977, Jerusalem: edited by Miriam Balaban (Hingham, Massachusetts: D. Reidel Publishing Company, 1978), pp. 341–342.

Designing and Delivering Presentations and Workshops

Ric Calabrese

Many professional people have to make oral presentations or conduct workshops on the job. This article presents guidelines for planning and organizing a presentation, arranging the room, delivering the presentation, talking to listeners later, and conducting workshops. Use the preparation checklist to avoid last-minute problems before a presentation and the evaluation sheet to assess your own style after a presentation.

Introduction: Knowledge Plus Risk Equals Growth

Today a significant responsibility included in the day-to-day life of many professionals includes making oral presentations and/or conducting workshops for groups of clients, customers, patients, other professionals, and the general public. Unfortunately, the skills needed to plan and deliver a message orally and articulately cannot be learned by reading alone; one must be prepared to digest the information and then have the courage to take risks.

While entire books are written on the broad and general subject of giving oral presentations, the purpose of this article is to develop and to discuss concisely the most salient principles occurring in the process of preparing and delivering oral presentations and workshops.

Following are suggestions for advance planning and organizing of a presentation, advance physical arrangements of the room, delivering the presentation, postpresentation actions, and conducting workshops.

Advance Planning and Organization for the Presentation

Less is more when giving oral presentations. The key to holding an audience, particularly one with limited time to absorb the speaker's ideas, is to be coherent and to communicate simply; otherwise, their minds will wander. The secret of conciseness and simplicity is for the speaker to know his or her objectives prior to planning the talk. What does he or she want to accomplish? What changes in the attitude or behavior of the audience are desired? Does the speaker want them to perform a task or recall some information? A common mistake made among untrained presenters is to attempt to cover too much in the time allotted. Inexperienced speakers often feel the need to parade their expertise and thus they overload the audience with information.

Unlike reading, where one can go back and reconsider an idea, listening requires ongoing concentration. When the brain begins to feel overloaded and saturation sets in, it protects itself by shutting down. Has not everyone had the experience of pretending to be listening to an overly meticulous speaker while the imagination went elsewhere on holiday? Oral presentations need to be limited to a few major points that can be clearly explained and reinforced through details, examples, and a variety of media. Too much information and too many different points defeat the purpose. When listeners know where the presentation is going and are able to follow the presenter's reasoning, grasping examples with ease, they are much more likely to give their full attention.

Although presentations are often as brief as 10 minutes or as long as 90, presenters would do well to remember when adapting their goals to their group, what the Reverend William Sloane Coffin said about the length of an effective sermon: "No souls are saved after twenty minutes." Once the presenter has tailored the general and specific goals to a particular group, the next task is to organize them.

Introduction Critical to Speaker Credibility

Each of the three generally accepted divisions of a presentation (Introduction, Body, and Conclusion) requires its own internal organization and serves a specific function. The Introduction serves as the speaker's opportunity to establish credibility, to link himself or herself in some way with the audience, to let them know in what ways they will be better off as a result of having attended to the speaker, and to describe for them in some serial fashion what will be covered in the presentation.

Often when speakers are introduced, their credentials are presented in advance; however, there will be times when no one is present to introduce the speaker or, when "introducers" are available, they may neglect to mention important items related to the speaker's credibility. While self-serving comments delivered in a braggadocio manner may have a negative effect on audiences, audiences do, nevertheless, want to know that the person talking to them is worthy and knowledgeable. Presenters, therefore, should subtly let them know during the introduction that they are qualified. For example, one might say, "In an article I wrote last year for the Journal XYZ . . . ," or "While I was completing my graduate work at . . . ," or "Of the several hundred accounts I have worked with in the past. . . ."

Audiences tend to be more attentive when they believe that the person speaking to them is able to relate to their circumstances. During the introduction, whenever it is possible, speakers would do well to underscore any connection they may have with a particular group. For example, the speaker might say, "I have lived in this community for fifteen years . . . ," or "I see we are all baby boomers and share many of the same sentiments." If one gives it some thought, almost all audiences have some traits with which the speaker can identify.

During the introduction the speaker needs to let the audience know how the topic relates to their needs; in other words, *answer* the unasked question in everyone's mind, "What is in this for me?" Abraham Maslow has synthesized the basic human needs into five areas: physiological, safety and security, belonging and social activity, esteem and status, self-realization and fulfillment. When the presenter announces that after listening to the speech, the listeners will in some way be better able to control their health, be more secure, be in a position where others think better of them or respect them more, think more of themselves, or feel they have the knowledge to develop latent potential in themselves, the audience "perks up" and prepares itself to attend to the forthcoming message. Not every human need can be related to every topic, but as many as are possible and appropriate should be suggested during the introduction. During the "body" and "conclusions" of the speech as well, the speaker should remind the audience how what is being discussed can be related to fulfilling their own needs.

Finally, the last critical component to be included in the introduction is a list of topics that will be discussed. It might be presented something like this: "Today I intend to discuss three specific points. Number one, I will discuss the relationship of diet to heart disease; number two, I will illustrate the fat and cholesterol content of foods; and thirdly, I will model how to order at a restaurant." As mentioned above, not only does this help the audience to listen to the talk with an expectation of what is to come, but the organization itself adds to the "halo effect" and increases the audience's perception of the speaker's credibility.

Three Key Objectives of the "Body" of the Presentation

The Body of the presentation is the second major division and is where the points mentioned in the introduction are actually developed. Presenters need to have a rationale for the way they decide to organize this major section. The overall objectives generally are threefold: that the audience fully understand the message, that the audience believe, and that they be comfortable enough with the speaker to share their objections in the event that they are confused or wish to challenge the presenter.

To safeguard the first objective, understanding, speakers need not only to construct their messages clearly and concisely but also need to design visual aids, handouts, and/or participative experiences to enhance the audience's understanding. The second objective, audience belief, needs continuously to be reinforced throughout the presentation. Credibility or lack of credibility continues to be inferred from the presenter's style and the quality of all aids associated with the presentation. The third objective, to develop a rapport with the audience to the extent that they dare challenge or question, is critical. When objections are unexpressed, presenters may infer falsely that the audience agrees and understands, causing them to move too quickly from one point to another, leaving confused members behind.

A Presentation Can Be Concluded in More Ways than One

Like the first two, the third division of the presentation, the Conclusion, may be handled successfully in many different ways. The key is that all the ingredients be included somewhere. Ingredient one is for the audience to "feel" that the presentation is "winding down" and is about to end. This needs to be done gradually and smoothly. It sounds nonprofessional and haphazard to end with remarks like "any questions," "that's all folks," "thank you for your attention." Remember the presenter's credibility is influenced by the audience's perceptions of how well he or she is organized. Clues such as "In conclusion . . . ," "To summarize," "Before concluding, I want to leave you with one more thought," are helpful in letting the audience know that the presentation is about to terminate. If a summary is warranted, it should be given; if a final plea or pitch is warranted, that should be given; if a final quote, anecdote, or joke makes the point one more time, then that is appropriate.

One last word about conclusions is be proactive in asking if there are any questions. Of course there are, but people are often hesitant to ask. The speaker should prepare a few for the audience, and then call on a participant who has been paying close attention for a reaction. One might say, for example, "I noticed you looked confused when I was discussing the formula to determine appropriate profit levels. What is it you would like further discussion of?" After responding to that first question generated by the presenter, it is often much simpler to get others to respond when the question is asked: "Are there any *other* questions or comments?"

Correlation Exists between "Form" and Speaker "Credibility"

Although the discussion of the presentation's design is generally divided into three areas of introduction, body, and conclusion, there is no one perfect way to organize. Speakers with different styles can be equally successful. There are, however, several points to consider when deciding upon a presentation's format. A fact that most people are not aware of is that the presentation's organization is critical primarily because it relates to the presenter's credibility.

For example, if a speaker were to deliver a talk to two different audiences, giving one audience ten slices of information in an organized way and giving the second audience the same ten slices of information in a disorganized way, both audiences would retain about the same amount of information after hearing the talk. But only one audience would consider it seriously or possibly change their attitude toward the subject matter.

When an audience hears an organized speaker, one who gives the audience a sense that the presenter knows exactly where he or she is going, with a defined beginning, middle, and end, they are more likely to infer that the speaker is competent in the area being addressed. On the other hand, many brilliant and qualified professionals are not taken seriously when giving presentations because they sound too "loose," too unprepared, too disorganized.

Media credibility is related to inferences regarding presenter's credibility. All media, the way they are designed, presented, and utilized, are an extension of the speaker and, consequently, reflect directly the speaker's credibility. Media include such things as handouts, blackboard, flip chart, pad or easel, overhead projector, 35 mm slides, motion pictures, sound recording, videocassettes, samples of products, models or mock-ups, and so forth. "Excellence" tends to be inferred from those that are obviously carefully put together. For example, a presenter whose overhead transparencies are executed with large print that can easily be read (as opposed to ordinary typewriter print, which cannot easily be read from a transparency) allows for the inference of an experienced and considerate presenter. Other signs include using stenciled letters in bold colors on posters, rather than sloppy printing or cursive writing. The condition of the poster itself leads either to positive or negative impressions of the speaker. Those that are discolored, bent, and old looking suggest that the presenter doesn't care enough to add fresh aids for his or her group.

Even the quality and color of the paper used in handouts can add or detract from the overall impression. If it is possible, the presenter is wise to try and coordinate all aids in colors that may be symbolic or meaningful. For example, a presenter giving a talk to a group on Saint Patrick's Day could make use of green; a talk to an Italian-American Club could be done in red, white, and green. This may sound superfluous, but audiences do respond on an unconscious level to the extra care and preparation the speaker has made in tailoring a presentation on their behalf.

An unfortunate judgment is often made regarding presenters who do not carefully proofread all their materials, including handouts, overhead transparencies, flip charts, posters, and the like, prior to the audience's exposure to them. While it is not rational, audience members do tend to make negative inferences about how credible a presenter is in his or her area of expertise if he or she can't spell, count, or interpret statistics. PROOFREAD CAREFULLY.

Whenever presenters are going to be using audiovisual equipment with which they are unfamiliar, it is critical that they arrive early enough to study it beforehand. The most competent presenters may lose their credibility entirely by not knowing how to turn on a machine or advance a slide.

Language and Style of Oral Presentations Differ from Those of Written Presentations

The written text and the oral presentation are entirely different. There is no objection to a presenter writing out the entire talk, carefully organizing it according to topics, causes and effects, chronology, or whatever else seems appropriate. Once the talk is written, however, the speaker needs to recognize that the written manuscript represents the "science" of a presentation; the actual delivery represents the "art." Each time it is delivered, it should be somewhat different, using different words, different examples, different anecdotes, and so forth to suit each particular audience and situation.

The word "choice" in the spoken language tends to be different from the word "choice" used in the written language. Sentences in oral speech tend to be simpler, shorter, and sound more conversational, including common words and contractions, while the written manuscript may be more erudite and academic. The only way for speakers to develop the "art" of delivery is to rehearse from a simple outline and not from a manuscript, and to rehearse in front of real people who will react and comment, not mirrors, walls, or car windshields.

Oral presentations ought never to be read or memorized. There are good reasons for not rehearsing from a manuscript. The speech tends eventually to become memorized, and that can be deadly. Once a speech is memorized, speakers tend to become more "speaker-centered" than "audience-centered," which means that they tend to become more concerned about whether or not they can remember each line exactly as it is written on the manuscript and less concerned about whether or not the audience is enjoying, learning, listening, and understanding. Another problem that arises from manuscript speaking is that it is DULL! Because the facial expressions and vocal intonations are not spontaneous, the monologue tends to sound memorized and can easily become boring to listeners.

ADVANCE ARRANGEMENTS

Take control of seating the listeners before everyone settles down. It may be their boardroom, gymnasium, or meeting hall, but it is the presenter's "show." Conscious decisions should be made about whether to pull the group into a circle, half circle, rows, around tables, and so forth. When it is possible to know beforehand which persons are the most influential, their seats should be reserved and placed in the best position to see, hear, and appreciate any visual aids, as well as the speaker. A final checklist for presentations is found in Figure 1.

Creating positive impressions begins the moment the presenter enters the room. The presenter is being "sized-up" and adding to the presentation's ambiance from the moment he or she enters the room. Therefore, the presenter should, whenever possible, arrive early and make an effort to meet people. The speaker's self-confidence, whether real or feigned, will relax the audience and increase their perceptions of his or her desire to share information. Presenters

Attend to details and prepare a final checklist. Listed below are a few questions that the presenter might consider in order to avoid last-minute problems.

Do I have my presentation notes?
Do I have all my supporting materials?
Have I enough copies for each of the attendees?
Will the facility be unlocked and open?
Are the tables and chairs arranged to suit my design?
Do I understand how to operate the lighting system?
Do I understand how to operate the ventilation system?
Do I know the location and operational condition of the electrical circuits?
Do I know how to work the projector?
Do I have an extra bulb, cassette, video, extension cord, etc.?
Will the projection screen be in place and adequate for this size group?
Do I have the type of sound system I require? Is it working?
Have arrangements been made to handle messages during the presentation?
Are there arrangements for hats and coats?
Will there be a sign to announce the place of the presentation?
Will someone be introducing me, and have I given him or her all the information I want shared with the group?

Figure 1 Final Checklist

should never volunteer any negative information regarding their own stress or fear of speaking. The audience wants to learn and enjoy, and when they are aware of the speaker's fragility or stage fright, they tend to become nervous themselves in sympathy.

The presenter, waiting to be introduced, seated among the audience or on stage, should be aware that those people who know he or she is the guest speaker will be watching his or her every move. That means that even before beginning the presentation, speakers must be careful to smile, look confident, and extend themselves to others. Once introduced, the way the speaker walks up to the podium is critical. During those first moments an initial impression is being created. The speaker should consciously walk confidently, looking and smiling toward the audience. Prior to uttering the first words to the audience, it is a good technique to spend a long three seconds just looking out at the audience, smiling and establishing eye contact with several individuals. This allows them to infer poise, confidence, and the speaker's desire to "connect" with them.

DELIVERING THE PRESENTATION

Never share internal feelings of fright or anxiety. Speakers experience themselves in the situation from the inside out; the audience experiences them from the outside in. Simply, that means that if asked if he or she is nervous, the speaker should always answer "NO!" The audience is picking up the "tip of the iceberg" from their observations of the "outside" of the speaker; they do not actually feel the intensity of the speaker's anxiety and will probably be totally unaware of it, unless it is brought to their attention through one's own confession. One must act confident, even when one may not feel it internally.

One of the worst things presenters can do is to admit to an audience that they are scared, ill prepared, missing material, or have done the presentation better in the past for other groups. The audience does not know what it might be missing, and are generally much less critical of speakers than they are of themselves.

"Stage fright" is a vestige from our ancient past. The feelings commonly referred to as "stage fright" may date back to the dawn of the human race, a time when our prehistoric ancestors had to survive by living in caves and sharing the food supply with other beasts. Faced by a predator, our ancestors had a genuine use for a sudden jolt of energy, which gave them the power to do battle or run, fight or take flight. The vestiges of this power, stemming from the secretions of the adrenal gland, still manifest themselves today when people sense danger. Who has not felt that ice block in the stomach while being reprimanded by the boss, or experienced the sweaty palms and racing heart while walking into a room full of strangers?

Occasionally one still reads in the newspaper accounts of individuals under conditions of fear or danger exhibiting superhuman strength; the father, for example, who lifts the car off his child, who has been pinned under the wheels. This is an example of the power that comes with the adrenaline jolt; however, when one is unable to fight, run, or in some other way utilize this surge, he or she may become overwhelmed by the internal feelings themselves. It is this feeling prior to and during a presentation that is commonly labeled "stage fright."

The best safeguard against stage fright is adequate preparation and rehearsal. The more one practices in front of *live others,* the less nervousness one will have. Other ways of dealing with this feeling include being active during the presentation and "acting" calm and confident. If one knows one is going to be full of extra energy because of one's adrenaline flow, one could plan to engage in demonstrations during the presentation, distribute handouts, use a pointer, or any other activity that involves motion. Motion is a release for the tension and anxiety and allows the audience to infer enthusiasm from the speaker's movement rather than fright, nervousness, or tension. It may not work for everyone, but many people can learn to control their public behavior if they visualize themselves as acting. When one is "on" giving a presentation, one is acting. Regardless of how one may feel, one can *act* confident, calm, and poised.

Movement should be intentional and meaningful. All movement should be meaningful. Do not pace. Presenters ought to look for opportunities to break the invisible barrier between themselves and their audience. Walking toward the audience, walking around the audience, walking in and out of the audience, are all acceptable ways of delivering a presentation. What is not acceptable is pacing back and forth, particularly with eyes down, pulling thoughts together before uttering them.

Movement into an audience is a communication vehicle in itself. When the speaker penetrates that invisible barrier between speaker and audience, he or she tells the audience of a desire to connect, to be close, to better "sense" what they are feeling about him or her and the content. In fact, as one walks among the audience, one can begin to be seen from a different perspective and may gain new insights into how to better clarify the presentation's points from this experience.

Unnecessary barriers between the presenter and audience should be omitted. Presenters do best when they omit all barriers between themselves and their audience. Avoid using a podium or lectern, even when one is provided by the sponsoring organization. Of course, there may be times when, because of the quantity of material, a place to store things may be required. Even under this condition, using a table to set handouts and other materials on is preferred to a podium.

A lectern should only be used when the speaker does not intend to move about; intends to lecture and needs a stand to rest against and place notes upon. Adults generally do not learn optimally through the lecture method and unless the speaker is extraordinarily good, straight lecture behind a lectern should be avoided. When one delivers the message standing in front of the group, without a barrier, one is more disposed to stop the talk to respond to their verbal or nonverbal feedback. Gestures and movement too can be more expansive and visible without the lectern barrier.

Voice cadence, pitch, variation, and rate are communication vehicles. Vocal inflection and variation add interest and add to the impression of speaker enthusiasm. For some people controlling this variation is simple and natural, but for others it is a challenge. Nevertheless, the presenter needs to attend to voice modulation. The goal when speaking in front of a group is to sound natural and conversational; however, what sounds natural and conversational when one is standing in front of a large group is not the same as what sounds conversational in a small face-to-face group.

"Natural and conversational" from the presenter's point of view is exaggerated. The highs need to be a bit higher and the lows need to be a bit lower. What may sound to the presenter's ear as "phony" and "theatrical" generally sounds far less so to the listener. In any case, a delivery that is of narrow range or monotone is difficult to attend to for more than a few minutes. The good news is that this trait can be fairly easily and quickly developed, even in those who recognize a problem in this area. It requires risking sounding foolish and exaggerated in front of trusted others, until adequate reinforcement has convinced the presenter that the increase is really an advantage that allows him or her to be attended to more easily.

When talking in front of a group, generally the speaker should attempt to speak more slowly than in ordinary conversation. What is an appropriate rate in a small face-to-face discussion is probably too fast for a group presentation. For some reason, there seems to be a correlation between the speech rate of the speaker and the size of the audience. What might be easily grasped at a more rapid rate in face-to-face conversation, isn't understood as quickly in large groups. Also, the speaker's slower rate allows him or her to scan the audience to see if the presentation is being understood, to see if some people need an opportunity to disagree, to see if he or she needs to talk louder or increase variation because some look bored.

Nonverbal behavior, the silent language, can be meaningful and controlled. It is good practice to keep hands away from the body and from one another. Allow them to be free to gesture, and avoid holding anything in them while talking, unless it is a useful prop like a pointer or a visual aid. After twenty-five years of teaching presentational speaking to college undergraduates and corporate professionals, this writer knows empirically that once speakers allow their hands to mesh together or to grasp one another behind their backs, there is only a slight chance of their being released to gesture. Often people feel awkward with their hands hanging loosely at their sides. Perhaps if they could see themselves on videotape in this posture, they would realize that it is natural looking, but even more importantly, they would probably see that one tends not to stay in that position. If speakers talk with their hands hanging loosely at their sides, eventually their hands begin to rise and gesture spontaneously to emphasize important points.

It is very dangerous to begin a presentation holding a pen, paper clip, rubber band, or other instrument not directly related to the presentation. Unconsciously, the fingers begin to play with the instrument, and the audience becomes fascinated in watching to see what the speaker will do. A former student actually straightened out a paper clip and began to stab himself while talking. Needless to say, for the remainder of his presentation the class couldn't take their eyes off the mutilation scene, and missed the speaker's concluding points.

As one speaks so will one be judged. Professional speakers attend to their diction, particularly when pronouncing words like "for," "can," "with," "picture," "going to," and "want to." In ordinary conversation one isn't likely to judge negatively a speaker who mispronounces common words and engages in sloppy diction saying, for example, words like "fer," "ken," "wit," "pitcher," "gunna," and "wanna," for the words listed above. When that speaker is in front of an audience, however, these mispronounced words often stand out and lead to negative inferences regarding the speaker. If one thinks of well-known television anchor men and women, one will be unable to recall any who have faulty diction.

Professionals who present themselves in front of groups must attend to their diction because they risk losing credibility if it is poor. Logical? No! True? Yes!

Once one has decided to become conscious of one's diction and decides to improve it, several steps are required. Step number one is for the speaker to inform trusted others who are often around him or her to listen critically, and to stop him or her each time a diction error occurs. Only after one is made aware of common errors, can one begin to train one's self to hear them. That is step two. Because the human mind operates generally at five times the rate of human speech, it is possible to listen critically to ourselves as we speak. So those who are attempting to rid themselves of poor diction or some other vocalized interference ("um," "an," "and a," etc.) can train themselves to listen for the errors and to correct them. Like learning to ride a bicycle or to use a computer, this learning and training task is uncomfortable at first, but improvement comes quickly. Working on one's diction is an ongoing task. Professional speakers never stop listening to the way their words are coming out and planning ahead to pronounce them correctly.

The visage itself is a communication vehicle. Presenters need to train themselves to keep their faces animated, using a variety of facial expressions. For many of the same reasons expressed above, it is important that speakers use all the communication vehicles available to maintain the audience's attention. One's facial expression is itself a communication vehicle. When it is lively, animated, expressive, and changing regularly while the speaker reacts to the feedback coming from the audience, it enhances the verbal message and allows the audience to go on unconsciously inferring the speaker's audience-centeredness. Because of a natural dispositional personality, ethnic background, life experience, and so forth, this is easier for some people to do than for others; however, everyone can improve! Because it isn't easy for one to "act" expressive does not mean that one can't grow considerably in the ability to look expressive. Indeed one can.

Eye contact is a vehicle of communication. It should be used to see the audience and respond to their nonverbal feedback. It is surprising how many people seem to remember having learned some rule about being able to fool the audience into thinking that the speaker is seeing them while he or she is actually looking over the heads of the people in the last row. The point is that when speakers have the opportunity to present themselves and their ideas to an audience, they want them to understand, to believe, and to follow their recommendations. Speakers have the best chance of being successful in achieving those goals when they are able to interpret the audience's ongoing reaction to what they are saying. Even though presentation speaking is generally considered a one-way communication situation, with the speaker talking at the audience as they listen, it is, in fact, a two-way situation with audience and speaker communicating with one another constantly and simultaneously.

Trained speakers see almost everything from their position in front of the room. If they are alert and looking, they see people who are beginning to fidget and interpret their feedback. They might decide from this observation to give the audience a short break; they might liven up their own movements to regain attention; they might engage in a new activity, one perhaps that involves audience participation, and so forth. The speaker might see some people coming in late, looking awkwardly for a seat. This provides the opportunity to publicly welcome them and ask others to move over to provide seating. The speaker might see people who look angry. This provides the opportunity to say, "You look angry. What have I said to offend you?" In other words, speakers who use their eyes to "connect" with the audience make them a part of the presentation. The audience knows it too, and will begin to send signals when they realize the speaker is sensitive to them.

Facial expressions are important. Smile, and look like you are having a good time. Smiling can be rehearsed and may feel phoney, but it needs to be built into the design of the presentation. One does not need to be constantly grinning, but one does need to maintain an expression of gentleness, approachableness, and nondefensiveness. The easiest way to convey these impressions is to smile often. Unfortunately, it isn't easy to smile when one is unsure of the material. All the principles above can only be worked at after the speaker has sufficiently mastered the content and consciously, through rehearsal, developed skills.

Although individual situations may make this difficult, a general rule to remember is that when speaking for an hour or less, always plan on at least ten minutes for some audience interaction. In talks of more than forty-five minutes, actual audience participation activities should be planned where possible and appropriate.

Postpresentation

Speakers should remember to bring business cards and to remain afterwards for people who may want to talk. If they have done a good job with eye contact, smiling, and prompting the inference of warmth, people will want to talk with them. This is frequently a source of repeat talks to other groups, and it is an opportunity for the best kind of public relations, face-to-face. When there is a long line of persons waiting to talk and time is limited, this writer passes out his business card and tells everyone they are entitled to one call for free consultation. It is amazing how many people follow up on the offer.

Workshops

Workshops differ from presentations. There are significant differences between a presentation and a workshop or training session. The differences are in the amount of time needed, the amount of audience participation recommended, and the goals of the leader. Presentations generally run no more than ninety minutes, while workshops may run from ninety minutes to several days. Audience participation and training are an integral part of workshop design, and the goals of workshops are generally to teach and train, while the goal of a presentation often is to persuade the audience to accept a specific proposal. The suggestions below are particularly appropriate when designing workshops.

Workshop leaders should plan to use techniques that involve all participants and encourage open communication between themselves and the participants and among the participants. Ways to encourage participation include: dividing the larger group into smaller groups, giving them time in their small group to get to know one another, and then giving them activities related to the workshop topic. These small group activities may include: case studies, either real or hypothetical; role playing; and questionnaire completion and discussion. Occasionally a group may be given a preworkshop assignment to complete a questionnaire, read material related to the workshop, or some other task. Reactions to these tasks can be processed among the participants in small groups.

Frequently, the most important point in determining the climate and direction of the workshop occurs during the first twenty minutes. Some suggestions for using the opening minutes to establish an open climate include using a group introductory activity to promote a relaxed and open atmosphere. When the group is small enough (generally twenty or less), spending time to allow each participant to identify major concerns and questions tends to encourage this involvement. Workshop leaders can relieve participant anxiety by introducing themselves, sharing both personal and professional information, giving a brief overview of the objectives for the workshop, the main topics to be covered, their sequence, and approximate time span. They should also reinforce often that they do have expectations in terms of participants' cooperation and participation. Even when such introductory activities take as long as an hour, they are justifiable, because of their importance in establishing a common frame of reference with shared goals in a relaxed and receptive setting.

Leaders should be aware of signals of fatigue or boredom from participants. Some techniques used to avoid these problems include: identifying two or three participants whose behavior provides some type of clue to group climate, providing a variety of activities to break-up the routine, and providing a change of pace. Most successful workshops are a blend of information-presenting activities with "hands-on," experiential types of activities.

Another way to safeguard understanding and attention is to summarize regularly what has occurred, especially before moving on to a new topic. Continuity of training is enhanced, if there are purposeful and periodic reviews and summaries during the workshop. Leaders should note that it is not essential that they do the summarizing; in fact, having the participants themselves do it provides feedback as to whether or not the group has grasped the important points.

Anticipate from among the alternatives the appropriate conclusion for the workshop. Just as the introductory period to the workshop should not be rushed, so too should closure be carefully attended to. This depends upon how much time is available for the entire workshop, of course. But in a day's workshop of from six to eight hours, allowing a full hour at the end for closure is appropriate. During this time loose ends are tied; the group's original expectations are reviewed and assessed as to how well these expectations have been met. Requiring the group to complete an evaluation of the workshop during the time provided for the workshop is the method most likely to get the largest return. These evaluations are most helpful to leaders who want to continue to grow. They will process the evaluations and make changes in subsequent workshops based on the responses. A sample evaluation sheet is available in the "Appendix." The emphasis in a workshop is always on quality not quantity. It isn't how much the audience has heard; it is how much they will learn, will remember, and will use in the future that is the final measure of the workshop's success.

APPENDIX

Evaluation Sheet

Presenter: ______________

Date: ________

Presentation (Check your response)

	Excellent	*Good*	*Satisfactory*	*Poor*	*Unsatisfactory*
1. How would you describe the presenter's knowledge of the subject matter?	____	____	____	____	____
2. How did you find the presenter's style of delivery?	____	____	____	____	____
3. Did the presenter encourage individual participation?	____	____	____	____	____
4. Did the presenter hold your interest?	____	____	____	____	____
5. How did you find the organization of the content material?	____	____	____	____	____
6. How useful did you find the visual aids?	____	____	____	____	____
7. To what degree did you feel that the content and exercises were relevant to your job?	____	____	____	____	____

8. Comments:

Astute DP Professionals Pay Attention to Business Etiquette

Jeffrey P. Davidson

Successful professionals should understand the unwritten rules of business etiquette. Every company has its own set of rules for appropriate behavior. This article discusses general rules for handling telephone calls, attending office parties, and dressing appropriately and suggests tips for making casual conversation, smoking, and drinking coffee. Taking the time to learn the accepted etiquette at your company will make you a team player.

YOU ARE BRIGHT and you do your job well. You complete your fair share of all assignments, come to work on time and smile at everyone you meet at the coffee pot.

But no one asks you to lunch. None of your co-workers drop by your desk to chat about the weather. No one remembers your birthday with a card.

More important than the slights listed above, no one in your office wants to work on a project with you.

You may have gotten tangled in (and nearly choked by) the ropes of "business etiquette." The rules, as strict as any of Emily Post's, are usually unwritten. They vary from office to office. They govern appropriate behavior in that particular office.

If you don't pay attention to these unwritten rules, you will be flouting, purposely or not, the agreed upon norms. Whether your reluctance to follow business etiquette is viewed as arrogance or naivete, the people you work with will let you know you are not being a team player. Tension can build over even the smallest matters, disrupting office productivity and creating unnecessary stress.

ETIQUETTE DEFINED

Business etiquette dictates when certain behaviors are appropriate or inappropriate. Make your life easier by paying attention to the rules at your office. Be aware of how the factors discussed below—general rules and safe talk—come into play where you work.

Remember that business etiquette differs from social etiquette. For example, in situation A, Hank blushes when he is praised for his work, includes his co-workers in the praise and changes the subject to the next football game. In situation B, however, Hank enumerates his accomplishments over the last year, emphasizing his value to the company as an individual.

Same person, same subject. But situation A takes place at a dinner party of neighbors; situation B takes place at a luncheon of the board of directors of

Hank's company. Those who don't recognize the differences between social and business manners can easily make their lives difficult. No matter how good your work, or how knowledgeable you are, you also must follow the prescribed routine.

Judith Martin, author of *Miss Manners' Guide to Excruciatingly Correct Behavior,* says "good business manners often dictate the opposite of good social manners." For instance, she points out that knowing what your neighbor is doing, seen as nosiness in the social world, is regarded as intelligence in the business world.

General Rules

Every office has its own set of rules. Taking an extra 15 minutes for lunch may be part of the routine at one office while the office next door insists on strict punctuality. Generally, the etiquette rules won't be written down. No section in the employee manual will tell you what you can get away with and what you can't.

The following is a list of topics subject to etiquette rules. All of these topics have the possibility of creating a great deal of tension in almost any office. Look around you to discover what etiquette dictates in your office. Observe people who have been at your office for a long time and who are well-respected. Follow their examples.

Coffee. If there is a container provided to pay for your coffee, do it. Trying to get away without putting your quarter (or dime) in the container will get you a bad reputation—quickly.

Coffee pot. If you take the last cup of coffee, make a new pot. It's amazing how many people either take the last cup and leave a dirty, empty pot for someone else to fill, or notice the pot is low and walk away, deciding to get their coffee later from a full pot. The same holds true for filling up any supply you use the last of, such as paper in the copying machine.

Cigarettes. Your office may have designated smoking and nonsmoking areas. People will notice, and will not appreciate, those who step over the bounds. If you must smoke, stay in the prescribed areas. If you smoke in your office, buy an air freshener.

Dress code. At an office where I previously worked, the dress code was clearly spelled out in the employee manual. Part of the statement read: "Blue jeans, tee shirts and shorts are not considered appropriate dress for the office." That was respected by everyone in the office until the vice president wore blue jeans one Friday. Although nothing was ever discussed (nor changed in the manual), it became standard procedure for some of our workers to wear blue jeans every Friday.

A new employee, seeing that jeans were acceptable on Friday, wore blue jeans during her second week at the office—on a Wednesday. Her supervisor was kind enough to point out the blunder to her directly. You may not be so lucky.

Let the higher-ups in the company start new trends. You can follow once the trend has won widespread acceptance.

Telephone calls. These can be tricky waters to navigate. Most companies have a receptionist to answer the phone, or, if not a specific person to handle that duty, then a schedule for switching it among others. However, an occasion may arise when the receptionist is away from his or her desk and you have to answer the phone. Answer graciously, using the same greeting the receptionist uses. I'm impressed when I call an office and ask for the president, only to find out he or she is the person who answered the phone. On the other hand, there is nothing worse than a person answering the telephone as if the task is beneath them.

Eating at your desk. Some offices provide lunchrooms and expect employees to eat only in those areas while others allow discreet snacking at one's desk. The amount allowed by your company will probably depend on the number of customers/clients who walk through the office, the type of business conducted and the amount of privacy each desk or office affords. In addition to whether it is considered polite to eat at your desk, the whole issue of eating at your desk may create friction with your co-workers. They may see you eating at your desk, while they all go out, as an attempt to win favor at their expense.

Lunch, anyone? Who goes to lunch with whom is, in some offices, an important indicator of power shifts and organizational changes. Advice in recent magazine articles ranges from "don't ever go out to lunch with anyone below your position" to "a woman should not invite another woman out to lunch." I don't agree with creating any hard and fast rules about who you invite to lunch. Here, as with the other topics covered, be aware of the generally accepted behavior in your office. Remember, a sure way to win gratitude is to ask a person to lunch who is normally not included in the invitations.

When parties are not parties. Remember that business occasions, such as cocktail receptions, celebratory luncheons and the infamous office Christmas party, are really business activities. Letting yourself go at these doesn't make good sense. Remind yourself that you will see these people the next morning.

The topics discussed above are just a few examples of etiquette rules to be aware of in your office. Others you should watch for include:

- What business-related discussions are held at business social occasions?
- How birthdays and holidays are handled.
- The amount of time allowed for non-project-related, informal chatting.
- The routine when paychecks are handed out.
- What personal effects are on desks or displayed on walls around desks?
- Are personal phone calls allowed? Discouraged?
- How are messages handled?
- What information about another employee is generally given over the phone, such as "he's not available right now" or "he's out of town."

Taking the time to learn the ropes, through careful observation and asking questions (if necessary), will pay off for you.

SAFE TALK

Jean Nicely, public relations director of a small engineering firm in Tennessee, says, "I can tell the people at work a funny story about how my daughter's mother-in-law got lost in New York City or how my own mother-in-law drives me batty.

"But I would never tell those people anything about problems with my husband or daughters. That would just be unthinkable." "Safe talk" refers to the everyday tidbits that fuel office conversation. Certain subjects, like mothers-in-law, have a universal reputation as fodder for jokes and light talk, so they become safe talk in the office. It's entirely possible to work nine-to-five every day and hear nothing but the following:

- Is it raining yet?
- What a hot one!
- Hey, how's it going?
- Good morning/good night.
- Have a good weekend/evening/vacation.
- Thank God it's Friday.
- Do you want to contribute to Lester's retirement gift?
- Please sign this card for Mary.
- Want to go for lunch/coffee?
- How about those Cubs/Bears?
- Where are you going for vacation?
- Thank you/pardon me.
- Do you have a red pen/yellow highlighter?
- Another day, another dollar.
- Going up? Going down?
- I'll only be a minute.
- Are you feeling all right?
- Are you going to be using the terminal long?
- Please call me when you're done.
- See you tomorrow.

Of course, if you listen carefully, you can hear other things in the office. Betty telling Debra about her latest affair or Norm bragging to Dennis about extra deductions he is taking on his tax return. Many people find the camaraderie of an office an ideal place for finding new friends.

However, many more people end up sorry that they took a business relationship and turned it into a friendship. Stories abound of office friendships destroyed because one person was promoted, because a power shift occurred or because a new alliance was formed.

Personal secrets you have spilled to a person in your office can become a weapon if the friendship fades or breaks. Err on the side of caution when divulging your personal life. Safe talk, which can be very friendly, also will protect you from office gossip.

Eliminating Gender Bias in Language

Effective tone in writing includes sensitivity to any language that is sexist. This article presents guidelines for avoiding problems in using pronouns and gender-specific nouns. Use the guidelines to check your own drafts.

CLEAR WRITING EXPERTS RECOGNIZE that part of writing understandable documents is understanding and responding to the needs of the intended audience. This might mean limiting the use of jargon or technical terms for a general audience, or using a larger typesize for an elderly audience. It is the writer's job to maintain the audience's willingness to go on reading the document. Readers who are continually stumped by big words or offended by a pompous tone are likely to stop reading and miss the intended message.

Today, part of striking the right tone is handling gender-linked terms sensitively. Use of gender terms is a controversial issue. Some writers use the generic masculine exclusively, but this offends many readers, because it seems to be based on the presumption that all people are male unless proven female. Other writers are experimenting with ways to make English more neutral. Some have even tried to create a gender-neutral pronoun for the third person singular, but this has not caught on.

Avoiding gender bias in writing involves two kinds of sensitivity: (1) being aware of potential bias in the kinds of observations and characterizations that it is appropriate to make about women and men, and (2) being aware of certain biases that are inherent in the language and of how you can avoid them. The second category includes using gender-specific nouns and pronouns appropriately, and this article presents guidelines for handling these problems. The best approach right now is to write around the problem when you can, to avoid offending your audience. Here are a few ways to do that:

1. Use a gender-neutral term when speaking generically of your fellow creatures. For example:

man	the human race
mankind	humankind, people
manpower	workforce, personnel
man on the street	average person

2. Avoid clearly gender-marked titles. Use neutral terms when good ones are available. For example:

chairman	chair
spokesman	speaker, representative
policeman	police officer
stewardess	flight attendant

   When you can, avoid awkward coinages like "spokesperson."

3. If you are speaking of the holder of a position and you know the gender

of the person who currently occupies it, use the appropriate gender pronoun. For example, suppose the "head nurse" is a man:

Before: The head nurse must file her report by September 1, 1982.
After: The head nurse must file his report by September 1, 1982.

4. Rewrite sentences to avoid using gender pronouns. Use the appropriate title instead.

   Before: You should see your doctor first, and he should call the pharmacist directly.

   After: You should see your doctor first, and the doctor should call the pharmacist directly.

5. Recast your statement in the plural, thus avoiding the third person singular pronoun.

   Before: Each student should bring his text to class.
   After: All students should bring their texts to class.

6. Address your readers directly in the second person if you can appropriately do so.

   Before: The student must send in his application by the final deadline date.

   After: Send in your application by the final deadline date.

7. Replace third person singular possessives with articles.

   Before: Every branch chief should draft his preliminary schedule by Friday.

   After: Every branch chief should draft a preliminary schedule by Friday.

8. Write your way out of the problem by using the passive voice.

   Before: Each department head should do his own projections.
   After: Projections should be done by each department head.

   We recommend this solution only if nothing else works, since the active voice is generally clearer and more effective than the passive.

9. Use a third person singular pronoun to refer to a third person singular antecedent, unless you are sure your audience is as willing as you are to break the rules of English grammar.

   Before: Every child should brush his teeth after meals.
   After: All children should brush their teeth after meals.
   Or: Every child should brush his or her teeth after meals.
   But not: Every child should brush their teeth after meals.

   Using "their" is becoming more acceptable in speech. But if you use it in writing, you are likely to distract your reader.

10. Avoid "s/he," "he/she," and "his/her." They look awkward, but even worse, they interfere when someone is trying to read a text aloud. If none of the other guidelines has been helpful, use the slightly less awkward forms "he or she," and "his or hers."

11. Remember, the goal is to avoid constructions that will offend your readers enough, because of their views on grammar or politics, to distract them from your text.

Strategies for Effective Presentations

Janet G. Elsea

The more you know about your audience, the better you can organize and deliver an oral presentation. This article classifies audiences as (1) people who view you and your topic favorably, (2) people who approach controversial topics objectively, (3) people who are not interested in the topic, and (4) people who are hostile to you and your topic. Notice the tips on handling a mixed audience.

IT'S LATE ON FRIDAY AFTERNOON. Your boss has scheduled a staff meeting at which you are to give a report on the latest benefits package. You know a snowball in hell has a better chance of being listened to than you unless you manage to be:

- Brief, to the point
- Animated and interesting
- Available after your talk (asking nothing of them during it).

Your analysis is accurate—this is a "could care less" audience. It has the attention span of a flea, a weekend-bound mentality, and no interest in the topic.

It is also a captive audience that won't forgive if you drone on, show lots of graphs and charts, ask for questions, or ignore the clock.

Take that same group, however, on a Wednesday morning with the topic geared to how they will benefit personally from the new package, and you may find them interested and involved. Why the difference?

An examination of the basic facts about human nature—facts that meeting planners and speakers often overlook—reveals several reasons.

Audiences are of four types, scattered along an attitude continuum:

- *Love You*
- *Think They're Impartial*
- *Could Care Less*
- *Love You Not.*

The more you learn about your audience ahead of time, the better you can tailor content and delivery to meet their needs and your objectives.

Time of day, day of week, order of appearance and physical setup affect people's response, as do your topic and who you represent.

These strategies can help prepare you and your remarks for a variety of audiences and situations.

AUDIENCES FALL INTO ONE OF FOUR CATEGORIES

Because communication is "the sharing of meaning" and not the transmission of information, it is important to find common ground and approach audiences from *their* perspective. There are four types of participants, and each demands adjustment of both the content and delivery of your presentation.

1. The Audience That "Loves You." Some groups are friendly and predisposed to view you and/or your topic favorably. Therefore, your delivery should be open and warm, using:

- Lots of eye contact
- Smiling facial expressions
- Ample gestures and movement
- Variation in rate and loudness
- Humor, anecdotes, examples and personal testimony as supporting material.

With a positive audience, you can try new ideas, ask for response midstream, and urge a specific action.

Most organizational patterns will work because these are active listeners who will track you throughout—as long as you don't take them for granted and ramble or fall into a monotonous delivery.

It is not necessary to establish your credibility directly, although it's nice if someone introduces you with some glowing remarks.

2. The Audience That "Thinks It's Impartial." Though studies indicate most people approach controversial topics with their minds made up, some *think* they are objective—therefore, you must honor that perception by giving both sides.

Because these people consider themselves calm and rational, your delivery must mirror that dispassion.

Do not try to entertain them with your speaking prowess: "just the facts, ma'am." The key is control: be even in your delivery but not monotonous.

Supporting material, therefore, might consist of facts and figures, expert testimony, comparison and contrasts. Stay away from humor (you may be seen as frivolous), personal stories (they're not interested in your anecdotes), and flashy visual aids (such as films, video or slide shows).

Organize your material in a precise, noncontroversial fashion with the sequence of main ideas readily apparent; the pro-con pattern works well with this audience.

Allow time for questions and response—they like to probe and will offer their point of view while listening to yours.

As a final note, have someone the group respects give your credentials in her or his introduction of you.

3. The Audience That "Could Care Less." In many ways, the crowd that could care less is the other side of the coin from the one above: delivery is the probable key to your success.

You must be dynamic to listen to and interesting to look at: vary rate and loudness levels, move and gesture, have facial expressions and make lots of eye contact.

Given their short attention span and the fact they don't want to be there, make your motto the words of Falstaff: "Brevity is the soul of wit." Entertain them while you present a set of brief and interesting remarks (don't call it a "speech").

Use supporting material that may draw them in:

- Humor, cartoons and anecdotes
- Interesting and colorful visual aids
- Metaphors
- Powerful quotations
- Startling statistics.

Do *not:*

- Darken the room
- Stand immobile behind a podium
- Pass out anything for them to read
- Make them look at boring overheads
- Expect them to ask questions and give response.

Do put the content into a concise organizational pattern with a built-in interest factor, such as cause-effect, problem-solution, or the sign-post (as long as you use fewer than three main points). Avoid the historical overview, pro-con and the topical patterns as they seem endless to a captive audience.

Finally, don't worry about establishing your credibility—not only does that take time but may focus their hostility upon you.

4. The Audience That "Loves You Not." This dangerous audience may be looking for chances to take control or ridicule you. It is aroused before you open your mouth, either by your topic or who you represent.

Your immediate goal, then, is to lower their thermostats (thus lessening their arousal levels) and present your opinions calmly.

Delivery and content must work together. Be calm and controlled, speak slowly and evenly; remember, rapid rate and loud tones arouse both you and your listeners. Force yourself to speak in a measured tone of voice and with purposeful gestures; avoid random movements as they will energize you.

The organizational pattern should be inherently noncontroversial, such as: historical, geographical, chronological or topical. Avoid the cause-effect, pro-con and problem-solution formats.

Likewise, select supporting material that seems objective, such as data or expert testimony from their sources or neutral ones. Avoid anecdotes and jokes since they may set off a hostile audience, but do be empathetic and concerned.

Give careful attention to the question-answer period ahead of time. If possible, don't have one because you risk losing control and giving the opposition center stage.

But, if you must answer questions, stage manage both the format (insist on a moderator or written questions, for example) and the time frame (never let the other side have the final word). Make it clear that you will give a final statement after you have taken the last question.

Finally, have your credentials given by a neutral introducer, but don't put much stock in that exercise. With a hostile audience your best hope and strategy is to comport yourself in such a manner that you will at least be seen as credible.

In addition to where audiences fall along the attitude continuum, time of day and day of week can affect their attention span and retention levels. For example, most people are at their mental best in the morning and are least effective after a meal.

Also consider the effect of sequence or order of appearance. Research tells us that those who speak first and/or last will be remembered more than those who speak in the middle (the "primacy/recency effect").

Room temperature, audience sight lines, and seating arrangements also have an effect upon people's abilities to attend to and retain information.

With the audience that "loves you," time of day is not as crucial as with the other three types—although if you want people to take action, they will be more alert in the morning.

Likewise, nearly any slot in the speaking order will work with this audience, with the middle one being the weakest (keep that in mind for the hostile audience!).

Have an open seating with few barriers between you and your "friendlies."

With an impartial audience, give thought to precision: start and end on time, regardless of the day or time you select (again, this is not crucial because these participants are also active listeners), and stick to the agenda.

Make certain the physical arrangements are orderly with equipment and visuals ready to go. Remember to have easy access to microphones for their Q and A session.

Timing is critical if you are to shift the "could care less" audience members toward the positive end of the continuum. Try a time frame not critical to their work, such as midafternoon midweek, or just before lunch (when their attention span is high and you have a natural adjourning time).

Arrange the meeting space with chairs and tables facing you—this may discourage side conversations. Keep lights bright, temperature low, and air moving.

In terms of sequence, get on the program early; the longer you and they wait, the shorter their listening curve.

With the hostile audience, use facts about attention span to your advantage: schedule meetings after a meal, late afternoon, on early Monday mornings or late Friday afternoons.

Another trick of the trade is to hold a meeting just before quitting time, carpools, or other "deadlines" that allow you to wrap things up—in their own best interests, of course.

In terms of setup, put some distance and/or objects between them and you (tables, flip charts, podium). Make it difficult for them to get to the few microphones or the moderator.

In short, you're better off controlling as much as you can from the outset: starting and stopping times, room arrangements, thermostat, equipment and materials.

You need not be at the mercy of an audience that "loves you not"; indeed, the more control you have the more confident you will feel.

Delivery Is the Key to a Mixed Audience

Assume you have analyzed your audience every which way: demographics, past experience, friends' experiences, available printed material (annual reports, trade journals, newspapers) and talking to the person who invited you to speak. What if it now appears there is a mixture of types likely to attend?

My advice is to plan the content as if the entire audience were an impartial one, then stick to your prepared organizational pattern and supporting material throughout your presentation. You may cut material but try not to add any.

Vary your delivery as your speech proceeds. Start with warm body language and a tone of voice that is controlled and calm. As you read feedback, adjust your voice and/or your nonverbal actions appropriately.

For example, if people nod, smile, lean forward and watch you—typical signs of interest—shift to a more energetic and involving style. But if the response seems restrained and thoughtful, tone down the nonverbal and/or vocal.

If there seems to be a negative response from several people, damp down your energy levels and consider cutting out some supporting material. With those who seem bored, look away or whisper to others, become more animated—perhaps by moving or gesturing, picking up the pace, or increasing your energy level.

If disinterest seems widespread, consider shortening your remarks and wrapping them up.

The way to deal with a mixed audience is with delivery, not content, for several reasons:

- In your anxiety you may not read audience response accurately.
- Unless you are an experienced speaker it is difficult to make impromptu changes in the fabric of your speech.
- There may be cultural differences you aren't aware of that may lead you to read nonverbal responses incorrectly.

It seems Aristotle was right thousands of years ago when he defined rhetoric as "finding the available means of persuasion in any given situation."

That's really what effective presentations are all about, whether they are at an employee briefing, a staff meeting, before stockholders, at a press conference, or at a local service club.

By using these audience-centered strategies to tailor both your message and how you deliver it to audience needs and experiences, your chances of establishing a positive first impression—and, at very least, getting a hearing—are much higher.

Remember, there really aren't any boring topics—just boring speakers.

A Good Résumé Counts Most

John D. Erdlen

Most personnel recruiters assume that the person without a résumé is not prepared for the job search. This article discusses the importance of having an effective résumé. Use the checklist of dos and don'ts when you write your résumé.

ADVICE BY SOME CAREER COUNSELORS to throw away résumés and rely on networking may actually be an obstacle to a successful employment search. This "beat the system" approach replaces résumés with "contacts" and substitutes "informational meetings" for job interviews. It is an attempt to outsmart prospective employers and can easily backfire by antagonizing recruiters.

Personnel recruiters and other hiring managers can see through the facade of a candidate who requests "just a few minutes of their time" for an informational meeting. Most employment and staffing managers agree that the lack of a résumé indicates poor preparation for a job search and invariably creates an unfavorable impression. In fact, over 95% of all hires are introduced by a résumé. It is the major factor in deciding whether to interview a candidate.

Corporate recruiters are often frustrated by "innovative" approaches because most of them provide no systematic way to obtain basic information and evaluate a candidate's qualifications. While networking and informational interviewing are indeed valuable job-seeking tools, they are intended to enhance a traditional employment search, not to replace it.

The problem in obtaining employment is usually not the résumé; it is with the information it contains or lacks. Many job opportunities are lost because the résumé omits pertinent data, contains too much irrelevant information, or is difficult to read or understand. A properly constructed résumé can be a powerful sales tool that opens career doors. Here are a few suggestions for résumé preparation:

- Indicate a job objective or summary of qualifications in the résumé or cover letter. The "I'll do anything" plea is not recommended.
- Emphasize accomplishments.
- Mention any education related to your job objective.
- Describe your role in outside activities or professional associations related to your career objective.

Résumé "don'ts" include:

- Don't state a specific salary requirement; it may not concur with the company's job evaluation and cause you to be eliminated from consideration.
- Don't leave gaps in the résumé; omissions can be interpreted as attempts to hide negative details about your background.

- Don't list reasons for leaving previous positions; however, be prepared to furnish this information at an interview.
- Don't be too brief or too lengthy. One-page is enough for new graduates; two pages for most senior candidates.
- Don't use an unusual résumé format; this tactic attracts negative attention.

Even your closest friends are seldom fully aware of all your working credentials. Their response to a request for job-hunting assistance usually includes, "Send me your résumé."

How to Deliver Dynamic Presentations: Use Visuals for Impact

Marya W. Holcombe and Judith K. Stein

Oral presentations are most effective when the speaker uses intriguing graphic aids. This article discusses how to choose graphics for certain audiences and purposes, how to prepare graphics, and how to use them effectively. Use the chart to select appropriate types of graphics for your oral presentations.

PRESENTATIONS ARE MORE PERSUASIVE when supported by visuals. But audience members frequently walk out of a presentation wondering what the visuals were all about. Why? Because presenters often don't take the time to develop and learn how to use them. A little common sense and a few guidelines can help produce effective visuals.

People remember more when they see *and* hear a message. And they remember images better than words. But few people can read and listen at the same time. Thus, for visuals to be effective, the audience must be able to see the visual and grasp its message quickly. To meet these criteria:

- Select the best kind of visual support,
- design visuals for the greatest clarity and impact, and
- use visuals correctly.

CHOOSING PROPER SUPPORT

What types of visual aids are available?

Managers most often use transparencies, flip charts or 33 mm slides, usually in that order. Although some conference rooms contain blackboards, most managers regard "chalk talk" as too informal or too academic for anything but pure problem solving.

Audience size influences the choice of medium. So do the purpose of your presentation, industry and organizational norms and available resources.

Each audience member must be able to see the visuals clearly. If the group exceeds 20 people, flip charts are not effective. Overhead transparencies, on the other hand, can be used with groups containing up to about 100 people. But those transparencies should be shown on a large screen. Slides are necessary with audiences of several hundred people. Whatever visual you use, ensure that it is legible from a distance.

The purpose of your presentation also affects the type of visual you use. If you're trying to persuade your boss to approve your budget, you wouldn't put on a slide show in his or her office. Yet, if you're trying to persuade your company's board of directors that a major expansion program is vital, you wouldn't appear in the boardroom with rough, hand-lettered flip charts.

The type of visuals required to make a persuasive argument helps determine the appropriate medium. Only the simplest charts and graphs are appropriate for flip charts. For complex material, use transparencies, which are enlarged when projected onto a screen. If photographs are essential to your point, use slides. But remember: Copying a photograph onto a transparency generally is unsatisfactory because the reproduction will be poor.

Industry and organizational norms also influence selection of visuals. If your company has the attitude that "we're all in this together," you might prefer to use creative, informal visuals. Presentations usually follow a detailed format, sometimes preserved in a bound book, in organizations with rigid hierarchies. In that case, professionally produced slides may be in order.

Time, money and the availability of equipment also will guide your decision. Slides, for example, are expensive and take at least a day to produce.

You can make overheads on an office copier instantly and inexpensively by using a typed or hand-drawn original on letter-size paper. Although professionally made overheads are superior, do-it-yourselfers usually are acceptable if they are neat.

Flip charts are easy to make and are the least expensive form of visual. But when drawn professionally, time and cost factors increase.

How to Prepare Visuals

Visuals can consist of words or graphics, or both. They reinforce the presentation's main points and help the audience follow the flow of arguments.

Regardless of the type of visual used, two guiding principles surface. Keep it simple, and ensure that it focuses attention on your message.

Text visuals, consisting of only words, primarily are used as road maps. They preview or summarize information and remind the audience where you are in the presentation. Although text visuals also may emphasize main points, most presenters consider pictures more powerful than words.

Simple visuals, which allow the audience to grasp the concept easily, are best. Because the human eye can take in about 40 characters in one glance, lines should be limited to that number. Use phrases rather than full sentences, expanding on each phrase as you speak.

To ensure that the visual is easy to read, use upper- and lowercase letters. Choose a simple typeface. Use bullets or Arabic numerals to set off items in a list: They are easier to read than letters or Roman numerals.

Consistency and white space help the audience focus on your message. Start text at the same place on all visuals—about one-fifth down from the top. Leave a space about the height of a capital letter between each line. Then audience members will not have to search for a starting place each time a new visual appears.

Using color, boldface, or large-size type for important concepts helps distinguish them from supporting ideas. Reinforce your message by using the same feature each time you make a similar point. For example, you can use the same text visual several times to remind the audience of your place in the argument. Highlight your location each time with one of those features as you move through the presentation.

GRAPHIC VISUALS

People usually assimilate visual forms more quickly than words. Thus, use graphic visuals to emphasize important relationships. To create effective graphics, select the most appropriate chart form, follow the rules of good design and reinforce your message with headings.

Most management presentations include numerical data. For those presentations, line, bar and pie charts are the most frequently used graphic forms. Yet diagrams, maps and photographs also are valuable. Tables are appropriate only when the audience must have exact numbers.

Choose the form that best fits the point you are making:

- Line charts show changes over time of one or several variables.
- Bar charts, whether vertical or horizontal, show the relationship between two or more variables at one point in time or at several points in time. Although line and bar charts are used frequently for similar purposes, line charts are more effective in depicting change over time. Bar charts highlight differences between distinct time periods or compare components of a whole at different times.
- Pie charts show the relationship among parts of a unit at a given period. Although pie charts are round, other shapes, such as bars, also can be used to show parts of the whole.
- Diagrams, such as organization charts, process charts or flowcharts, help the audience visualize relationships and processes.
- Maps focus attention on one or several locations and indicate spatial relations and distances.
- Drawings show the details of a design. And cartoons can be used to make a point with humor.
- Tables provide data in a precise form. Complex tables, however, are inappropriate for presentations where the audience is expected to catch the visual message quickly. In addition, anyone seated behind the first row usually cannot see long columns of numbers. If you must show all the numbers, limit them to three or four columns and three rows. Or use handouts. Avoid using as an overhead or slide a full-page table copied from a written document.

DESIGN RULES

When preparing visuals, follow the rules of good design. Visuals are intended to be snapshots of your data, not detailed blueprints. Once you have selected an appropriate form for your chart, include only the data needed to get your message across.

If you want the audience to focus on declining sales, for example, leave out earnings and dividends. If you want to emphasize that two-thirds of corporate earnings went to pay the energy bill, a pie chart showing all other expenditures in detail will detract from your message.

Color, when used in moderation, is an effective way to focus the audience's attention. Shade a part of a graph, highlight a number, or use a colored arrow to point to details that specifically support your argument. Be consistent when using color so as not to confuse the audience. Remember these points when choosing colors for your visuals:

- Use colors in moderation;
- use bright colors to attract attention to your main point, but use red sparingly;
- use contrasting colors to contrast concepts or to signify a major change; and
- use two shades of the same color to point out minor changes.

Reinforce your message with headings. A good heading explains your visual explicitly. If an audience member is distracted while you are speaking, he or she still can determine significant points by looking at the heading. Even an attentive listener appreciates the reinforcement.

Useful headings contain a noun and a verb and tell the audience what is important. This heading, for example, makes a precise point that the audience will remember: "Earnings did not exceed 1972 levels until 1981." If the heading had been "Earnings: 1972 to 1983," audience members who had not heard the message would have been left to draw their own conclusions—possibly one the presenter didn't intend. Or, they might not bother to draw any conclusion at all.

The placement, size and kind of lettering also are important considerations. Headings always should be above the chart. And, in general, heading type should be twice as large as other type. Text identifying the mechanical proportions of the visual should be lightface, while text identifying the data lines, bars or parts in the graphic representation can be boldface. Again, for reasons of simplicity, run all lettering in the same typeface.

How to Use Visuals

Visual means "perceptible by sight; visible." Visuals are worthless if the audience can't see them because the speaker blocks their view, or the screen is out of place, or the focus is poor, or the projector is not working. Follow these simple rules when using slides, overhead projectors and flip charts:

Slides. Effective slide presentations require careful orchestration; conscientious planning and rehearsing are important. Check that all slides in the carousel or tray are right side up and in the correct order. Use the slides as your notes; a lectern light is disconcerting to the audience. If the slide projector doesn't have a remote-control feature, rehearse the slide presentation with the projectionist. Saying "next slide, please" interrupts the flow of thought.

Use a pointer to focus attention. Some presenters use a "light" pointer, which allows them to point to the screen with a ray of light. However, the novelty of this tool may cause your audience to focus more on the pointer than on the point. Plus, your dependence on yet another battery and a bulb increases the chances of something going wrong. Unless you have a very large, high screen, use a conventional pointer.

Overheads. When using overhead projectors professionally, your equipment must be in good working order. And you must know how to operate it

Guidelines for Selecting Visuals

Considerations	Flip charts	Overheads	Slides
Audience size	Fewer than 20 people	About 100 people	Several hundred people
Degree of formality	Informal	Formal or informal	Formal
Design complexity	Simple	Simple; can be made on office copier	Anything that can be photographed
Equipment and room requirements	Easel and chart	Projector and screen; shades to block light	Projector and screen; dim lighting
Production time	Drawing time only	Drawing or typing time; may be copied instantly	Design and photographing time plus at least 24 hours production time
Cost	Inexpensive unless professionally drawn	Inexpensive unless professionally designed or typset	Relatively inexpensive

properly. Before anyone arrives, turn the projector on and off, check the focus and make sure you have an extra bulb. Also, know how to change the bulb, if necessary.

The visual on the screen should amplify the point you are making. If you finish the topic and do not have a visual for your next point, turn off the projector. A glowing, empty screen is a dreadful distraction, and a covered transparency with light leaking around the edges is almost as bad. Furthermore, by turning off the projector, you signal audience members to return their attention to you.

Where you position yourself throughout the presentation is important. Don't block anyone's view. When a transparency is projected on the screen, move to one side.

When you want to point to something on the screen, don't turn your back to the audience. Instead, point with your hand nearest the screen. Or use a pointer if the screen is large.

If you are near the projector, use a pencil, not a finger, to point to something on the transparency. Your finger would look enormous. Or, use a specially designed pen to circle the concept you want to emphasize.

Some presenters add information to a transparency while it's projected on the screen. Although writing on the transparency during the presentation adds action, it requires good handwriting and self-assurance. It is difficult to write on a transparency without having your hand shake (the projector will magnify the movement many times over) and without moving the transparency. If you do write on the transparency, always stand to one side of the machine as you write. Then move away so everyone in the audience can read what you have written.

Flip charts. Because they require no electrical equipment, flip charts are easy to use. Easels should be anchored securely on the floor, and flip charts should be attached firmly to the easel. If using an adjustable floor stand, set the height so you can comfortably grasp the page in the middle and flip it over the pad. Moreover, ensure that the chart braces are fixed in advance so that your charts don't slide to the floor while you're speaking.

Before the audience arrives, open the chart to the first blank page. As you proceed from one visual to the next during the presentation, grasp and flip over the completed visual to the following blank page. Flip the visual with your hand nearest the chart as you face the audience. If you are not ambidextrous, stand on the side most convenient for you.

If you do not have a visual for the next section of your presentation, skip to a blank page, and move to another part of the room. And, as in any presentation, always face the audience.

Occasionally, for external presentations, managers use professionally made boards the size of flip charts. However, those are expensive and bulky to transport. Furthermore, they're unwieldy when showing and removing: Presenters usually need assistance to change the boards, which distracts the audience.

Making effective presentations is easy when you use a little common sense and follow these guidelines. Text and graphic visuals should be simple and should focus attention on your message. And remember: Review all visuals carefully. Ask an outsider or a disinterested co-worker to review them for typographical or conceptual errors, inconsistencies or points of confusion. Then set them aside until you are ready to rehearse the presentation.

Writing Online Documentation

William Horton

User-friendly online documentation helps readers understand commands and use computer equipment efficiently. This article offers guidelines for converting computer hardware and software terminology into messages that are useful and nonthreatening. Use these guidelines when you prepare online documentation.

Secrets of User-Friendly Messages

Online writing is a second language to most writers. It embodies a new grammar, new idioms, and new rhetorical strategies. Many online documents fail because the writer, through ignorance or arrogance or mental inertia, fails to account for the differences between online and paper documents. Such writers produce the worst of both worlds, a document with all the weaknesses of paper and none of the strengths of online technology. The writing challenge is especially acute for people who produce documents that will appear both on paper and on a screen.

This column reviews some differences between paper and online documents and what these differences imply for the way online documents are written. It offers guidelines to aid writers of online messages and documents. It does not replace your familiar University of Chicago or Government Printing Office style guide. Rather, it supplements them by adding guidelines specifically for online documents.

Write Messages in the User's Terms

Little is communicated when the writer and reader speak different languages. A similar lack of communication occurs when messages are written with computer hardware and software terminology instead of terms the user should understand. Express messages in terms of the data objects the user is manipulating, not in terms of the computer's internal representation of those objects. For example, in an error message, don't say

INPUT DATA VALUE OUT OF RANGE

Say

Font value is out of range.
Fonts can range from 10 pt to 24 pt.

Unless you can guarantee that most of your users have extensive knowledge of the architecture and operation of computers, you should avoid computer terminology in messages and prompts. Check proposed messages to see if you are relying on computer terminology instead of terms from the user's world. Terms to avoid include these:

Output	Directory	Root	Data	Response
Byte	Pathname	System	Parameter	Input

Avoid Hostile Terminology

No one sets out to be rude, but have you heard what your computer is saying? Terse, concise error messages sometimes offend users. Writers must be sensitive to how messages could be misinterpreted.

Use Positive Words. Avoid words and phrases that leave a bad taste in the mouth. Beware of words with negative connotations, for example:

Illegal input	Error
Invalid command	Bad data specification

Instead, tell the user what would be legal, valid, or correct. For instance, instead of saying

Illegal command

say

Use COPY, MOVE, or DELETE.

Forgo Metaphorical Mayhem. The computer screen is more violent than TV. Seems like most programs are scripted like a B-grade disaster movie:

Catastrophic error.
Print request aborted.
All running processes will now be killed.
Fatal error.

These messages can frighten a novice user. Inexperienced users sometimes think they can break the computer and the cost will come out of their paychecks. These messages don't help.

Write No-Fault Messages

Many messages unintentionally belittle the user. Error messages that imply that the user is at fault can increase anxiety or hostility. Consider this rude message:

Account: 4413A
Password: *******
INVALID PASSWORD. ACCESS DENIED.

One has the image of a supercilious maître d' refusing admittance to a fine restaurant. See how much friendlier the message seems when a neutral tone is employed:

Account: 4413A
Password: *******

Password did not match the one for this account.
Please try again.

Never accuse or blame the user for the problem. Notice the sense of blame placed on the reader in this accusatory message:

The data *you* entered is invalid. Reenter *your* data.

Consider instead this no-fault message:

We're sorry, but we were unable to complete your call as dialed. Please hang up, check your number, or consult the operator for assistance.

Here, there is no fault, or if there is fault, it is the writer's, not the reader's.

Let the User Set the Pace

Users like to feel that they are in control. Many users react negatively to computer-driven dialogs in which the computer demands information from them, much like a lawyer conducting a hostile cross examination.

Don't demand input:

Enter value now:

Instead, reveal the state of the system:

Ready:

Or show what response is expected:

Command: Filename: Typeface:

Make Messages Pertinent, Not Impertinent

There is no evidence that cuteness works and some evidence that it annoys or interferes with communication. Writers often employ humorous messages in the expectation that they will relieve tension and make using the system more enjoyable. What writers often forget is when and how often those messages will be seen. What is funny the first time is boring the 10th time and positively annoying the 27th time it is seen. Not only that, but some users won't get the joke, or won't find it funny even the first time. People's senses of humor differ—and some people don't even have one.

The reader's state of mind must be considered also. A user who has just lost four hours of work is not likely to find anything funny in a message that makes light of the predicament. Avoid messages like these, which were spotted on the supposedly user-friendly Macintosh:

Drat an error occurred while reading or writing!

Nearly Fatal Error ID = −4.5

Illegal or Rude Command!

The memory allocator is in deep, deep, trouble.
Close a file immediately!

If you must use humor in messages,

- Never poke fun at the user.
- Never make jokes about the product.

Not unless you want to join the ranks of unemployed comedians.

The Secret?

The secret, then, of user-friendly messages is simple, direct sincerity. It is treating the reader of the message as a friend and a fellow professional. It is speaking truth in simple words. Not a bad idea for all writers.

Is It Worth a Thousand Words?

Graphics help readers understand a text by illustrating objects and concepts. This article discusses when and why to use graphics in designing a document. Use these guidelines as you plan your documents.

RESEARCH HAS SHOWN that graphics can help readers understand the information that is presented in a text by showing them what it looks like. You can include graphics in your text more easily than ever before by using new desk-top publishing packages and other computer graphics software.

But before you begin to take advantage of this technology, stop and think about why you are using graphics, when to use them, and how to use them. Here are some guidelines to help you decide why, when, and how to use graphics in technical documents.

THINKING ABOUT WHY AND WHEN

In addition to using graphics to show examples or to illustrate an object you are discussing, you can use graphics for these purposes:

- Use graphics to *guide your readers* through the text. For example, you can use a flowchart like a map, highlighting each section as you describe it. You can help readers find information by using icons in the section headings.

- Use graphics to *show the relationship* between numbers. People are accustomed to seeing numerical relationships explained in charts or tables, but you must set up the chart or table carefully so that the relationships are clear.
- Use graphics as *metaphors* to help your readers relate the concepts they are learning to concepts they already know. For example, you can use a picture of a file cabinet with file folders in the drawers to help readers learn about directories and subdirectories on a disk.

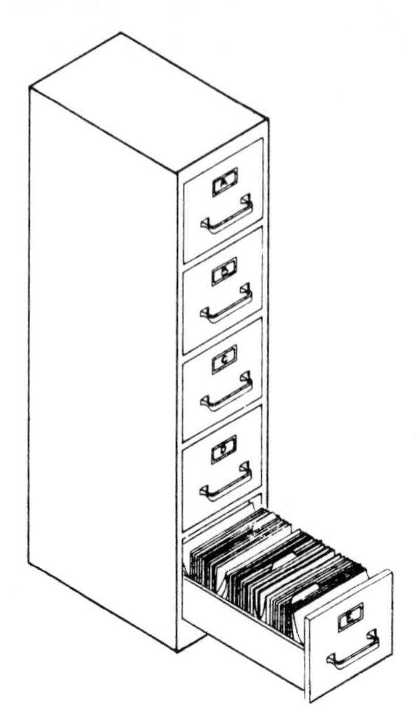

Thinking About How

It's easy to over-complicate graphics. Graphics have to be simple and well thought out to make them useful to the reader. Too many colors, segments, or legends can make a graphic cluttered and hard to understand. To make your graphics easier to use and understand, keep these ideas in mind:

- Always *label* your graphics. Readers shouldn't have to guess what the illustration is about. It is especially important to include the units and magnitude of any numbers in your graphics.
- Use *callouts* to point out important features of the graphics. Don't use too many callouts. If you need more than four or five callouts, use two pictures. For example, you can show a piece of machinery from two perspectives or show part of it in an "exploded" view.

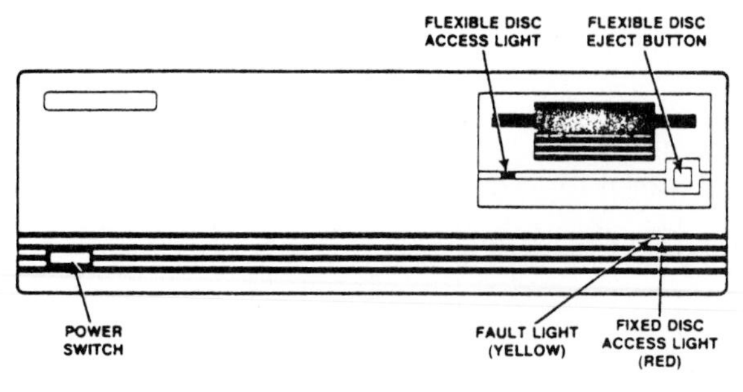

The HP 9153 Disc

- Gear your graphics to the level of the *audience*. Just as you don't expect novice readers to understand technical jargon, you shouldn't expect them to decipher complicated flowcharts or technical drawings. Keep the drawing simple.
- Take advantage of the *universal symbols* that most people are familiar with—for example, the circle and slash to mean "don't"—and use them correctly.

A Survey of the Frequency, Types and Importance of Writing Tasks in Four Career Areas

Mary K. Kirtz and Diana C. Reep

Effective writing on the job is important for job performance and for promotion. This article reports the results of a survey of people working in four different career areas. Look for your career area in the three tables and check (1) the amount of time you are likely to spend writing on the job, (2) how important writing is for your career area, and (3) the kinds of documents you can expect to write on the job.

RESULTS OF A SURVEY OF PEOPLE employed in four different career areas indicate that effective writing on the job is important both for job performance and for promotion. The survey, conducted in 1987 and 1988, asked 118 participants in writing seminars held in companies or on university campuses about the types of writing they did and how important writing was to their careers. Participants included 38 management people, 27 people in technical fields, 24 in clerical positions, and 29 in social service occupations.

Table 1 shows the percentage of time the participants said they spent writing while on the job. Although none of those surveyed was employed as a professional writer, writing clearly occupied a major part of the work day. Nearly 50% of managers said they spent from 21% to 60% of their day writing. In social services, 51% said they spent between 21% and 60% of their time writing. Over 40% of the technical people and 29% of the clerical people also wrote that much.

Table 1 Percentage of Time Spent in Writing Tasks

	Management (N = 38)	Technical (N = 27)	Clerical (N = 24)	Social Services (N = 29)
20% or less	17 (44.7%)	13 (48.1%)	10 (41.6%)	9 (31.0%)
21%–40%	15 (39.4%)	6 (22.2%)	5 (20.8%)	9 (31.0%)
41%–60%	4 (10.5%)	5 (18.5%)	2 (8.3%)	6 (20.6%)
61%–80%	2 (5.2%)	1 (3.7%)	4 (16.6%)	3 (10.3%)
81%–100%		1 (3.7%)		

Not all respondents answered, so totals may not be 100%.

Table 2 reports the answers to questions about the importance of writing. Most said effective writing was either critically important or very important to job performance: 84% of the managers, 81% of the technical people, 86% of the clerical people, and 89% of the social services people. For promotion, 70% of the technical people and 57% of the managers said writing well had an important effect on their opportunities for advancement.

Table 3 lists the kinds of documents written daily or often. Managers most frequently wrote letters, memos, short reports, and instructions or procedures. Technical people wrote fewer letters than other groups did, but they frequently wrote memos, short reports, and instructions or procedures. Clerical people wrote primarily letters and memos, while those in social services wrote primarily letters and short reports. In addition, individuals named a variety of documents that they wrote frequently in their work, such as policy statements, minutes, and proposals. Obviously, people on the job write often and they need to do it well to succeed.

Table 2 Importance of Communication to the Job

	***Management* (N = 38)**	***Technical* (N = 27)**	***Clerical* (N = 24)**	***Social Services* (N = 29)**
How Important Is the Ability to Write Effectively in Your Present Job?				
Critically Important	11 (28.9%)	7 (25.6%)	6 (25%)	10 (34.4%)
Very Important	21 (55.2%)	15 (55.5%)	10 (41.6%)	16 (55.1%)
Somewhat Important	6 (15.7%)	5 (18.5%)	7 (29.1%)	3 (10.3%)
Unimportant				
Does the Ability to Write Effectively Have Any Effect on Promotion?				
Critical Effect	2 (5.2%)		4 (16.6%)	2 (6.8%)
Important Effect	22 (57.8%)	19 (70.3%)	8 (33.3%)	10 (34.4%)
Minor Effect	7 (18.4%)	5 (18.5%)	9 (37.5%)	11 (37.9%)
No Effect at All	5 (13.1%)	2 (7.4%)	1 (4.1%)	3 (10.3%)
Don't Know	2 (5.2%)	1 (3.7%)	1 (4.1%)	2 (6.8%)

Not all respondents answered all questions, so totals may not be 100%.

Table 3 Kinds of Communication Written Daily or Often

	***Management* (N = 38)**	***Technical* (N = 27)**	***Clerical* (N = 24)**	***Social Services* (N = 29)**
Letters	24 (63.1%)	5 (18.5%)	14 (58.3%)	16 (55.1%)
Memorandums	29 (76.3%)	13 (48.1%)	13 (54.1%)	5 (17.2%)
Short Reports (1–5 pages)	19 (50%)	9 (33.3%)	8 (33.3%)	22 (75.8%)
Long Reports (6+ pages)	4 (10.5%)	2 (7.4%)	1 (4.1%)	4 (13.7%)
Proposals (internal)	5 (13.1%)	3 (11.1%)	1 (4.1%)	
Proposals (external)	4 (10.5%)	3 (11.1%)	4 (16.6%)	2 (6.8%)
Instructions or Procedures	15 (39.4%)	11 (40.7%)	7 (29.1%)	3 (10.3%)
Policy Statement or Guidelines	13 (34.2%)	3 (11.1%)	3 (12.5%)	1 (3.4%)
Minutes	6 (15.7%)	5 (18.5%)	2 (8.3%)	1 (3.4%)
Journal Articles	2 (5.2%)	2 (7.4%)		
Speeches	4 (10.5%)	1 (3.7%)		
Advertising/ Promotion Materials	4 (10.5%)		2 (8.3%)	
Other: Manuals	1 (2.6%)			
User Documentation		1 (3.7%)		
Patient Log		1 (3.7%)		
Case Log				7 (24.1%)

Writing Technical Procedures

Ralph L. Kliem

Procedures communicate information about performing specific tasks or handling company systems. This article discusses organizing procedures in one of three patterns: step-by-step, playscript, and item-by-item. Notice that procedures must be reviewed and updated to remain effective and useful.

DOCUMENTING ADMINISTRATIVE AND TECHNICAL OPERATIONS is very good business. And developing effective procedures is a relatively easy process and an inexpensive one, especially if a logical and systematic approach is followed.

A procedure is nothing more than a set of instructions for completing a series of operations. It specifies what is done, who does it, where done, how done, and when done.

An entire procedure is composed either of a single operation or a process. An operation is a distinctive step in the flow of work or the completion of an action that is necessary to produce a single unit of work. A process is a series of related operations that when completed will result in the production of one or more units of work.

Basically, two generic kinds of procedures exist in the business environment. The first are operational procedures. They provide information that is essential for the completion of a single specific task. Administrative procedures are second. These procedures do not directly contribute to the accomplishment of a work effort. Instead, they play a supplemental or complementary role.

ORGANIZING THE PROCEDURE

The organization of a procedure is somewhat similar to that of writing an essay or article. The procedure contains, however, only an introduction and body; it does not have a conclusion.

The introduction contains general information that is not incorporated within the body. This information is usually supplemental or complementary material; it may even refer to other related procedures. Occasionally, the introduction will include the scope of the procedure; that is, an overview of the entire process and how it relates to other procedures. In addition, the introduction might contain a list of individuals or positions involved in the process.

The body of the procedure contains the bulk of the information. It provides the information necessary to complete a particular action or to produce a specific unit of work. It also provides information on the functions performed and who does them.

BASIC TEXT PATTERNS

Procedures usually take one of three text patterns. Each pattern serves a specific purpose and function.

The step-by-step text pattern for a procedure is selected whenever a sequential operation must be recorded. For instance, a procedure may be needed for completing a form, performing calculations, or conducting balancing operations. This text pattern usually contains section headings and subheadings that serve as functional blocks of operations throughout the procedure.

A playscript text pattern, like the step-by-step one, is used for sequential descriptions. However, this text pattern is used whenever more than one user is participating in a procedure. For instance, a procedure may involve two or more individuals. Each person has a specific function. In the playscript pattern, the operations by each individual are recorded in a manner that reflects the actual flow of the procedure. Section headings and subheadings serve as main functional blocks of tasks throughout the procedure.

An item-by-item text pattern, unlike the previous two, is for procedures that do not follow a sequential pattern. These involve incorporating blocks of information that do not follow a logical flow but are related in some particular way. Generally, these procedures are organized by unit or position. The section headings and subheadings indicate a major responsibility, unit, or position. Within each section, the text can take the step-by-step text pattern.

ENHANCING THE PROCEDURE

To make procedures more effective and comprehensible, displays can be included either for clarifying, supplementing, or complementing material in the text. The list of displays is endless. For instance, it can include charts, graphs, drawings, tables, and forms. All displays must be clear, concise and relevant. And above all else, they must be referenced within the text. Generally, displays serve as either exhibits or appendices. Exhibits are directly used with the text; appendices are indirectly mentioned and provide supplemental material to the reader. Materials in the appendices are usually mentioned in the introduction only.

PREPARING THE PROCEDURE

The information for a procedure can be gathered in countless ways. Interviewing, observation and literature reviews are the most common techniques.

Whatever means is adopted, the information should reveal the nature of the tasks, the exact steps to accomplish the procedure and the proper sequences within the procedure.

The information should be inscribed within the first draft. The document should follow all accepted documentation standards that exist within the firm, and it should be written in a style, format, and level that is appropriate for the readers.

Once the draft is completed, the writer should ask himself or herself some very important questions, including:

- Is the procedure organized; that is, does it follow a logical pattern throughout its contents?
- Is the procedure functional; that is, is it something that the reader can use during the course of his or her operations?
- Is the procedure clear; that is, is it a product that a reader can understand immediately without having to leaf from page to page to what is needed?
- Is the procedure concise; that is, does it contain only the main essential words and operations?

The answer to these questions, of course, must be yes. It makes little sense to develop a procedure that is not organized, functional, clear or concise. However, many firms fail to develop procedures with those very attributes. But excellent procedures can be developed if a systematic means is used for having procedures reviewed prior to their publication.

Having the Procedures Reviewed

The first task is to determine who should have the opportunity to review drafts. Initially, the individuals performing the operations mentioned in the procedure should review it to inspect for accuracy and relevancy of the content. Thereafter, the procedure *writer* should use his or her discretion. Usually, a supervisor has the ultimate say as to whether the procedure is correct and accurate. Occasionally, however, a supervisor does not want to be bothered with this step and leaves the approval in the hands of the employee performing the operation. Whatever the case, the writer should ensure that all those individuals concerned with the procedure have the opportunity to review it. Not only will this process ensure the accuracy of the procedure, it also protects the writer from criticism, particularly from individuals who did not originally agree to the development of a procedure or to some specifics within it.

When *final* approval is required, the writer should ensure that the final document is circulated for approval and that a slip is attached to record the reviewer's signature and the date of his or her approval or disapproval. This will protect the writer from being the target of any future criticisms.

Distributing the Procedure

Once approved, the procedure must be distributed to those who need it. To accomplish that task, answers to the following questions are necessary:

- Who will need the procedure?
- Is the procedure replacing another one?
- How many copies are needed?
- When is the procedure needed?
- What manuals does it belong in?

Maintaining Procedures

Once procedures have been distributed, a means must be available to update them. Usually, that entails devising a process which requires a periodic review, whether annually or semiannually. This periodic review should provide opportunities to rewrite or to even obsolete procedures.

The systematic review of procedures should entail, at a minimum, steps for determining:

- Who should review the procedure
- When the procedure should be reviewed
- How the review process should occur
- How the reviewer's signature and comments can be recorded

Conclusion

Clear procedures offer many advantages to a firm. First, they are an effective means for communicating operational and administrative information. Second, they enable business activities to occur in a standardized manner. Third, they serve as a repository of information which can alleviate the disastrous consequences of employee turnover.

But just writing procedures is not enough. Procedures must be developed in a systematic way that will provide the necessary information to the user. In addition, this means must include the review and maintenance of procedures. Without a systematic process, many procedures will ultimately serve no purpose, much like an unread novel sitting on a bookshelf.

For Writers: Understanding the Art of Layout

James L. Marra

Writers and artists often have difficulty understanding each other's interests in designing manuals. This article explains the general principles of design and also describes how a group of technical writing interns designed a training manual. Notice the differences in visual-balance emphasis created by various cover designs.

OFTENTIMES, THE DIFFICULTIES ENCOUNTERED in the preparation of technical manuals are due to a lack of understanding between writers and artists. In effect, one hand doesn't know quite what the other hand is doing or why it's doing it. One reason for the problem is that the language of the two specialists is often full of esoteric, field-oriented jargon. For instance, the writer may well be mystified by the procedures involved in halftone or duotone reproduction. The artist, on the other hand, may be equally mystified by parallel or pyramidal sentence construction. And the danger, of course, is that such barriers hinder progress and strain work efforts and cooperation between two groups whose harmonious collaboration is a must if the "new look" in technical manuals is to be successful.

These barriers need not exist. If the writer understands the main principles of design, he or she will be better able to understand the artist's motivation and rationale. Conversely, such understanding will evoke an attentive ear from the artist when the writer enters the design field. All in all, more harmony between the two will become a distinct probability. And more effective manuals will result.

DESIGN PRINCIPLES

Virtually any good design book will elaborate upon a multitude of design principles which one should consider in creating layouts. Naturally, artists are schooled in such principles, which basically boil down to four fundamental concerns: unity, balance, sequence, and proportion.

Unity

The main principle, unity, is the relationship of all graphic elements, white or negative space included, on a page. It is the binding agent to the layout. In Gestalt thinking, the modes of perception patterns lead to the result that the whole of a visual image is greater than the sum of its parts. This is unity. It is the principle whereby type style and size, art (line or halftone), borders and frames, and colors all lend themselves harmoniously and efficiently to a unified presentation.

To achieve unity the layout artist develops axis points, or points around which the elements (type, art, etc.) revolve. As a result, the vibrating visual effect a reader experiences from widely spaced, random groupings is eliminated and a whole, unified image results. This idea of wholeness is unity. And in many ways it is the umbrella principle to balance, sequence, and proportion.

Balance

Balance is the distribution of optical weight represented in sizes of art and type, tones, and colors. It may be symmetrical (formal) or asymmetrical (informal). If symmetrical, the weight of the figures or elements on the ground of one-half of the layout page will be repeated on the ground of the other half. Thus, if one is working with balance left to right and places a mirror down the center of the page, one-half of the design will look like the other. The optical weights will be the same. In asymmetrical design, optical weight is still important, but the weight is not equally distributed. Rather, there is a shifting of its emphasis to one side or the other.

Sequence

Whereas balance involves weight, sequence involves rhythm. It is the guiding principle which emphasizes control of a reader's eye movement. Normally, the primary visual area of a page will be its top left. The eye will then move in a zigzag, or Z pattern, to the bottom right. Consequently, many layout artists are fond of placing the design elements within the path of the Z. However, eye movement can be created in any direction by the choice of position, size, or color of the elements.

Proportion

The final principle, proportion, is the relationship of sizes and tones. The layout artist who is considering how much width of an element to use compared to its depth is working in proportion. Likewise, when consideration is given to the amount of light area as opposed to the amount of dark area, proportion is the principle at work. The main idea to bear in mind when working with proportion is the need to employ the 3/5ths rule. The 3/5ths rule, or golden oblong as it is commonly referred to in design circles, simply says that a page layout is more dramatic, more appealing to the eye when the major element (photo, copy block, etc.) occupies 3/5ths rather than half of the available space. In fact, any odd-fraction dimension is more suitable, avoiding the pitfall of humdrum which is a result of page layout in halves.

THE INTERN PROGRAM

And perhaps that is what layout design in action seeks to do—avoid the humdrum, the unevocative. In manuals the quest is under way. For instance, a select group of technical writing interns from the Redstone Arsenal in Huntsville, Alabama, recently underwent an intensive training program in technical communication at Texas Tech University in Lubbock, Texas. Organized by the Department of Mass Communications, the program was designed to bring the interns into contact with the layout possibilities of technical manuals. A key word here is "possibilities," for the instruction was geared to creative thinking in layout and design. First, the principles of design had to be understood. Then, by the actual employment of the principles, the interns were thrust into the hot seat of the layout artist. Possibilities emerged. Alternative methods of presentation were realized firsthand. And a number of workable, though rough, layouts were produced as alternatives to the training manual cover shown in Figure 1.

The Procedures

Prior to these rough layouts, which originated with quick and plentiful thumbnail sketches (the crude, initial drafts of design possibilities), a critique of the existing manual cover was offered. For instance, as regards the design principles, the cover was found to be lacking axis points around which the elements of type and art should revolve. The lack of contrasting sizes for greater points of emphasis (proportion) was also noted, particularly in the typesize. In addition, there appeared to be no definitive attempt at balance, either symmetrical or asymmetrical. And overall, it was felt that the design gave the appearance of random selection and positioning.

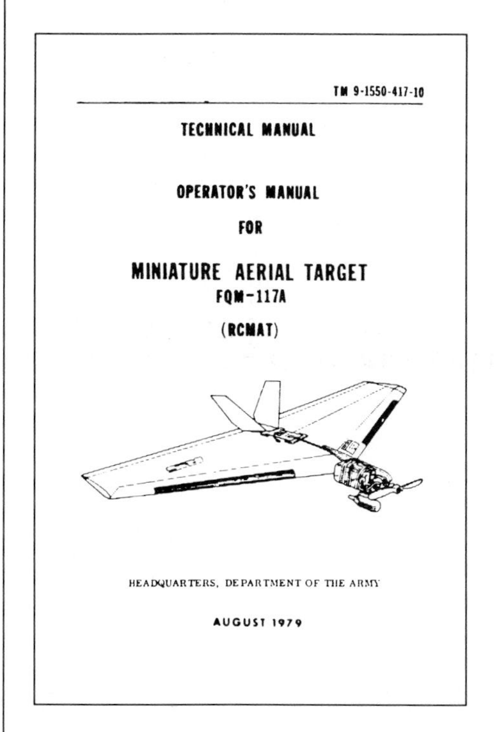

TM 9-1550-417-10

TECHNICAL MANUAL

OPERATOR'S MANUAL

FOR

MINIATURE AERIAL TARGET

FQM-117A

(RCMAT)

HEADQUARTERS, DEPARTMENT OF THE ARMY

AUGUST 1979

Figure 1 The Cover of an Existing Manual

Following the critique, the interns began thinking with their pencils in the form of thumbnails. Working with existing copy and art, they were encouraged to make size and position changes and to incorporate singly or in combination the four main design principles. A random selection of nine quick thumbnails, re-created for this article, is represented as Figure 2, A to I. At this stage the layout artist was simply attempting to imagine as many alternatives as possible, deferring judgment so as to allow the visual imagination free rein. Likewise, the writers, by entering into this initial phase of artistic production, experienced the problem as it would be attacked at an artist's drawing board.

The Results

Following the thumbnail stage, the writers undertook execution of rough layouts. The second phase is generally the stepping stone to the comprehensive layout, which, in turn, evolves into the finished mechanical used in print production. In effect, the rough layout is that which the artist imagines to have the most promise of the many thumbnails produced. Consequently, it is given more detail and elaboration than the thumbnails and represents a more vivid picture of what the end product will be. Again recreated for reproduction here, a selection of rough layouts produced by the interns is represented by Figures 3, 4, and 5. These evolved from the thumbnails in Figure 2: A, B, and C, respectively (the remaining thumbnails are offered as alternative approaches). And in the rough layouts the design principles are well exemplified.

For example, in Figure 3 the intern sought to capture symmetrical balance in order to give the layout a dignified and formal look. The weight of the elements to the left equals the weight of the elements to the right. Also, the typesize has been altered to create contrast and thereby highlight the key parts of the message.

In Figure 4, the reader's eye is nicely directed through the principle of sequencing. By contrasting the weights of type and using reason in positioning the elements, the intern developed the zigzagging or Z pattern of eye flow. The reader's eye begins reading at the upper left because of the highlighting of "Miniature Aerial Target." It then moves down through the target to the propeller, which directs the eye back to the closing message, "Headquarters, Department of the Army."

In Figure 5, the abstracted view of the aerial target has been magnified, giving it a larger proportional weight and creating a dramatic effect as it becomes a dominant part of the layout. It is also consistent with the odd-fraction-dimension principle in that the line art now occupies the great percentage of the page. In addition, two more interesting visual effects are created. First, the lines of the art move off the page. This produces closure, a phenomenon which allows the reader's imagination to complete the drawing. More active involvement of the reader is the end result. Second, the terms "Operator's Manual For . . ." are placed in the wing span. From this device it is possible to create a dimensional effect which encourages the reader to move into the page, again creating more active involvement.

It is also interesting to note that in each of the figures represented there is a unified composition of the elements. The contrasting of weights establishes emphatic parts of the message. At the same time, the elements are tightly grouped in units and appear to flare out from various axis points while not giving the appear-

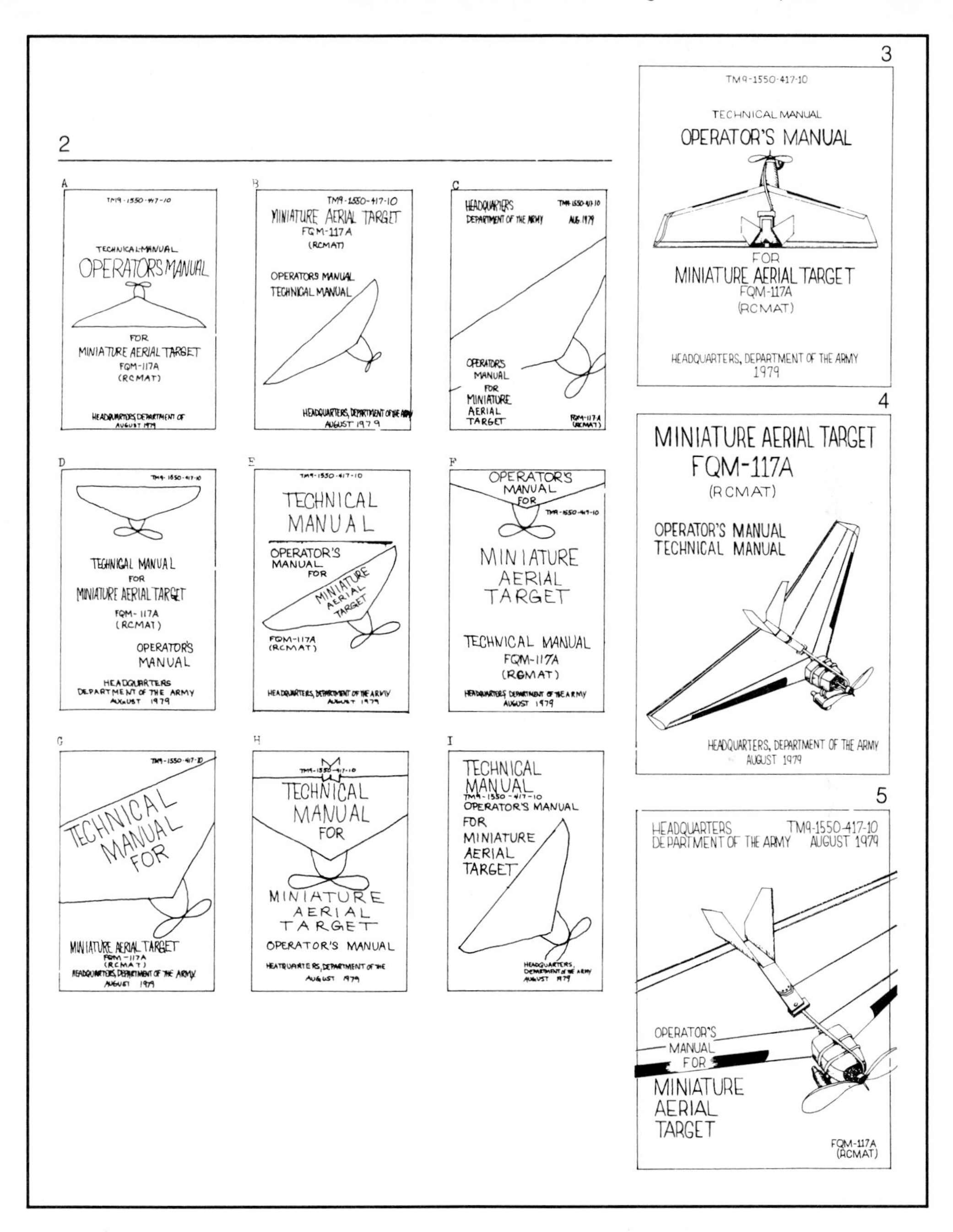

Figure 2A–I Thumbnail sketches of possible variations.

Figure 3 Symmetrical balance and tightening of the elements give the layout a dignified look.

Figure 4 A zigzag (Z) pattern of eye movement is created by the positioning of the elements.

Figure 5 Added weight is given to the illustration by its large size.

ance of random scattering about the page. Consequently, the amount of vibrating space between the elements is diminished, and the reader's eye is able to fix on key parts of the message.

CONCLUSION

In conclusion, by employing one principle, then another, and by considering combinations of each, the interns were able to imagine a variety of possibilities. In addition, they were able to experience dilemmas which confront layout artists on a daily basis. And moreover, they were working with the language and processes of design which are the daily bread of those artists. The likely result in such a situation is that writers gain a better understanding of design and the people who create it. The language barrier is overcome. More harmony ensues. And the collaborative effort, relieved of the strain of communication problems, pursues a common direction toward more effective "new looks" in technical manuals.

Combating Speech Anxiety

Myles Martel

Most speakers suffer from anxiety before making an oral presentation. This article provides guidelines for controlling and deemphasizing fear of public speaking. Read the guidelines before making an oral presentation.

CALL IT STAGE FRIGHT, speech fright, speech or communicator anxiety or apprehension, its symptoms are unmistakable and its significance as a barrier to effective presentations cannot be overstressed.

A quaking voice, trembling knees, erratic pacing or rocking, stomach butterflies, pronounced monotone, a deadpan facial expression, and lack of eye contact are common cues that the speaker is experiencing the malady I prefer to call speech anxiety—a tension-laden self-consciousness which dominates the speaker's concern for communicating ideas.

Few speakers are immune from this phenomenon. Leaders in all professions in all ages have experienced speech anxiety, regardless of status, sex, intelligence, age or income. Sir Winston Churchill remarked that when he began to speak he felt as if there were a nine-inch block of ice in the pit of his stomach. And Cicero said, "I turn pale at the outset of a speech and quake in every limb and in all my soul."

Communication experts and psychologists are not certain about the precise causes of speech anxiety. The most popular explanation, however, calls it a form of approach-avoidance conflict. On the one hand, the speaker may feel impelled to represent the organization, while on the other, feel pulled in the opposite direction by another dominant force, often by some fear of failure.

Speech anxiety has no sure-fire cures, but it can be alleviated by a proper attitude toward (1) the anxiety, (2) the self, (3) the audience, (4) preparation and practice, and (5) delivery.

ATTITUDE TOWARD SPEECH ANXIETY

1. Recognize that most if not all speakers experience some form of normal nervous anxiety. If you feel a little anxious, don't panic. Say to yourself that you have normal nervous anxiety, which you can control—not eliminate.
2. Consider the normal nervous anxiety to be potentially helpful. Your anxiety may motivate you to speak with greater forcefulness, and result in a more animated presentation.
3. Realize that your anxiety state is often much more obvious to you than to your audience.

4. Don't be afraid to admit to your speech anxiety in discussions with friends or colleagues.

Attitude Toward Self

1. Think "expert." With few exceptions you would not be making an appearance if the host did not feel you had something of value to share with your audience.
2. Assume a 100 percent attitude regarding your engagement. Many speakers have a tendency to take on a "less than" attitude toward the engagement, often not realizing that this attitude may reflect their own fear of coping with the challenge to communicate. The ability to "wing it" is a talent presumed by many but possessed by all too few.

Attitude Toward the Audience

1. Regard the audience as being composed mainly of friendly listeners with whom you will be engaging in an enlarged conversation.
2. Don't overreact to hostility, disagreement, lack of interest or apparent confusion.

Attitude Toward Preparation

1. Understand what a speech is, the specific means by which it can be prepared and practiced.
2. Avoid memorization. Fear of forgetting normally generates tension, self-consciousness and other unnatural delivery characteristics.
3. Learn how to speak extemporaneously from a carefully prepared outline consisting of key words and phrases. Conscientiously practice expressing each idea without "overscripting" yourself into memorized passages. Practicing with audio or videotapes while standing in front of one or more constructive critics can be especially helpful. Practicing in front of a mirror, despite the popularity of this advice, can be counterproductive; it can make you more concerned with your body than with your message.
4. Analyze the audience, occasion and setting well in advance. Learning about the demographic and attitudinal composition of your audience, the layout of the room and the specifics of the program can diminish your normal fear of the unknown.
5. Prepare a neat, well-spaced and well-selected set of notes with delivery advice and notations.
6. Devote extra efforts to creating within the speech itself an impactful introduction that captures the audience's attention, enhances your credibility and orients the audience toward your subject.
7. Do your best to be well-introduced. Send a suggested introduction to the host and always have a copy with you in case the host has lost the original.

8. Avoid major last-minute changes in your outline. The natural anxiety produced by the engagement may prevent you from handling such changes well.
9. Arrive early and mingle with the audience. This will help you to regard them as individuals with whom you can hold an enlarged conversation rather than as an unknown, fear-provoking mass of people.
10. Try relaxation exercises shortly before your appearance. Isometrics and deep-breathing exercises can be especially helpful.
11. To relieve your television or radio talk-show jitters it may help to have a friend ask you immediately before your appearance a few warm-up questions.

Attitude Toward Delivery

1. Gesture and move around purposefully (when a lectern is not present) to reinforce the message physically, while you simultaneously create a natural outlet for your nervous anxiety.
2. Establish and maintain direct and distributed eye contact with the audience. The more you establish eye contact, the less you will be struck with fear of the unknown. Also, you will be better able to respond to their feedback with an enlarged-conversation frame of mind. The popular advice "pick a friendly face" may be helpful when you need visual reassurance, but don't concentrate on this person for too long.
3. Consider using visual aids. Flip charts in particular are effective means of inducing natural bodily movement.

Probably no one has ever conquered speech anxiety by just reading about it. Reading merely provides the insights to be applied in real speaking situations. The more you seek these situations and systematically prepare for them, the more confident—and compelling—you should become.

Technical Writing for Computer Software

Polly Perryman

Computer user manuals must show readers what the system can do and help readers use the system efficiently. This article discusses the difficulties in producing effective computer manuals and identifies the needed writing strategies, such as definition and description. Before writing a user manual, a writer must thoroughly understand the computer software and its capabilities.

WRITERS BECOME GOOD WRITERS when they work at their craft. That is, they write. The more they write, the more they can write. The more they can write, the better their work becomes. Just as any other profession requires experience to attain expert status, writing requires years of practice to become effective. Discipline is necessary to produce material on demand. Technical writing is a specialized area of writing, and the technical writer, just like novelists and journalists, needs to consistently produce quality material for his audience.

The material must be comprehendable and serve the purpose for which it was written. A novelist writes stories that have a beginning, middle, and end. Readers must be able to determine what the story is about. They must be able to identify characters and see a definite plot. A newspaper or magazine journalist must also write material that has a beginning, middle, and end. It must be informative, thereby providing the reader with new information or a new slant on old information. The reader must be able to understand the vocabulary and structure of the article. A technical writer must meet those same requirements in writing a manual. There must be a beginning, middle, and end to even the most technical manuscripts written. It is here, among the very rudiments of writing, that the documentation produced for today's computer software is lost.

System user documentation is not, contrary to some interpretations, needed to explain the programmer's hard work in creating a system users should be thankful for. It is meant to provide explanations, descriptions, and instructions of what the system can do for the user. This fact is recognized by many software developers. Still, guidelines for writing documentation are sketched out and scarce. Qualified technical writers for computer software are hard to find.

IDENTIFY SPECIFIC TASKS

From the very beginning, the technical writer must identify the specific tasks and activities of the user. This identification process should be apparent from the first page the user reads. It must be reflected in the table of contents. It is here, in the table of contents, that the user will look for the information he needs to learn and reference. Technical writers must not only write well and interpret technical capabilities, they must also understand the intended application function in a user's environment.

The current trend is to use the menu options from the system's main menu as chapter headings. This is a good method for dividing the manual; however, the manual must provide the user with additional information. Information such as: the unique use of characters or symbols programmed for special use throughout the system, required use of control keys, and "paging" instructions that explain how to get from one screen to the next. This information must also be listed in the table of contents. It is all too often swallowed within the pages of the manual. The user is left on his own to try to find the information or guess what it is.

For novice users this becomes a special problem. Novice users are unaware that such capabilities can be programmed into the system. Many times a user will use a system for years before he fully utilizes it to its full potential. This situation can be avoided. The writer can and should think through the material that must be included in the manual and start at the beginning. This thought process carried out over and over again by the writer will enable him to better organize the manuals he writes.

The user will expect the page number in the table of contents to be correct. This won't be a problem when proper editing has been completed. The problem is that paging a manual is a tedious job often overlooked in the editing process. With more manuals being printed directly from letter-quality word processing systems, direct correlation between the page number listed next to the chapter headings in the table of contents and actual pages of information is often incorrect. This happens when the table of contents is completed and not modified as revisions to the manual are made. When a user meets with frustration in the very beginning of the manual it is safe to assume he won't make much use of the rest of the book. However, when the writer has taken care to identify information properly, giving the user a good beginning, the user will move on with enthusiasm. Learning is basic to human nature.

The middle of the manual must also be meaningful. In the middle of the manuscript certain things must take place. In a novel, for instance, characters become more colorful and the plot thickens. In a newspaper or magazine article, details are told. In a technical document or user manual explanations must be given, descriptions must be complete, and instructions must be accurate. All manuscripts must include the information for which the manuscript was written or it becomes of little value to anyone.

Some novelists capture the reader with greater finesse than others. Some reporters dig out stories with a deeper interest. Publishers demand quality, and authors/journalists respect their authority to recognize good work. With technical manuals, software publishers seem to lack the knowledge necessary to recognize whether the requirements in documentation have been met. Their primary concern appears to be with making their product more marketable, not necessarily usable. The users of today's systems have become wise to the need for usable documentation. Published software manuals will have to begin meeting and mastering the requirements set forth for their intended use.

Review Requirements in Detail

It may be necessary at this point to review those requirements in some detail. The requirements for quality user manuals are: first, that they have a beginning, middle, and end; second, that the table of contents list all necessary information in logical user terms; and third, that definitions, explanations, descriptions, illustrations, and instructions are included in their appropriate places throughout the manual.

Definitions are one or two sentences which provide the specific meaning of a capability, menu, menu option, or field prompt. Definitions should be used at the beginning of a section to let the user know what he is about to learn. Explanations are written to interpret specific information. They provide the meaning of information and express the reasons for it. In a technical manual explanations must tell the user exactly what the purpose of the system is. If it is an application system for accounts payable, it must say that, and then proceed to methodically tell what the system will or will not do. This process must be written in words that the user will understand. For instance, the word invoice should be used to define a field required for billing input. Replacement words used solely to "dress-up" the narrative nullify the original objective in writing an explanation. Some commonly misused replacement words for invoice are: voucher, statement, bill, payment, or ticket number. While some may be acceptable terms they are not clear. Clarity is essential in explanations. Some other common errors are: reports mislabeled as lists, lists mislabeled as edits. Precision in the interpretation process leaves no room for misnomers.

Descriptions are also a required source of information in a user manual. Descriptions represent the quality of the system functions. They are needed to tell the user about the system capabilities, relate a system function to a previous manual operation, and provide a conceptual understanding for implementing instructions. Verbal descriptions must be concise and simply stated to quickly provide a visual sense of what is happening.

Instructions must above all else be accurate. Typos, spelling errors, verb misusage, and inconsistent word usage must be eliminated to build a foundation for accuracy. On that foundation tediously accumulated details must be given to the user. A step-by-step account of each action the user must perform has to be given. Objectivity is important. Assumptions cannot be allowed. The obvious must be recorded in writing along with the remote. Instructions do not necessarily require many words, but each word must be chosen for its value in conveying the exact connotation intended. A frequently abused word in instructional documentation is the word *depress* rather than the proper word *press*.

When each of the requirements has been met, a writer may add enhancements to the user manual just as a programmer adds "whistles and bells" to a system to make it more attractive. The enhancements might include sections providing examples, illustrations, or samples of screen displays, printed lists, edits, or reports. Other enhancements might include glossaries, tutorials, or suggested sample data. The possibilities for improving a user manual to attract attention and make it sell the product better are endless; but, an *improvement or enhancement can never be substituted for required material.*

Writing a manual that includes all of the requirements takes time. Writing manual after manual takes patience. Time and patience are the most valuable assets a writer can have in producing any manuscript. They are assets that make good writers great. Writing a user manual for publication demands the skills of a writer who works at his craft.

Logical Approach

In order to meet the requirements set down for a user manual, the technical writer must logically approach the task in much the same way any author would. A novelist couldn't very well produce a manuscript weaving a plot through the Russian revolution without having a complete understanding of the geological, political, and social structure that existed at the time. A journalist can't report on a scientific advancement without finding out what was innovative about it. So, too, the technical writer must thoroughly understand the basic concepts and subject matter a system is designed and developed to do.

A novelist must methodically build the characteristics and personalities of the hero or heroine. A technical writer must identify the capabilities of the system —effective use of system functions makes the user the hero of the manual. There are probably several ways that would work in approaching this task. There is one way that works best.

It begins by assuming the technical writer knows nothing about the system. He becomes as innocent to the computer environment as a novice user in a small computerized shop. By doing this, he broadens his base for questioning the system. He thinks about what he needs to have happen rather than what is happening. This willful act of ignorance is the beginning of *system research*; the first step a writer must take when a quality manual is to accompany a software system.

Subject and Function Studies

System research is a deliberate act of education. It is a two-fold process. The first part is subject study. A writer must thoroughly comprehend the manual procedures and basic skills necessary to perform a task before he can begin to tell users how to accomplish the task on a computer. The novelist, when beginning, must be familiar with the architecture of a given period of time in order to have his characters weave in and out of their surroundings in a realistic, meaningful manner. So it is with the technical writer providing meaningful and realistic explanations and instructions to the user. A manual designed to assist a user in learning and operating a general ledger system or financial forecasting system must be written with the correct foundation and structure. The terminology must be in tune with the environment the system is being used in rather than the shop it came from.

The second phase of system research is function study. Function study is the process of learning exactly what the computer software will or will not do. It is the translation of computer function into manual tasks in accordance with the software design. Notes must be taken to insure accuracy in interpreting the steps taken to access, enter, review and print data. When writing a user manual the internal flow of the software is not as important as the exterior flow of the system. What's happening on the display screen? What has to be done next? What's going to happen when that is done? A writer must think of all the problems users might encounter. He must define each message, each option, and each prompt request field. He must remember that the user is, in fact, reading his work to become a competent master of the system. He must realize that the words that are being committed to paper for this manual are critical to the on-going operations of a business no less important than his own.

System research is, then, an accumulation of the subject and software specifications developed to provide users with a faster, simpler method of recording, storing, and processing data. It is the basis for writing, learning, and referencing material that will enhance the system design from start to finish.

Outline Is Important

When sufficient information has been gathered about the subject and software, a writer must plan the best presentation of the material. He begins an outline. The outline must be detailed enough to include each of the major functions, all of the minor functions, notations of special features, and primary facts about the individual messages that will be or might be displayed.

The outline can be written in any format the writer chooses. It should be comprehensive enough that it can be used to record and track the progress of the manual as it is being written. The outline is important for other reasons, too. Once written in complete detail, the outline may be presented to the programmers and system designers providing them with a visual means for reviewing the writer's interpretation of what is occurring in the system. At this point, the wording of a menu option may be changed to represent more clearly what the option is used for. Additional modifications may be noted and refined. When an outline is presented the writer provides the system development team with an objective user viewpoint of the software capsulized in a manageable report. It is from this outline that a system begins to have a leading edge on its competitors.

The outline should also be used throughout the development of the software. Notes should be kept indicating where changes or enhancements are being made. The outline should be written as the system is being designed. Writing should start at the same time as the programming. The final copy may then be prepared as the programmers "clean up" the system getting it finalized for the marketplace.

Starting the First Draft

As soon as the outline has been reviewed with the persons responsible for the original design and development of the software, the first draft should be written. At this time putting facts on paper is more important than actual sentence structure and good writing. The flow of the system should be evident from this first draft. It must be accurate. It must be clear and precise. It also must be remembered that it is merely a first draft and should be edited a minimum of three more times before it can be called a user manual. It is at this point that many companies who provide users with excellent software systems begin to falter.

The editing process is long, tiring, and time consuming. It is expensive, too. Writers, editors, and system personnel must be paid while the product is held back 2 to 4 weeks. In the rush to hit the marketplace, drafts are deemed fit for user training. The commitment to documentation has been sideswiped again.

Anytime the user manual is pulled out of production prematurely, the user loses. He loses the real reason the manual was provided in the first place. It is not reasonable to assume that users can or should learn about the system from uncompleted manuals.

THE SMOOTH COURSE TO SUCCESS

The only way to avoid the problem of premature documentation release is to become as committed to the user's learning needs as the computer industry is to the productive needs met by a software system. This commitment is critical to eliminating the problem of incomplete documentation and staying on a smooth course with success.

The smooth course to success means less technical support will be required and more satisfaction will be realized by the consumer. Keep in mind that in many instances it is the competently trained secretary who will be a primary user of the system—a person whose professional skills demand a thorough understanding of language form and usage. This is the time to think about how the manual will stand up to an on-site edit. On-site edits can be successful only when the proper in-house editing has been completed prior to releasing the system.

The editing process is different from a proofreading project. Editing is the process of modifying, enhancing, and omitting information from the user manual. Proofreading is the process of seeking spelling or typographical errors. Editing is crucial to comprehension; proofreading is crucial to cosmetics.

THE PRODUCTION PHASE

Long before the final copy of a user manual is completed, the outer design including the page formatting, typesetting, and binding style must be decided. When all of the editing is complete, this last step in publishing a user manual that shows off a system, must allow swift and smooth production. The selection of a printer is important.

A writer, though perhaps not directly responsible for the printing of his book, should have some knowledge of what is going to happen with his written manuscript upon completion. Novelists who want to be taken seriously in the world of journalism don't work for the gossip papers given to printing scandalous information that may or may not be factual. Technical writers should be equally concerned about the publication of their work. Software publishers and developers need to become educated in the techniques needed to produce quality documentation. This will not come about if writers don't demand sufficient time to write quality work.

Not all the blame can be placed on software publishers. Often the writer of a user manual is a programmer, system analyst, or data processing clerk. The assignment of the user manual project is decided by highly competent, generally successful business people. The decision is sometimes based on economics. The reasoning behind that decision is that the programmer already knows the system, thus eliminating the need for a writer to complete system research that would delay the marketing of the product.

Sometimes the reason is that writers are thought of as glorified clerks, not as sophisticated as programmers, therefore not essential in producing a manual.

Sometimes the reason may be the lack of qualified technical writers available in the computer industry.

Whatever the reason, these business people are beginning to realize the importance of quality documentation. They have been trained and conditioned to have a "feel" for the mood of consumers. They know the consumers of computer software are rapidly gaining insight through the technical doors of data processing. The users are demanding quality documentation.

It would be wise to have technical editors review any work submitted by programmers writing user manuals. This alone would be a step toward progress.

The Editor's Nightmare: Formatting Lists Within the Text

Gene A. Plunka

Lists in technical writing emphasize specific information for readers. This article provides guidelines for preparing stacked lists and lists within sentences. The article includes samples of how to format lists effectively and tips on how to punctuate lists and when to use bullets. Take the quiz in the article to test your knowledge of how to format lists effectively.

EVERY EXPERIENCED EDITOR has been confronted with the task of editing lists within the text. Almost daily, editors encounter writers who do shoddy work and turn in sloppy copy. However, even quite capable writers often do not know how to handle large blocks of information that must be compiled in the form of a list. Because there are no standard rules for the preparation of lists, writers must invariably create their own system (or systems) of notation.

Such "creativity" causes major problems for the editor, especially when the writer is not even consistent in preparing the same type of list. In addition, the editor, who usually has only a vague notion of how to edit lists properly, must make arbitrary decisions based on personal preferences. Such capricious marking raises doubts about the validity of the editorial changes, especially when there are many lists to edit, as is often true of scientific or technical material.

In this article, I provide guidelines for editors and writers when preparing or editing either vertical lists or lists within a sentence.

MISCELLANEOUS ERRORS

As a former editor, I can recall some of the most common errors that made the editing of lists a tedious and frustrating task. First, many writers create their own notation system, using dashes, asterisks, daggers, or other artistic "designs" to introduce listed items. When preparing lists, writers could help the editor considerably if they used only the standard punctuation marks that will be discussed later in this article.

The second major headache-producer for the editor is the shoddy writer's insistence on using "For example" to introduce a list, thereby writing in phrases rather than sentences. I am not suggesting that "For example" must be eliminated from a writer's vocabulary; instead, just be sure that when "For example" begins a statement, the two words will be followed by a sentence and not by a colon.

The third major fault is the long list that is not grammatically parallel—e.g., participial phrases mixed with noun phrases, infinitive phrases with adverbial phrases, etc. The following example indicates why many editors cringe at the thought of editing technical manuscripts full of lists.

Our expenses included the following costs:
1. Mail—$500
2. Telephone—$230
3. Using the Printer—$200
4. Copier Fees—$75

Obviously, lines 3 and 4 need to be changed to "Printer" and "Copier" to maintain parallel structure. In short, when the editor has to worry about both formatting the lists for consistency and correcting for parallel structure, the editing of lists becomes that much more demanding, time consuming, and frustrating.

WHEN TO CREATE LISTS

First, try this quiz to test your knowledge of how to format lists. Choose the column number that represents the proper method for formatting the list in the following sentence:

The elements in the test included

I	II	III
1. zinc	(1) zinc	1) zinc
2. lead	(2) lead	2) lead
3. tin	(3) tin	3) tin
IV	**V**	**VI**
a) zinc	a. zinc	zinc
b) lead	b. lead	lead
c) tin	c. tin	tin

This example illustrates the types of decisions editors must make when correcting lists. The solution to the puzzle is that none of the six columns is the best choice. The writer should not have created a list at all. Instead, the sentence should read, "The elements in the test included zinc, lead, and tin." Most lists are superfluous. Scientists, engineers, and technicians frequently create lists when they could have otherwise saved space and avoided excessive formatting decisions. Thus, writers should first ask themselves whether there is a significant need for the vertical list that is being considered.

Lists within a sentence are often preceded by a colon. Most style manuals or grammar texts recite a rule that should be memorized by "list-happy" authors: **Do not use a colon between a verb or preposition and its direct object.** Often, however, writers assume that colons must automatically precede long lists, even when a verb introduces the list. To avoid violating the rule above, writers may merely add the words "as follows" or a similar variation after the verb. For example, instead of writing *The criteria were: (1) cost, (2) quality, (3) reputation, and (4) service,* the author should be consistent in applying the rule: *The criteria were the following factors: (1) cost, (2) quality, (3) reputation, and (4) service.*

Similarly, writers who use a colon after a preposition (e.g., *He supplied the scientists with: pipettes, bunsen burners, and ether*) should instead write *He supplied the scientists with the following materials: pipettes, bunsen burners, and ether,* or more simply, without the colon, *He supplied the scientists with pipettes, bunsen burners, and ether.*

Some writers insist on following the old rule that the first word after a colon is capitalized if a complete sentence follows the colon. Today, however, writers have the option of capitalizing or lowercasing the first word of a complete sentence following a colon; they must, of course, be consistent in following the chosen option throughout the text.

Elements in a list can be specifically enumerated to prevent misreading, to clarify a complex or lengthy series, or to highlight the listed items. Enumerations within a sentence should be introduced and identified only by numbers or lowercase letters, not by capital letters or roman numerals. In *A Manual for Writers of Term Papers, Theses, and Dissertations,* Kate L. Turabian notes that numbers or letters in a list stand out better when they are surrounded by parentheses rather than when they are followed by periods **[1]**. *The Chicago Manual of Style* concurs with Turabian's assessment and suggests double parentheses surrounding numbers or lowercase letters **[2]**.

Following the colon that introduces the list, a comma or semicolon may be used to separate enumerated entries; again, the most important factor is consistency of style throughout the manuscript. Of course, when there are commas within the items of the list, semicolons must be used for clarification.

Thus, models to follow for enumerations within a sentence would be these four examples:

> The report consisted of four sections: (1) introduction, (2) discussion, (3) results, and (4) conclusion.
>
> The report consisted of four sections: (1) introduction; (2) discussion; (3) results; and (4) conclusion.
>
> The report consisted of four sections: (a) introduction, (b) discussion, (c) results, and (d) conclusion.
>
> The report consisted of four sections: (a) introduction; (b) discussion; (c) results; and (d) conclusion.

When writers use letters or numbers in a sentence list, they should keep in mind that a line that ends in (1) or (a) is considered a poor line break. Many editors will, by habit, mark such errors, even in draft copy, as carry-over lines to run over the letter or number to the rest of the list on the following line. By paying closer attention to inappropriate line breaks in the middle of lists, writers could simplify the editor's tedious and time-consuming job of correcting and marking errors concerning typography.

Vertical Lists

Lists that contain four or more items in a series may be prepared vertically, set off from the sentence that introduces the list. Again, the rule to follow is that a short list should be incorporated into the sentence of which it is a part, not into a vertical list. Vertical lists are usually introduced by a colon; however, if the list is preceded by a complete sentence, a period may follow the sentence and thereby introduce the list. Dashes or other punctuation marks should not introduce vertical lists.

Introducing and Enumerating Vertical Lists

Writers who seek clarification with regard to the rules for enumerating vertical lists will discover a potpourri of contradictory information. *The Chicago Manual of Style* uses numbers and periods in its examples of such lists, yet no guidelines are provided. The *Publication Manual of the American Psychological Association* suggests the use of arabic numerals followed by a period but not enclosed or followed by parentheses **[3]**. The editors of *Words Into Type* also support the use of numerals with periods **[4]**. *The Council of Biology Editors Style Manual* states that lowercase roman numerals, arabic numerals, or lowercase letters in single or double parentheses may be used, or the writer may prepare the list "in accordance with the style of the journal" **[5]**.

The other style manuals, including the *U.S. Government Printing Office Style Manual*, the *MLA Handbook*, *The Associated Press Stylebook and Libel Manual*, and *The New York Times Manual of Style*, offer no advice on the preparation of vertical lists. The consensus is that the format adopted for vertical lists should be arabic numerals or lowercase letters (be consistent as to which you use) followed by periods. The style therefore differs from the formatting of lists *within* sentences where double parentheses are suggested to make the items easier to differentiate for the reader.

Sometimes vertical lists may be introduced by phrases instead of sentences, or the items in the list may be needed to complete the sentence. In such instances, no colon is used to introduce the list items, thereby precipitating different rules to follow with regard to punctuation and numbering. The following sentence is an example of such a list:

> The hydroponic flooding system included
> a sealed reservoir,
> an aquarium vibrator air pump,
> an automatic household timer, and
> a plant container

Such lists cause numerous problems with regard to punctuation, numbering sequence, and capitalization. In addition, the editor must worry about whether or not the list's format is consistent with similar vertical lists displayed elsewhere in the text.

A good rule is to use the colon to introduce all lists (i.e., use complete sentences preceding vertical lists) so as to develop a degree of consistency in formatting vertical lists throughout the manuscript. When the colon is used to precede a list, rules can be followed. However, the lack of a colon usually leads to arbitrary rules with regard to style and formatting.

In vertical lists, numerals or lowercase letters are aligned by periods and can be set flush left or indented. Any run-over lines in the list should be aligned with the first word following the numeral, rather than aligned with the number, lowercase letter, or period. The writer has the option of either capitalizing or lowercasing the first word of each list item, provided that the style is adhered to consistently throughout the manuscript. An editor's work will be substantially increased when a writer capitalizes first-word list items in one sentence and then lowercases the next time around, and if the manuscript is of book length, obviously, errors of this sort will be more difficult for the editor to catch.

Punctuating Vertical Lists

Punctuating vertical lists is often a major source of contention among editors and writers. If the listed items are complete sentences, periods may follow each separate entry in the list. Otherwise, no punctuation marks should be included in the vertical list. The advantage of this simplified method of punctuation is that it eliminates semicolons or commas between entry items. The disadvantage is that no period ends the sentence, which is annoying to many writers and editors [*including this one, who ends lists with a firm period.-FRS*]. An example of the suggested format is found below.

Chapter Four includes the following information on flight dynamics:

1. Regional defenses against multiple ICBM attacks
2. Minimum energy and nonoptimum ICBM elliptic trajectories
3. Dynamic equations of interceptor rockets
4. Free-flight phases of ballistic missile trajectories

In this example, lowercase letters could be used to replace arabic numerals. In addition, each entry in the list could conceivably begin with a lowercase letter instead of a capital letter.

Using Bullets in Vertical Lists

During the past 25 or 30 years, bullets have been used increasingly to enumerate vertical lists. A bullet is a typographical mark, a solid dot, used to distinguish elements in a list. I have been unsuccessful in tracing the exact origin of the term, although I am fairly certain that it is not of British derivation. *The Oxford English Dictionary* does not refer to "bullet" in relationship to printing or proofreaders' marks. In addition, Judith Butcher, chief subeditor of Cambridge University Press and author of *Copy-Editing*, an important reference work for editors, does not mention "bullets" in her list of editorial marks.

In the United States, "bullet" did not appear as a typographical mark in *Webster's Second International Dictionary*, first printed in 1934 and reprinted in 1949. However, the first printing of *Webster's Third New International Dictionary* in 1961 defines the term as "a large solid dot so placed in printed matter as to call attention to a particular passage **[6]**." Thus, the term probably evolved in the United States sometime between 1949 and 1961.

The exact origin of the bullet is a mystery. In his humorous approach to the subject, Frederick P. Szydlik amused us with possible origins that he speculated upon in his 1986 article in *Technical Communication* **[7]**. In July 1983, the editors of *The Editorial Eye* suggested that bullets evolved from printers' center dots, but no specific date of origin was mentioned **[8]**. Robert Compton, writing in the October 1983 issue of *The Editorial Eye*, stated that representatives of the Burger writing course refer to bullets as "Burger dots" in honor of author/teacher Robert S. Burger, who claims to have invented them **[9]**. Burger has been teaching businessmen how to write since 1958, so he could be a likely candidate for authorship.

Style manuals and editing texts do not help much since most of them, including the *U.S. Government Printing Office Style Manual, The Chicago Manual of Style,* and Karen Judd's *Copyediting: A Practical Guide*, do not even list the bullet as a typographical symbol. Even *The Associated Press Stylebook and Libel Manual* refers the reader to "Weapons" when one looks up "Bullets" in the body of the sourcebook.

Bullets are generally used as flagging devices, for they call attention to, and readily distinguish, items in a list. Writers choose bullets instead of numbers or letters when the items in the list have no special rank, sequence, or significance—in other words, when all of the elements in the list are equally important. In addition, bullets should be included when the listed information in each entry is composed of more than one line, e.g., one long sentence, two or more sentences, or a paragraph. In short, bullets are not used in lieu of numbers or letters; instead, they serve a distinct and different purpose in the enumeration process and should be used sparingly. As Szydlik wryly points out in his précis, poor writers are apt to substitute bullets for complete sentences, thereby turning some reports into long bulleted lists that put facts together in piecemeal fashion. In effect, what we are witnessing in private industry is that writers who have trouble constructing sentences are using bullets to convey ideas in the form of phrases.

Writers should take the time to format bulleted lists properly. There are essentially four ways to use bullets. First, bullets can be flush left with the margin, with the text also running over flush left. Second, bullets may be flush left, with the text running over only to the beginning of the first word in the copy (hanging style). Third, bullets may be indented from the left margin, yet when the copy runs over, it is flush left. In the last form, bullets start flush left, the copy extends flush left, but the bullets are run-on in such a way that they follow each sentence in the middle of the paragraph. The following examples differentiate the four methods of using bullets.

```
• XXXXXXXXXXXX          • XXXXXXXXXXXX
XXXXXXXXXXXXXX            XXXXXXXXXXXX
XXXXXXXXXXXXXX          • XXXXXXXXXXXX
• XXXXXXXXXXXX            XXXXXXXXXXXX
XXXXXXXXXXXXXX          • XXXXXXXXXXXX
Flush                   Hanging

  • XXXXXXXX            • XXXXXXXXXXXX
XXXXXXXXXXXX            XXXXXXXXXXXX
XXXXXXXXXXXX            XXXXX • XXXXX
  • XXXXXXXX            XXXXXXXXXXXX
XXXXXXXXXXXX            XX • XXXXXXXXX
Indented                Run-On
```

Writers should decide which style is most appropriate to save space and make for easier reading; then they should avoid combining the various styles, thereby pleasing editors as well as attentive readers.

Concluding Remarks

As an increasing amount of our literature is becoming more scientific and technical, writers are including more lists in their manuscripts. Until there are standardized rules for the compilation of such lists, inconsistencies will increase in the formatting of scientific and technical articles and books. Writers could assist editors in reducing the number of formatting errors in lists by asking themselves these questions:

1. Is there a need for a vertical list, or will a list within a sentence be sufficient?
2. How should the list be punctuated?
3. Should numbers, letters, or bullets be used to format the list?
4. Have I been consistent in preparing all similar lists in the text the same way?
5. Are all of the items in the list parallel in structure?

Astute writers who pay attention to all five questions will improve the quality of their writing, develop an eye for detail, and spare their editors the extra antacid needed to get them through yet another manuscript.

References

1. Kate L. Turabian, *A Manual for Writers of Term Papers, Theses, and Dissertations,* 4th ed. (Chicago: The University of Chicago Press, 1973), 28.

2. *The Chicago Manual of Style,* 13th ed. (Chicago: The University of Chicago Press, 1982), 245.

3. *Publication Manual of the American Psychological Association,* 3d ed. (Washington, D.C.: American Psychological Association, 1983), 68.

4. Marjorie E. Skillin and Robert M. Gay, *Words Into Type,* 3d ed. (Englewood Cliffs, NJ: Prentice-Hall, Inc., 1974), 175.

5. *Council of Biology Editors Style Manual,* 4th ed. (Arlington, VA: Council of Biology Editors, Inc., 1978), 111.

6. *Webster's Third New International Dictionary* (Springfield, MA: G. and C. Merriam Co., 1961), 294.

7. Frederick P. Szydlik, "The Bulleted List in Technical Communication: A High-Caliber Paradigm," *Technical Communication* 33, no. 2 (Second Quarter 1986):93.

8. "Ask the Eye," *The Editorial Eye,* July 1983, 7.

9. Robert A. Compton, "About Those Bullets," *The Editorial Eye,* Late October 1983, 4.

The Practical Writer

Written documents often fail to achieve their purpose because they confuse or irritate readers. This article discusses how to achieve effective, clear style through careful word choice, using active voice instead of passive and concrete words instead of abstract ones whenever possible.

FROM TIME TO TIME most educated people are called upon to act as writers. They might not think of themselves as such as they dash off a personal note or dictate a memo, but that is what they are. They are practising a difficult and demanding craft, and facing its inborn challenge. This is to find the right words and to put them in the right order so that the thoughts they represent can be understood.

Some writers deliberately muddy the meaning of their words, if indeed they meant anything to begin with. When most people write, however, it is to get a message across. This is especially so in business and institutions, where written words carry much of the load of communications. The written traffic of any well-ordered organization is thick and varied—letters, memos, reports, policy statements, manuals, sales literature, and what-have-you. The purpose of it all is to use words in a way that serves the organization's aims.

Unfortunately, written communications often fail to accomplish this purpose. Some organizational writing gives rise to confusion, inefficiency, and ill-will. This is almost always because the intended message did not get through to the receiving end. Why? Because the message was inadequately prepared.

An irresistible comparison arises between writing and another craft which most people have to practise sometimes, namely cooking. In both fields there is a wide range of competence, from the great chefs and authors to the occasional practitioners who must do the job whether they like it or not. In both, care in preparation is of the essence. Shakespeare wrote that it is an ill cook who does not lick his own fingers; it is an ill writer who does not work at it hard enough to be reasonably satisfied with the results.

Unlike bachelor cooks, however, casual writers are rarely the sole consumers of their own offerings. Reclusive philosophers and schoolgirls keeping diaries are about the only writers whose work is not intended for other eyes. If a piece of writing turns out to be an indigestible half-baked mess, those on the receiving end are usually the ones to suffer. This might be all right in literature, because the reader of a bad book can always toss it aside. But in organizations, where written communications command attention, it is up to the recipient of a sloppy writing job to figure out what it means.

The reader is thus put in the position of doing the thinking the writer failed to do. To make others do your work for you is, of course, an uncivil act. In a recent magazine advertisement on the printed word, one of a commendable series published by International Paper Company, novelist Kurt Vonnegut touched on the social aspect of writing: "Why should you examine your writing style with the idea of improving it? Do so as a mark of respect for your readers. If you scribble your thoughts any which way, your readers will surely feel that you care nothing for them."

In the working world, bad writing is not only bad manners, it is bad business. The victim of an incomprehensible letter will at best be annoyed and at worst decide that people who can't say what they mean aren't worth doing business with. Write a sloppy letter, and it might rebound on you when the recipient calls for clarification. Where one carefully worded letter would have sufficed, you might have to write two or more.

Muddled messages can cause havoc within an organization. Instructions that are misunderstood can set people off in the wrong directions or put them to work in vain. Written policies that are open to misinterpretation can throw sand in the gears of an entire operation. Ill-considered language in communications with employees can torpedo morale.

A Careful Writer Must Be a Careful Thinker

In the early 1950s the British Treasury grew so concerned with the inefficiency resulting from poor writing that it called in a noted man of letters, Sir Ernest Gowers, to work on the problem. Out of this Gowers wrote an invaluable book, *The Complete Plain Words,* for the benefit of British civil servants and anyone else who must put English to practical use. (Her Majesty's Stationery Office, London, 1954.)

Gowers took as his touchstone a quotation from Robert Louis Stevenson: "The difficulty is not to write, but to write what you mean, not to affect your reader, but to affect him precisely as you wish." To affect your reader precisely as you wish obviously calls for precision in the handling of language. And to achieve precision in anything takes time.

Gowers suggested that the time spent pursuing precision more than cancels out the time wasted by imprecision. People in administrative jobs might well protest that they were not hired as writers, and that their schedules are crammed enough without having to fuss over the niceties of grammar and the like. The answer to this is that it is an important part of their work to put words on paper. It should be done just as thoroughly and conscientiously as anything else for which they get paid.

No one should be led to believe writing is easy. As great a genius as Dr. Samuel Johnson described composition as "an effort of slow diligence and steady perseverance to which the mind is dragged by necessity or resolution." Writing is hard work because *thinking* is hard work; the two are inseparable. But there is some compensation for the effort invested in trying to write well.

The intellectual discipline required to make thoughts come through intelligibly on paper pays off in clarifying your thoughts in general. When you start writing about a subject, you will often find that your knowledge of it and your thinking about it leave something to be desired. The question that should be foremost in the writer's mind, "What am I really trying to say?" will raise the related questions, "What do I really know about this? What do I really think about it?" A careful writer has to be a careful thinker—and in the long run careful thinking saves time and trouble for the writer, the reader, and everybody else concerned.

The problem is that many people believe that they *have* thought out ideas and expressed them competently on paper when they actually haven't. This is because they use nebulous multi-purpose words that may mean one thing to them and something quite different to someone else. Gowers gave the example of the verb "involve," which is used variously to mean "entail," "include," "contain," "imply," "implicate," "influence," etc., etc. "It has . . . developed a vagueness that makes it the delight of those who dislike the effort of searching for the right word," he wrote. "It is consequently much used, generally where some more specific word would be better and sometimes where it is merely superfluous."

The Right Word Will Almost Tell You Where It Should Go

There are plenty of other lazy man's words lurking about, threatening to set the writer up beside Humpty Dumpty, who boasted: "When I use a word, it means just what I want it to mean." It is therefore wise to avoid words that can be taken in more than one way in a given context. This ties in with the first commandment of practical writing, which is: "Be Specific." "Specify, be accurate, give exact details—and forget about fine writing and original style," Rudolph Flesch says in his book, *How to Be Brief.*

Style tends to take care of itself if you select the right words and put them in the most logical order; so, to a large extent, do grammar and syntax. Find the right word, and it will almost tell you where in a sentence it should go.

This is not to say that grammar and syntax are not important. Words scattered on a page at the discretion of the writer simply would not be comprehensible. The rules of language usage also exert a degree of discipline over your thinking about a subject by forcing you to put your thoughts in logical order. Many grammatical conventions are intended to eliminate ambiguity, so that you don't start out saying one thing and end up saying something else.

Most literate people, however, have an instinctive grasp of grammar and syntax that is adequate for all ordinary purposes. The rules of usage (in English more so than in French) are in any case flexible, changeable, and debatable: new words are invented as the language lives and grows, and a solecism in one generation becomes respectable in the next. So while grammar and syntax have their roles to play in written communications, they must not be adhered to so slavishly that they interfere with intelligible expression. Gowers quoted Lord MacAulay with approval on this score: "After all, the first law of writing, that law to which all other laws are subordinate, is this: that the words employed should be such as to convey to the reader the meaning of the writer."

Vocabulary Is Usually the Least of a Writer's Problems

Since words come first, an ample vocabulary is an asset in conveying meaning. Oddly enough, though, people who have difficulty getting their written messages across rarely lack the vocabulary required. They know the apt words, but they don't use them. They go in for sonorous but more or less meaningless language instead.

People who are perfectly able to express themselves in plain spoken language somehow get the idea that the short, simple words they use in everyday conversation are unworthy to be committed to paper. Thus where they would say, "We have closed the deal," they will write, "We have finalized the transaction." In writing, they "utilize available non-rail ground mode transportation resources" instead of loading trucks. They get caught in "prevailing precipitant climatic conditions" instead of in the rain. They "utilize a manual earth removal implement" instead of digging with a shovel. When so many words with so many meanings are being slung about, nobody can be quite sure of just what is being said.

The guiding principle for the practical writer should be that common words should always be used unless more exact words are needed for definition. The reason for this is so plain that it is all but invisible. It is that if you use words that everybody knows, everybody can understand what you want to say.

A common touch with language has always distinguished great leaders. Winston Churchill comes immediately to mind; he "mobilized the English language and sent it into battle," as John F. Kennedy said. Churchill mobilized the language in more ways than in his inspiring speeches. As Prime Minister of Great Britain, he was that nation's chief administrator at a time when governmental efficiency was a matter of life and death for the democratic world. In August, 1940, while the Battle of Britain was at its peak, Churchill took the time to write a memo about excess verbiage in inter-departmental correspondence. It read:

> Let us have an end to such phrases as these: 'It is also of importance to bear in mind the following considerations . . .' or 'Consideration should be given to carrying into effect . . .' Most of these woolly phrases are mere padding, which can be left out altogether or replaced by a single word. Let us not shrink from the short expressive word even if it is conversational.

Churchill's own wartime letters and memos, reproduced in his memoirs, are models of effective English. It is interesting to speculate on how much his clarity of expression, and his insistence upon it in others, helped to win the war. He was, of course, a professional writer who had earned a living from his pen since he was in his early twenties. He was something of a literary genius. In the light of this, it may seem ridiculous to exhort modern white-collar workers to write like Winston Churchill. Nevertheless, the principles of writing which Churchill followed are not at all hard to grasp.

Churchill was an admirer of H. W. Fowler's *A Dictionary of English Usage*, to which he would direct his generals when he caught them mangling the language. Fowler set five criteria for good writing—that it be direct, simple, brief, vigorous and lucid. Any writer who tries to live up to these is on the right track.

By keeping in mind two basic techniques you can go some way towards meeting Fowler's requirements. These are:

Prefer the active voice to the passive. It will make your writing more direct and vigorous. It's a matter of putting the verb in your sentence up front so that it pulls along the rest of the words. In the active voice you would say, "The carpenter built the house;" in the passive, "The house was built by the carpenter." Though it is not always possible to do so in the context of a sentence, use the active whenever you can.

Prefer the concrete to the abstract. A concrete word stands for something tangible or particular; an abstract word is "separated from matter, practice, or particular example." Churchill used concrete terms: "We have not journeyed all this way, across the centuries, across the oceans, across the mountains, across the prairies, because we are made of sugar candy." If he had couched that in the abstract, he might have said: "We have not proved ourselves capable of traversing time spans and geographical phenomena due to a deficiency in fortitude." Again, there are times when abstractions are called for by the context because there are no better concrete words, but try not to use them unless you must.

Sticking to the concrete will tend to keep you clear of one of the great pitfalls of modern practical writing, the use of "buzz words." These are words and expressions that come into currency not because they mean anything in particular, but merely because they sound impressive. It is difficult to give examples of them because they have such short lives; the "buzz words" of today are the laughing stocks of tomorrow. They are mostly abstract terms (ending, in English, in *-ion, -ance, -osity, -ive, -ize, -al,* and *-ate*), but they sometimes take the form of concrete words that have been sapped of their original meaning. The reason for giving them a wide berth is that their meaning is seldom clear.

Jargon presents a similar pitfall. It has its place as the in-house language of occupational groups, and that is where it should be kept. It too consists mostly of abstract words, and by keeping to the concrete you can shut out much of it. But jargon is contagious, so it should be consciously avoided. Never use a word of it unless you are certain that it means the same to your reader as it does to you.

The combination of the active and the concrete will help to make your prose direct, simple, vigorous, and lucid. There is no special technique for making it brief; that is up to you.

The first step to conciseness is to scorn the notion that length is a measure of thoroughness. It isn't. Emulate Blaise Pascal, who wrote to a friend: "I have made this letter a little longer than usual because I lack the time to make it shorter."

Use your pen or pencil as a cutting tool. No piece of writing, no matter what its purpose or length, should leave your desk until you have examined it intensely with a view to taking the fat out of it. Strike out anything that does not add directly to your reader's understanding of the subject. While doing this, try to put yourself in his or her shoes.

Be hard on yourself; writing is not called a discipline for nothing. It is tough, wearing, brain-racking work. But when you finally get it right, you have done a service to others. And, like Shakespeare's cook, you can lick your metaphorical fingers and feel that it was all worthwhile.

Creating Computer Menus That People Understand Easily

Janice (Ginny) Redish

Menus in computer software help users select the options they want to use for a specific task. This article offers ten guidelines for writing effective on-screen menus for computer software. Notice the two versions of a computer menu, and identify the changes that make the second version more effective.

A MENU IN A COMPUTER SOFTWARE PROGRAM is like a menu in a restaurant. It's a list of choices. From the Main Menu in my word processing program, I can choose to begin any one of several tasks, including:

C Create a new file

E Edit an existing file

P Print a file

S Save a file

just by typing the letter that represents the task I want. The computer responds in a way that is appropriate for the choice I make. If I choose C, it asks me for a name for my new file. If I choose P, it brings up another menu with more choices to help me decide how I want to print the file.

Menus for computer users are relatively new. They've come in with the development of microcomputers and programs for new and casual users.

Traditionally, a computer user had to remember all the commands that make a program work. The user had to remember both which command to use to do each task and exactly how the computer expected the command to be typed. And the commands weren't always mnemonic nor was the syntax always simple and consistent.

Menus make life much easier for users. With a menu, the user only has to recognize the task he or she wants to do. The letter or number that the user types replaces the command that had to be remembered.

(Expert computer users still often prefer command-driven programs because they are faster. With menus, you may have to make several choices in succession to get to exactly the task you want to do. With commands, you can often accomplish several steps at once—for example, tell the system you want to create a new file and name that file all in one command.)

Menus for a computer program come into the realms of both the software developer and the document designer. The software developer usually creates the menu because it is part of the interface—the way that the program works for the user.

But menus are also a form of documentation, and, like other documentation, they can be well designed or poorly designed, clear or confusing, easy to use or difficult to use.

Dr. Joseph Dumas of AIR's New England Office has been studying menu design as he helps clients create software programs and online documentation for them.

The following examples and guidelines are adapted from Dr. Dumas' forthcoming book, *Designing Effective Software for Users.*

You'll find several types of menus in different software programs. Sometimes one line of choices is shown across the top or bottom of the screen you are working on. In some systems, you can "pull down" or "pop up" a menu that covers only part of the screen.

The most common type of menu, however, is still a vertical list of choices that sits on the screen by itself (or in a window on top of the same or another program).

The guidelines in this article focus on the full-screen stand-alone menu, but the guidelines are relevant to other types of menus—and to other aspects of screen design and documentation, as well.

Here are two versions of the same menu.

```
**MENU**

01 ADD INFO
02 DELETE A FILE
03 UPDATE INFO
04 DISPLAY PROJECT INFO
05 MODIFY REPORTS
06 DISPLAY REPORTS
07 QUIT

ENTER OPTION: 06
```

Example 1 A Poor Menu

```
Managing Project Files and Reports

Options:
   1 Create a new file
   2 Delete a file
   3 Look at a file
   4 Change a file

   5 Look at a report
   6 Change a report

   7 Return to the previous menu

Type option number: __
```

Example 2 A Better Menu

We used these guidelines to change the first example into the second:

Guideline 1. Put a descriptive title on every menu. The title MENU doesn't help much. It doesn't distinguish this menu from any other in the program. The second menu has a title that brings the choices together and helps users know where they are in the program.

Guideline 2. Use the center of the screen. Legibility decreases on the periphery of the computer screen. There's no need to jam everything into the left corner.

Guideline 3. Organize the options logically. Users want to scan the list on a menu quickly to find the one they need. On many menus, options seem to be listed in the order in which they occurred to the developer—not in an order that makes sense to the user.

Useful organizing principles are the logical sequence of tasks, expected frequency of use, and alphabetical order.

Guideline 4. Group the options if they cover different topics. Leave a space between groups. Our sample menu really covers three topics—project files, project reports, and "quitting." The second menu makes the structure clearer by leaving spaces.

Guideline 5. Be consistent in the way you present options. The goal of all these guidelines is to make the user's task as easy as possible. Ambiguity creates confusion. To be consistent:

- Use the same word for the same action.
- Use verbs for actions.

- Use common English words, not jargon.
- Spell the words out; don't abbreviate.

The first menu uses both "update" and "modify" to mean the same task. In the second, we've substituted the plain English verb "change" for both.

The first menu uses INFO, but not everyone will realize that means "information." The first menu gives different information about tasks 1 to 4 when they are really parallel tasks—they all refer to project files.

Guideline 6. Use explicit words that make the options clear. What does "quit" mean? Leave the menu? And go where? Leave the program entirely? Users may be reluctant to choose this option because they don't know what will happen.

Guideline 7. Always provide a way out. A "quit" option is important. The user may have gotten to this menu by mistake and may not want to do any of the tasks. The second menu explicitly says, "return to the previous menu."

Guideline 8. Use letters or numbers. If you use numbers, start with 1, not 0, not 01. More research is needed on whether numbers or letters are better. Both have advantages. Letters work well if you can find a one-to-one correspondence of letter and action throughout the program, not just for this menu. But the lack of standardization makes letter choices hard for users to remember. (Does K mean kill a file or keep a file? Does D mean display a file or delete a file?)

Numbers work well, but they refer only to a particular menu. No intrinsic connection exists between the number and the option.

If you use numbers, start with 1. Why ask the user to type two numbers, 0 and 1, when one will do? The number range on the keyboard starts with 1, not with 0.

Guideline 9. Prompt the user for the correct action and set the cursor so the user types the option in the correct place. This guideline relates to the guidelines on being explicit and on using common English. The action verbs "type" or "press" are more appropriate for what you want users to do than the harsh computer jargon "hit" or the ambiguous "ENTER." Does ENTER on the first menu mean the two-step process, type a letter and press the ENTER key, or just type the letter?

Make the correct action consistent through all the menus and consistent with the way the program works. If users have to press ENTER elsewhere, but not on menus, they may do it without thinking and find themselves two steps from where they wanted to be.

Guideline 10. Don't use all capital letters. People read upper- and lowercase letters faster and more easily than all capital letters because the shape of lowercase letters helps in sorting out the words.

Finally, an important question that developers ask is, How many items should a menu have? Researchers don't know what the magic number is—if there is one. Recent research finds that users get to their task more accurately and faster if there are more items per menu and fewer menus (4–8 items on 2–3 menus are better than only 2 items if you need 4 menus to get to the task). More than 10 items on a menu is too many. But don't use two screens for one menu. If you have too many choices for one screen, make a hierarchy like the menu in the first paragraph of this article. There, "Print a file" is one choice, but all the options for how to print a file are on a second menu that I only get to if I choose P for "Print."

The Writer as Market Researcher

Lee Ridgway

Writers who prepare instruction and reference manuals for products must be fully aware of the readers and their needs. This article explains how to interview product users and determine how they use the product in their work and what they need in manuals. Use the sample interview questions to generate your own set of appropriate questions for interviewing potential readers.

AS TECHNICAL WRITERS, we have often defined our audience of readers. An article written for *Scientific American*, for example, is read by a well-defined audience with known characteristics and known variation. Any successful technical publication must know its readers, either through conventional market research techniques (such as reader surveys), association and contact (particularly for specialized trade or hobby magazines), or inspired guesswork by the editor (pretty risky). When we write for such a publication, therefore, we have a good audience description when we start.

But those of us who write instruction and reference manuals for manufacturers have been in a somewhat dangerous position. The people who bought the products (from can openers to software systems) received the information without any effort on their part. Furthermore, if they needed the information, they had to use the material, whether or not it was fun to read, whether or not it was useful, whether or not they understood it. In fact, the primary (buying) audience was the technical people who had built the product. If they didn't like the material, it didn't reach the customers. Everyone assumed that the users were just like the developers, and that developers' reviews were appropriate. So we wrote for an audience of developers.

For complex products, this approach often worked very well. Because only experts were expected to use the products, the developers were in fact very like the users of the products.

For simple products, people simply ignored the information. They used reference material as a last resort, when experimenting was not successful. Sometimes they drew inferences from pictures of the product. Frequently, such logic was enough.

But increasingly, ordinary people are being expected to master complex equipment and learn complex concepts. For many products, logic and past experience do not provide a basis for intuitive understanding. So we need to provide palatable information that ordinary people can cope with.

The Normal Writing Process

Normally, at the beginning of a project someone tells us who the buyers of the product will be. This identification is made by marketing and sales people and is based on some form of crystal-ball gazing. Somehow, given the predicted buyers, we deduce who the actual users will be. During the original writing, reviewing, and testing stages of the manuals, we assume that the identified audience is the real one. We base our information design on that audience, and we use various techniques to see whether the design is acceptable.

Writers use their skills to write truthfully and in a way that is useful to the specified reader. We review for accuracy and completeness, we test for readability, we test for usability, and the test results help us present appropriate information in a helpful way for the assumed audience. When the information is complete, it has already been exposed to all sorts of checking. The information is fine for the audience as we or the product planners defined it.

But did we define it correctly?

Interviewing the Audience

When the products have actually become available and there are people actually using them, you have the opportunity and the duty to see whether the users are as you defined them and whether the tasks they perform are the ones that you supported. To do this, you must take on another role. You must become an interviewer, visiting customers and finding out how they do their work using the product.

Interview Techniques

Being a successful visitor requires a good deal of tact. Frequently visits are very structured, with you and your readers sitting in a conference room together. They are aware that they are criticizing your work and may be quite apologetic about their comments. You have to project genuine interest and acceptance and avoid defensiveness. If you argue, the users will stop volunteering opinions, or they will become angry and uncooperative. You may ask for clarification but must take care that the request doesn't sound challenging. ("I dare you to show me a place where that is true.")

When you ask questions, you should phrase them in such a way that the users do not know what answer you want (which may affect the reply). If it is quite obvious what the "right" answer would be, you may phrase a question in a way that invites another answer. ("I wasn't sure about the presentation of topic. How could it have been made clearer?" Or "I thought about putting in a diagram of the phlisbug. Do you think that would be a good idea?")

Especially during early use of the product, interviews give you the only feedback available, so you want to cover the right ground and talk to the right people. It often helps to have an outline or questionnaire to refer to. If you are asking about a specific manual, questions to ask are like these:

- How frequently do you use this manual?
- Under what circumstances do you use it?
- When did you last use it?
- What tasks are you doing when you use it?
- How do you use it? Do you sit down and read it, or do you refer to it while you are doing something else?
- How would you rate it (excellent, good, fair, poor, awful)?
- What do you like best about it?
- What would you change about it?
- Do you have marginal notations in it?
- Who else uses the manual?
- Do you know what they use it for?

When there is a set of manuals for a product, you may be interested in the library structure and the placement of information within manuals. You may ask questions like these:

- How easy was it to find the information you wanted? Did you
  1. always find the right manual right away
  2. sometimes try another manual before the right one
  3. almost always try the wrong manual first
  4. hunt through several manuals to find the information
  5. mostly not find it
- When you had the right manual, how easy was it to find information in it?
  1. very easy
  2. pretty easy
  3. took more than three tries
  4. had to leaf through looking
  5. never found it
- Were the titles of the manuals
  1. very descriptive
  2. descriptive
  3. all right
  4. not very descriptive
  5. not at all descriptive

If people admit to making marginal comments, and if you can persuade them to share them, you will find them invaluable. Information that people have added for themselves is obviously important to them. The way they organize the information and the level of detail that they add tell you how expert they are and how they want information to be presented. But people are often shy about their notes, because they are hastily or sloppily written. The notes often contain simple information that they feel they should not need, and they are embarrassed to show it to you.

Uses and Advantages of Interviews

During interviews, you are finding out something about the people who are using your material and about the tasks and responsibilities they have. But during these interviews, you see only a few of your readers and you must not generalize too much from the ones you meet and talk to. You can, however, learn about the mental set that at least a few readers have, and try to find out whether the writer and reader share the same assumptions.

Readers are usually not able to pinpoint what they dislike or would change. They know that they are dissatisfied with something, but they often do not know why they are dissatisfied. They are uncomfortable because their expectations (which may not have been conscious) have in some way not been met.

You have to listen carefully to deduce their unfulfilled expectations. Perhaps they thought that the product would be "like a typewriter," meaning that they could sit down and use it immediately. The 300-page user's guide makes them suspect that this is not true. They may then react by saying "The guide is too big," when what they mean is that product requires too much information and it really isn't like a typewriter at all.

Often people have some analogy in their minds when they deal with something new ("It's a blender, really" or "It's an overgrown adding machine"). When the analogy fails in some crucial way, they feel that they have been misled. You need to understand what analogy they were making, whether you encouraged and supported it, and whether it failed or succeeded. If it failed, you need to discourage other readers from thinking along the same lines.

You are also trying to find out whether your understanding of the users' environment and duties is correct. If possible, you should observe users in their work setting, actually using information. You have to be very careful not to interfere with normal activities or to make people feel self-conscious by watching them too closely. But you might ask to see where the manuals are kept, or where the product is located, and see some of the environment that way. During a discussion, you might try to get a description of how things are done, establishing a scenario, as it were, of a day in the life of a terminal operator, or what the procedures are for running the payroll, to help you understand the real tasks.

To sum up, visits enable you to meet real readers, and to get a detailed picture of some readers and some work environments. You find out where your information fits and does not fit a few users. You may get new ideas from your reader sample. However, you have only seen a small sample of your readers. You must decide whether the sample is representative and whether it is complete enough to be useful.

SURVEYING THE AUDIENCE

Many products depend on mass markets, so that a few users may or may not be representative. What would suit one customer perfectly would suit only that one customer. To find out about the whole customer group, you have to prepare questionnaires and survey representative customers. You don't personally need all the skills of statisticians and market researchers, which include how to select samples and how to analyze the statistical significance of data. You do need to be advised by someone who has these skills, if you don't have them.

The selection of the sample is very important because it determines whether the results of the survey apply to the whole customer set. A sample can be the whole set, it can be customers chosen at random, or it can be representative customers. You need advice on sample selection to balance cost, time, and reliability. An appropriately chosen representative sample can yield more reliable results than a random sample of the same size, for example.

Surveys can be done by mail, by telephone, or in person. Mail is the most convenient but the least reliable and the least likely to get responses. You get few responses (sometimes only a few percent), and the people who reply are mostly people with very strong opinions (not necessarily representative). The sample is self-selecting, because every user chooses whether to respond and become a part of the sample. However, mail questionnaires can be longer and more detailed than telephone surveys, and may bring good results if used appropriately.

Telephone surveys have a much higher response rate, and the sample is chosen instead of choosing itself. The results are more reliable, but telephone survey questionnaires have to be written carefully, because the interview may be conducted by someone who does not know the topic at all. Questions have to be simple enough for answerers to understand them without explanation. Any list of choices must be short enough for the answerer to remember on one hearing.

However, you can be comforted by two facts about telephone surveys. First, people (surprisingly) are very cooperative about being interviewed over the telephone. Even so, they probably have an upper limit of about 20 minutes for question answering, except by special arrangement. (Many people do not like to talk on the telephone for long periods, so that their patience wears thin. They might also be very busy.) Second, there are professional market research companies that specialize in telephone surveys. They can help you with the questionnaire, do the actual telephone interviews, and gather and tabulate replies.

Personal-interview surveys are the most expensive and most difficult, but they can also be the most fruitful. The advantage is that a skillful interviewer can find out a great deal in a personal interview. The disadvantages are the cost and the need for a very skillful interviewer. The same considerations of personal interaction that affect visits also affect face-to-face surveys to a smaller degree. Interviewers are less likely than you to signal what they would like to hear. But sending trained interviewers to users can be very expensive, if a large sample is needed or users are spread thinly.

A variation of the face-to-face interview is the round-table discussion, where a number of customers discuss questions in the presence of an interviewer. The group interaction makes it easier for people to express opinions that they feel may not be flattering, but group pressure may give a false impression of unanimity. The interviewer tries to make sure that no one is dominating or dominated to a misleading degree.

Uses and Advantages of Surveys

From surveys, you can get a broad picture of who the real users of the product are, what environment the product is used in, and how satisfied the users are with the product. Although you do not get an in-depth knowledge of any single user (unless the survey was done by personal interviews), you have a lot of sketches of users, and a high degree of confidence that the users sketched are a realistic picture of the various user types.

In addition, surveys can verify task definitions. Although you learn more about actual tasks in detail from visits and long interviews, you can check that those tasks are realistic by survey questions. As a result, task-oriented instructions can be refined if necessary to reflect the way tasks are really done.

Don't expect surveys to give you facts about how long it takes a user to do a task, or find a piece of information, or correct a mistake. People's perception of time depends on how anxious they are, how interested they are, how impatient, or worried, or tired. Their time sense is affected by how long they expected something to take. Five minutes can be a very long time or a very short time, and depending on circumstances will be seen by the user as much too long or very quick. Time estimates will be unreliable. People can tell you whether they were happy or unhappy with the time they spent doing a task. You can then deduce what tasks need to be redesigned so that they can be done more quickly.

At this point, you may go back to the testing lab to find the facts that users could not give you. You can compare how long a task takes in testing with the users' evaluation (too long, very short, etc.) to uncover problem areas and to provide a basis for measuring improved performance. This testing may not test the same tasks that were tested during development. Task descriptions may have been redefined as a result of customer visits and surveys.

You may change users' expectations by changing the description of the task. One description may make the task sound trivial, while a different description can make it seem reasonable that the task should take longer. You may break the task into two parts, describe checkpoints, or give the user a distraction. Tell the user that there will be 10 messages before the task is done, and the user may count the messages instead of being impatient.

In short, you use survey results to verify definitions of tasks and users and to gather user opinions about the information. These opinions, although highly subjective, can be shown to be valid if the sample is large and representative. By structuring the survey questions carefully, you or your statistics expert can find out what factors affected users' judgments and deduce what improvements will have most effect.

SUMMARY

Collection and analysis of feedback are new skills for the technical writer when they are undertaken in a systematic and formal manner. Technical writers working for manufacturers need to be very much aware of their audience and its needs. Since the usual indicators of acceptance do not apply very well to manufacturer-supplied material, technical writers must depend on customer interviews and surveys to gauge their success. They therefore must develop interviewing skills and an understanding of statistical methods—another accomplishment of the Renaissance technical writer.

Six Graphics Guidelines

Typographic elements can help readers understand and use information efficiently. This two-part article explains how to use specific typographic elements, such as type size, white space, and capital letters. Notice the samples of type style and size.

IN THE FIRST PART OF THIS ARTICLE, we present six graphics guidelines that the Document Design Center has developed to help writers and graphic artists produce documents that are both visually appealing and easy to read. Good graphics can help readers to overcome resistance to a text or even help them to understand the material more easily.

1. Use highlighting techniques, but don't overuse them.

Highlighting techniques are a way of emphasizing important aspects of your document by calling attention to them visually. Some highlighting techniques are: *boldface, italics, and white space.* The six guidelines are printed in boldface. The list of highlighting techniques in this paragraph is printed in italics. We have used all of those techniques in *Simply Stated.*

In addition to emphasizing a part of your document, highlighting techniques generally contribute to the attractiveness of your document by providing visual relief in what might otherwise be a uniform page of text.

You can use highlighting techniques to emphasize important points, set off examples from the rest of the text, or set off sections of text by calling attention to headings.

Be careful, however, not to overuse any highlighting technique. If too much of your text is highlighted, the purpose of highlighting will be defeated. For example, if a whole section is in italics, it will not stand out; in fact, it will be more difficult to read than text set in regular type. If too much is in boldface, your text will also become more difficult to read, because the page will look too black and heavy. Also be careful not to use too many different highlighting techniques in one document. This can make your text look cluttered and can lead to confusion. Finally, make sure you are consistent in the way you use a particular technique. For example, if you are going to use boldface for a certain level of heading, be sure always to use it for that level.

2. Use 8 to 10 point type for text.

For most documents, 8 to 10 point type is the most readable size. If your type is too small, your document can look crowded and uninviting. Your readers may skip over text, or they may get eyestrain.

If your type is too large, your text will use more space than it needs to. If you have constraints on space, you may have to cut down on important information. If you choose to add pages to accommodate the large type, it can get expensive. Inappropriately large type can even make a text more difficult to read.

In addition to considering point size (see "Measuring Type"), you must consider the look of the type you choose. Some typefaces look larger than others of the same type size. For example:

This is an example of 10 point Souvenir type.

This is an example of 10 point Avant Garde type.

Simply Stated is set in 10 point Century Schoolbook. See how much more difficult it would be to read if it were set in 6 point type of the same style.

Simply Stated is set in 10 point Century Schoolbook. See how much more difficult it would be to read if it were set in 6 point type of the same style.

3. Avoid making your lines of type too long or too short.

The best line length for most text is 50–70 characters. Experience has shown that this length is less tiring to the eye than either short lines, which make the eye jump back and forth, or long lines, which strain the eye as it tries to stay on course. With either extreme, the eye is likely to jump to another line.

The lines in *Simply Stated* average about 48 characters. If we were to keep the same type size but lay out *Simply Stated* in four columns, the line length would be too short. The short lines would make awkward breaks in the text and cause the eye to jerk back and forth. The change would make for uncomfortable reading.

If we were to keep the same type size and lay out *Simply Stated* all the way across the page, with no column breaks, the long lines would be difficult to follow, and the text would look too densely packed to make pleasant reading. It would even appear to have less leading between the lines. (See "Leading.")

4. Use white space in margins and between sections.

If you use white space well, you can make your document look better and easier to read. The term "white space" refers to any of the blank space on a document, such as the margins and the space between sections, or the space that sets off an example. A text with too little white space can look cramped.

Use white space as an integral element in designing your document, and use it functionally. For example, the white space between sections helps the reader to see how the document is organized. The white space surrounding a title or an example isolates it and emphasizes its importance.

In *Simply Stated* we use white space to isolate the masthead and the titles of the articles. We leave adequate margins and separate each paragraph from the next with white space. If the margins were narrower and there was no space between paragraphs, the text would look cramped and less inviting.

5. Use ragged right margins.

When all the lines of text begin at a left margin but end at different points on the right, the right margin is "ragged." *Simply Stated* is set with a ragged right margin.

When the lines all end in the same place, the right margin is even. The text is "justified." This paragraph is justified, or set "flush left" and "flush right." To get the lines to come out even on the right, the typesetter had to put extra spaces between words and letters all along the line.

Many document designers today prefer ragged right margins, because they are less formal than justified text and create a more relaxed, contemporary look that many readers find inviting. Tastes do change, however, and ragged right margins may go out of fashion.

There are practical advantages to ragged right margins, however. Some printers have found that they reduce production costs, because it is easier to make corrections on unjustified type. It is not necessary to change every line in a paragraph in order to add or take out a word in one line. Some readers find that ragged right margins make a text easier to read. When lines end at different places, it is easier for readers to keep their place in the text. They are less likely to go back and reread the same line twice, because the right profile distinguishes one line from another. Also, the eye does not have to adjust to different spacing between letters, as it does with justified type.

6. Avoid using all capital letters.

Some documents use ALL CAPS, or uppercase letters, for emphasis. But ALL CAPS interferes with the legibility of text. We recommend other highlighting techniques instead, such as boldface, italics, or color.

All caps make text harder to read because the shapes of the letters do not vary very much. One of the ways that readers differentiate among letters is by their shape. Lowercase letters have more distinctive shapes than uppercase. Compare the shape of these two sets of words:

FLASHLIGHT BATTERY

Flashlight battery

When a whole block of text is printed in ALL CAPS, it takes up more space and takes longer to read. Also, the highlighting effect of the uppercase letters is lost because it is overused. See how this paragraph would look in ALL CAPS.

WHEN A WHOLE BLOCK OF TEXT IS PRINTED IN ALL CAPS, IT TAKES UP MORE SPACE AND TAKES LONGER TO READ. ALSO, THE HIGHLIGHT-

ING EFFECT OF THE UPPERCASE LETTERS IS LOST BECAUSE IT IS OVER-USED. SEE HOW THIS PARAGRAPH WOULD LOOK IN ALL CAPS.

A Typography Primer

In this section, we describe the basic features used in typesetting.

Serif and Sans Serif Type

Type comes in many styles, but there are two basic ones: serif and sans serif. Serif type has short, horizontal strokes that project from the tops and bottoms of the letters. *Simply Stated* is printed in a serif typeface. Serifs enhance the horizontal flow of a line of type, making it easier to read across a line. Serifs also make the individual letters easier to distinguish. Ancient Roman stone cutters used serifs to make their hand cut lettering look more even and horizontal. Serif typefaces are generally considered to be formal and traditional.

Sans serif type is type without serifs. It has a clean, contemporary look, and is generally considered to give an informal feeling. The clean design of sans serif type makes it easy to read, but because it does not have serifs to enhance the horizontal flow, a sans serif typeface requires more space between lines of type than a serif typeface of the same size.

serif Typography
sans serif Typography

Families of Type

Most styles of type are available in six different versions. The *style* is the same, but the letters vary in thickness, width, or slant. For example, here are six versions of Helvetica, a sans serif typeface.

Light	typography
Regular	typography
Bold	**typography**
Condensed	typography
Extended	typography
Italic	*typography*

These six versions form a "family" of type. If you want to emphasize individual words or short passages, you can use bold or italic type of the same family as the rest of your text. Titles can be in a larger size of the same face. (See "Measuring Type.") You can mix light, regular, bold, and italic type, as long as they are in the same family. But do not mix condensed or extended type with regular type; the shapes of the letters differ too much to look good together. In addition, condensed and extended faces have their own families of light, bold, and italic.

Measuring Type

Graphic artists work with two basic measurements: *picas* and *points*. They use picas to measure the length of a line or the width of a column. The are six picas to an inch, twelve points to a pica, and 72 points to an inch.

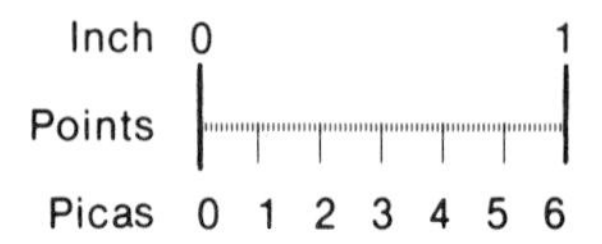

Points are used to measure the size of the type itself—the individual letters. Typefaces are identified by *point size* and *style;* for example, *Simply Stated* is printed in 10 point Century Schoolbook. The point size measures the whole letter, including *ascenders* (the tops of the "T" and the "h") and *descenders* (the tails of the "y" and the "p").

Graphic artists also speak of the "x-height" of a letter. This is the height of the body of the letter *without* ascenders or descenders, or the height of a small letter like an "x." When you are choosing a type size, it is important to consider the "x" height of your typeface. Although two different styles of type may both be 9 point, one may have a smaller "x-height" than the other. If you choose that face, you might need to go up to a 10 point size to get as readable a text as you would with a 9 point size and a larger "x-height." "X-height" will also influence how much space or "leading" you need between lines of type. (See "Leading.")

Leading

The space between lines of type is called *leading*. This term derives from the practice of setting hot metal type. When the type is set, strips of lead are laid between the lines of type. Although phototypesetting is more common than hot metal today, graphic artists still use the term "leading" to describe the space between lines. The leading, like the type itself, is measured in points. A text set in 10 point type with 11 point leading is said to be set "10 on 11," or "10/11." *Simply Stated* is set 10/11.

The amount of leading you use can affect the legibility of your text. If your text is *set solid*, with no extra space between lines, it will look too heavy and difficult to read. If you put in too much leading, your lines of type will appear to drift apart, making it difficult for the reader to follow the text.

Features of the type will affect your leading. Very small type and bold type require extra leading to relieve the density of the text. Sans serif type requires more leading than serif type of the same size, because it does not have serifs to carry the horizontal flow of the line and establish a clean break between the lines. Long lines require more leading than short lines in the same typesize; the extra leading helps the reader to follow across the long lines.

Look how much easier it is to read this passage in 7 point type when it is set with 8 point leading, as opposed to 7 (set solid). But see how it floats apart when it is set 7/11. (This is set 7/8)

Look how much easier it is to read this passage in 7 point type when it is set with 8 point leading, as opposed to 7 (set solid). But see how it floats apart when it is set 7/11. (This is set solid)

Look how much easier it is to read this passage in 7 point type when it is set with 8 point leading, as opposed to 7 (set solid). But see how it floats apart when it is set 7/11. (This is set 7/11)

When You're Not the Whole Staff

When a writing project is a team effort, all team members should be fully informed at all times about progress on the project. This article identifies the typical team members on a large writing project, such as the project director, senior editor, managing editor, and writer.

ARE YOU FACED WITH THE PROSPECT of running a large project in technical documentation? Over the years, we at the Document Design Center have completed projects ranging from a two-hour consultation on a client's letter to a multiyear project in which we produced forty books to document several pieces of software.

The principles of good document design are flexible enough that we have been able to adapt them to different types of projects. Principles of project management also must be adapted. On small projects, management usually comes easily. But in a project with a large staff, multiple contacts at the client's office, and a great number of documents to produce, managing the project becomes a major task.

Now that we've done large projects for several years, we'd like to share with our readers some of the techniques we've developed for keeping these projects running smoothly. There are two keys to a successful large project: open communication and defined project roles.

COMMUNICATING WITH THE CLIENT

The most important principle in managing any size project is to keep the client informed. On a small project in which you are the project director *and* the entire staff, it is easy to pick up the phone and talk with the client's contact person. But on a large project in which many activities are happening at the same time, you need to do more.

At the Document Design Center, we have evolved the "Doctrine of No Surprise." Keep the client—and, we might add, the project staff and your own company's management and administrative staff—fully informed. This includes letting people know beforehand when there is a potential problem. If the problem is averted, then you have good news. If the problem happens as a surprise, that's bad news.

Procedures for keeping the client informed can be as simple as a weekly status conference conducted by phone. For a more complicated project, you might be writing a detailed monthly report that includes project milestones—perhaps by book—and budget information including projected versus actual expenses; current expenses; and cumulative expenses. For some projects, you might have periodic status meetings with the client, or even be required to make a formal presentation with visual aids.

At the beginning of a DDC project, we sit down with the client to jointly work out a format for the monthly report. We structure the report based on how the project is structured, what the schedule is like, what kinds of information the client needs, and what kinds of information we are set up to generate.

Establishing Project Roles

In a small project, the roles staff members take can be flexible. In a large project, it is important to establish formal project roles. When the project involves several books and several writers, there must be a project director whose primary responsibility it is to keep the project on track and communications with the client running smoothly.

Our *project directors* are experienced editors as well as managers. They supervise the project team and provide staff with any resources they need to get the job done, including equipment, software, contact with the client, and clerical support. They may make arrangements for the writers to speak directly to the client's technical experts and to get hands-on experience with the product they are documenting.

Project directors monitor progress on the project and make sure that schedules are met. They review drafts before they are sent out to the client. They keep in touch with the client and let the client know if it looks like the schedules or resources are in jeopardy. They write the monthly report and keep the lines of communication open. In short, they are in charge of eliminating surprises.

Our *senior editors* develop a style sheet and production specifications for each set of books, working with the writers, the project director, and the client. They work individually with each writer in developing the approach to the books and their content and structure. A senior editor establishes a consistent voice or tone for all of the books, and makes sure that the books are consistent in style, content, and the way that specific subjects are treated.

Senior editors develop and monitor schedules and serve as *assistant project directors.* In this role, they supervise the day-to-day business of writing the books. If the project is large enough to have junior editors, they are supervised by the senior editors, as is the copy editing staff.

Some projects are large enough to have more than one senior editor. One of these will then be designated as a *managing editor,* whose job it is to maintain consistency among all of the groups of books.

All of our projects have *copy editors,* who review each draft for grammar, punctuation, adherence to the project's style guide, and correct formatting.

A large project generally includes a *word-processing specialist* or an *administrative assistant,* who helps the writers to produce drafts and keeps monthly backups of each book. This person usually also acts as the project director's assistant and is often responsible for maintaining project archives, tracking any confidential documents that the client has provided, and taking care of routine correspondence with the client.

Last—but by no means least—comes the *writer,* whose work all of these other people support. The writer's job is perhaps the least sharply defined of all.

The writer must learn how to use the product, elicit information from technical experts, translate the technical language into clear English prose, respond to editorial guidance and to the client's needs, type, proofread, and print drafts as necessary to meet deadlines, and serve as a peer reviewer for the other writers and as the representative of the user's point of view.

It is the writer's job to start out ignorant and end up as an expert user. The writer can then write the book that would have made the learning process easier, had it been available before.

Ethical Reasoning in Technical Communication: A Practical Framework

Mark R. Wicclair and David K. Farkas

Technical writers often confront ethical problems just as lawyers and other professionals do. This article discusses three types of ethical principles—goal-based, duty-based, and rights-based—and then describes and analyzes two cases in which a writer had to face an ethical dilemma. Notice that the third ethical case, which appears in a box, has no suggested solutions. Consider what you might do in these circumstances.

PROFESSIONALS IN TECHNICAL COMMUNICATION confront ethical problems at times, just as in law, medicine, engineering, and other fields. In recent years STC has increased its effort to generate a greater awareness and understanding of the ethical dimension of the profession.[1] This article is intended to contribute to that effort.

To clarify the nature of ethical problems, we first distinguish between the ethical perspective and several other perspectives. We then discuss three types of ethical principles. Together these principles make up a conceptual framework that will help illuminate almost any ethical problem. Finally, we demonstrate the application of these principles using hypothetical case studies.

THE NATURE OF ETHICAL PROBLEMS

Typically, when faced with an ethical problem, we ask, "What should I do?" But it is important to recognize that this question can be asked from a number of perspectives. One perspective is an attempt to discover the course of action that will best promote a person's own interests. This is not, however, a question of ethics. Indeed, as most of us have discovered, there is often a conflict between ethical requirements and considerations of self-interest. Who hasn't had the experience of being tempted to do something enjoyable or profitable even while knowing it would be wrong?

A second non-ethical perspective is associated with the law. When someone asks, "What should I do?" he may want to know whether a course of action is required or prohibited by law or is subject to legal sanction. There is often a connection between the perspective of law and the perspective of self-interest, for the desire to discover what the law requires is often motivated by the desire to avoid punishment and other legal sanctions. But between the perspectives of law and ethics there are several significant differences.

First, although many laws correspond to moral rules (laws against murder, rape, and kidnapping, for instance), other laws do not (such as technicalities of corporate law and various provisions of the tax code). Furthermore, almost every legal system has at one time or another included some unjust laws—such as laws in this country that institutionalized racism or laws passed in Nazi Germany. Finally, unethical actions are not necessarily illegal. For example, though it is morally wrong to lie or break a promise, only certain instances of lying (e.g., lying under oath) or promise breaking (e.g., breach of contract) are punishable under law. For all these reasons, then, it is important to recognize that law and ethics represent significantly different perspectives.

One additional perspective should be mentioned: religion. Religious doctrines, like laws, often correspond to moral rules, and for many people religion is an important motive for ethical behavior. But religion is ultimately distinct from ethics. When a Catholic, Protestant, or Jew asks, "What should I do?" he or she may want to know how to act as a good Catholic, Protestant, or Jew. A question of this type is significantly different from the corresponding ethical question: "How do I behave as a good human being?" In answering this latter question, one cannot refer to principles that would be accepted by a member of one faith and rejected by a member of another or a non-believer. Consequently, unlike religiously based rules of conduct, ethical principles cannot be derived from or justified by the doctrines or teachings of a particular religious faith.

THREE TYPES OF ETHICAL PRINCIPLES

To resolve ethical problems, then, we must employ ethical principles. We discuss here three types of ethical principles: goal-based, duty-based, and rights-based. Although these do not provide a simple formula for instantly resolving ethical problems, they do offer a means to reason about ethical problems in a systematic and sophisticated manner.[2]

Goal-Based Principles

Public policies, corporate decisions, and the actions of individuals all produce certain changes in the world. Directly or indirectly they affect the lives of human beings. These effects can be good or bad or a combination of both. According to goal-based principles, the rightness or wrongness of an action is a function of the goodness or badness of its consequences.

Goal-based principles vary according to the particular standard of value that is used to evaluate consequences. But the most widely known goal-based principle is probably the principle of utility. Utilitarians claim that we should assess the rightness of an action according to the degree to which it promotes the general welfare. We should, in other words, select the course of action that produces the greatest amount of aggregate good (the greatest good for the greatest number of people) or the least amount of aggregate harm. Public-policy decisions are often evaluated on the basis of this principle.

Duty-Based Principles

In the case of duty-based principles, the focus shifts from the consequences of our actions to the actions themselves. Some actions are wrong, it is claimed, just for what they are and not because of their bad consequences. Many moral judgments about sexual behavior are in part duty-based. From this perspective, if patronizing a prostitute is wrong, it is not because of harm that might come to the patron, the prostitute, or society, but simply because a moral duty is violated. Likewise, an individual might make a duty-based assertion that it is inherently wrong to lie or break a promise even if no harm would result or even if these actions would produce good consequences.

Rights-Based Principles

A right is an entitlement that creates corresponding obligations. For example, the right to free speech—a right that is valued and protected in our society—entitles people to say what they want when they want to and imposes an obligation on others to let them speak. If what a person says would be likely to offend and upset people, there would be a goal-based reason for not permitting the speech. But from a rights-based perspective, the person is entitled to speak regardless of these negative consequences. In this respect, rights-based principles are like duty-based principles.

While many people believe strongly in the right to free speech, few would argue that this or other rights can never be overridden by considerations of likely consequences. To cite a classic example, the consequence of needless injury and death overrides anyone's right to stand up and cry "Fire!" in a crowded theater. Nevertheless, if there is a right to free speech, it is not permissible to impose restrictions on speech every time there is a goal-based reason for doing so.

Applying These Ethical Principles

Together these three types of ethical principles provide technical communicators with a means of identifying and then resolving ethical problems associated with their work. We should begin by asking if a situation has an ethical dimension. To do this, we ask whether the situation involves any relevant goals, duties, or rights. Next, we should make sure that ethical considerations are not being confused with considerations of self-interest, law, or religion. If we choose to allow non-ethical considerations to affect our decisions, we should at least recognize that we are doing so. Finally, we should see what course, or courses, of action the relevant ethical principles point to. Sometimes, all point unequivocally to one course of action. However, in some cases these principles will conflict, some pointing in one direction, others in another. This is termed an "ethical dilemma." When faced with an ethical dilemma, we must assign priorities to the various conflicting ethical considerations—often a difficult and demanding process.

We now present and discuss two hypothetical case studies. In our discussion we refer to several goal- , duty- , and rights-based principles. We believe that the principles we cite are uncontroversial and generally accepted. In saying this, we do not mean to suggest that there are no significant disagreements in ethics. But it is important to recognize that disputes about specific moral issues often do not emanate from disagreements about ethical principles. For example, although there is much controversy about the morality of abortion, the disagreement is not over the acceptability of the ethical principle that all persons have a right to life; the disagreement is over the non-ethical question of whether fetuses are "persons."

We believe, then, that there is a broad consensus about many important ethical principles; and it is such uncontroversial principles that we cite in our discussion of the following two cases.

Case 1

Martin Yost is employed as a staff writer by Montgomery Kitchens, a highly reputable processed foods corporation. He works under Dr. Justin Zarkoff, a brilliant organic chemist who has had a series of major successes as Director of Section A of the New Products Division. Dr. Zarkoff is currently working on a formula for an improved salad dressing. The company is quite interested in this project and has requested that the lab work be completed by the end of the year.

It is time to write Section A's third-quarter progress report. However, for the first time in his career with Montgomery Kitchens, Zarkoff is having difficulty finishing a major project. Unexpectedly, the new dressing has turned out to have an inadequate shelf life.

When Yost receives Zarkoff's notes for the report, he sees that Zarkoff is claiming that the shelf-life problem was not discovered earlier because a group of cultures prepared by Section C was formulated improperly. Section C, Yost realizes, is a good target for Zarkoff, because it has a history of problems and because its most recent director, Dr. Rebecca Ross, is very new with the company and has not yet established any sort of "track record." It is no longer possible to establish whether the cultures were good or bad, but since Zarkoff's opinion will carry a great deal of weight, Dr. Ross and her subordinates will surely be held responsible.

Yost mentions Zarkoff's claim about the cultures to his very close and trustworthy friend Bob Smithson, Senior Chemist in Section A. Smithson tells Yost that he personally examined the cultures when they were brought in from Section C and that he is absolutely certain they were OK. Since he knows that Zarkoff also realizes that the cultures were OK, he strongly suspects that Zarkoff must have made some sort of miscalculation that he is now trying to cover up.

Yost tries to defuse the issue by talking to Zarkoff, suggesting to him that it is unprofessional and unwise to accuse Ross and Section C on the basis of mere speculation. Showing irritation, Zarkoff reminds Yost that his job is simply to write up the notes clearly and effectively. Yost leaves Zarkoff's office wondering whether he should write the report.

Analysis. If Yost prepares the report, he will not suffer in any way. If he refuses, he will damage his relationship with Zarkoff, perhaps irreparably, and he may lose his job. But these are matters of self-interest rather than ethics. The relevant ethical question is this: "Would it be *morally wrong* to write the report?"

From the perspective of goal-based principles, one would want to know whether the report would give rise to any bad consequences. It is obvious that it would, for Dr. Ross and her subordinates would be wrongly blamed for the foul-up. Thus, since there do not appear to be any overriding good consequences, one would conclude that writing the report would contribute to the violation of a goal-based ethical principle that prohibits actions that produce more bad than good.

Turning next to the duty-based perspective, we recognize that there is a duty not to harm people. A goal-based principle might permit writing the report if some good would follow that would outweigh the harm to Dr. Ross and her subordinates. But duty-based principles operate differently: Producing more good than bad wouldn't justify violating the duty not to harm individuals.

This duty would make it wrong to write the report unless some overriding duty could be identified. There may be a duty to obey one's boss, but neither this nor any other duty would be strong enough to override the duty not to harm others. In fact, there is another duty that favors not writing the report: the duty not to knowingly communicate false information.

Finally, there is the rights-based perspective. There appear to be two relevant rights: the right of Dr. Ross and her subordinates not to have their reputations wrongly tarnished and the company's right to know what is actually going on in its labs. Unless there is some overriding right, it would be morally wrong, from a rights-based perspective, to prepare the report.

In this case, then, goal- , duty- , and rights-based principles all support the same conclusion: Yost should not write the report. Since none of the principles furnishes a strong argument for writing the report, Yost is not faced with an ethical dilemma. But he is faced with another type of dilemma: Since refusing to write the report will anger Zarkoff and possibly bring about his own dismissal, Yost has to decide whether to act ethically or to protect his self-interest. If he writes the report, he will have to recognize that he is violating important goal- , duty- , and rights-based ethical principles.

Case 2

Susan Donovan works for Acme Power Equipment preparing manuals that instruct consumers on the safe operation and maintenance of power tools. For the first time in her career, one of her draft manuals has been returned with extensive changes: Numerous complex cautions and safety considerations have been added. The manual now stipulates an extensive list of conditions under which the piece of equipment should not be used and includes elaborate procedures for its use and maintenance. Because the manual is now much longer and more complex, the really important safety information is lost amid the expanded list of cautions. Moreover, after looking at all the overly elaborate procedures in the manual, the average consumer is apt to ignore the manual altogether. If it is prepared in this way, Donovan is convinced, the manual will actually lead to increased numbers of accidents and injuries.

Donovan expresses her concern in a meeting with her boss, Joe Hollingwood, Manager for Technical Information Services. Hollingwood responds that the revisions reflect a new policy initiated by the Legal Department in order to reduce the number of successful accident-related claims against the company. Almost any accident that could occur now would be in direct violation of stipulations and procedures described in the manual. Hollingwood acknowledges that the new style of manual will probably cause some people not to use the manuals at all, but he points out that most people can use the equipment safely without even looking at a manual. He adds that he too is concerned about the safety of consumers, but that the company needs to protect itself against costly lawsuits. Donovan responds that easily readable manuals lead to fewer accidents and, hence, fewer lawsuits. Hollingwood replies that in the expert opinion of the Legal Department the total cost to the company would be less if the manuals provided the extra legal protection they recommend. He then instructs Donovan to use the Legal Department's revisions as a model for all subsequent manuals. Troubled by Hollingwood's response, Donovan wants to know whether it is ethically permissible to follow his instructions.

Analysis. From the perspective of goal-based principles, writing manuals that will lead to increased injuries is ethically wrong. A possible good consequence is that a reduction in the number of successful lawsuits could result in lower prices. But neither this nor any other evident good consequence can justify injuries that might have been easily prevented.

A similar conclusion is arrived at from a duty-based perspective. Preparing these manuals would violate the important duty to prevent unnecessary and easily avoidable harm. Moreover, there appear to be no overriding duties that would justify violating this duty.

Finally, from a rights-based perspective, it is apparent that important rights such as the right to life and the right to health are at stake. It might also be claimed that people have a right to manuals that are designed to maximize their safety. Thus, ethical principles of each of the three types indicate it would be morally wrong to prepare the manuals according to Hollingwood's instructions.

Because Case 2 involves moral wrongs that are quite a bit more serious than those in Case 1, we now go on to consider an additional question, one not raised in the first case: What course of action should the technical communicator follow?

An obvious first step is to go over Hollingwood's head and speak to higher-level people in the company. This entails some risk, but might enable Donovan to reverse the new policy.

But what if this step fails? Donovan could look for a position with a more ethically responsible company. On the other hand, there is a goal-based reason for not quitting: By staying, Donovan could attempt to prepare manuals that would be safer than those that might be prepared by a less ethically sensitive successor. But very often this argument is merely a rationalization that masks the real motive of self-interest. The company, after all, is still engaging in an unethical practice, and the writer is participating in that activity.

Would leaving the company be a fully adequate step? While this would end the writer's involvement, the company's manuals would still be prepared in an unethical manner. There may indeed be an obligation to take further steps to have the practice stopped. To this end, Donovan might approach the media, or a government agency, or a consumer group. The obvious problem is that successively stronger steps usually entail greater degrees of risk and sacrifice. Taking a complaint outside the organization for which one works can jeopardize a technical communicator's entire career, since many organizations are reluctant to hire "whistle blowers."

Just how much can be reasonably asked of an individual in response to an immoral situation? To this question there is no clear answer, except to say that the greater the moral wrong, the greater the obligation to take strong—and perhaps risky—action against it.

CONCLUSION

The analyses offered here may strike the reader as very demanding. Naturally, we are all very reluctant to refuse assigned work, quit our jobs, or make complaints outside the organizations that employ us. No one wishes to be confronted by circumstances that would call for these kinds of responses, and many people simply would not respond ethically if significant risk and sacrifice were called for. This article provides a means of identifying and analyzing ethical problems, but deciding to make the appropriate ethical response to a situation is still a matter of individual conscience and will. It appears to be a condition of human existence that to live a highly ethical life usually exacts from us a certain price.[3]

Case 3

This case is presented without analysis so that readers can resolve it for themselves using the conceptual framework described in this essay.

A technical writer works for a government agency preparing instructional materials on fighting fires in industrial settings. Technical inaccuracies in these materials could lead to serious injury or death. The technical writer is primarily updating and expanding older, unreliable material that was published 30 years ago. He is trying to incorporate recently published material into the older material, but much of the recent information is highly technical and some of it is contradictory.

The technical writer has developed some familiarity with firefighting through his work, but has no special training in this field or in such related fields as chemistry. He was hired with the understanding that firefighting specialists in the agency as well as paid outside consultants would review drafts of all the materials in order to catch and correct any technical inaccuracies. He has come to realize, however, that neither the agency specialists nor the outside consultants do more than skim the drafts. Moreover, when he calls attention to special problems in the drafts, he receives replies that are hasty and sometimes evasive. In effect, whatever he writes will be printed and distributed to municipal fire departments, safety departments of industrial corporations, and other groups throughout the United States.

Is there an ethical problem here? If so, what is it, what ethical principles are involved, and what kinds of responses are called for?

NOTES

1. Significant activities include the re-establishment of the STC Committee on Ethics, the preparation of the STC "Code for Communicators," the group of articles on ethics published in the third-quarter 1980 issue of *Technical Communication* (as well as several articles published in other places), and the continuing series of cases and reader responses that have appeared in *Intercom*.

2. The arguments justifying ethical principles as well as ethics itself are beyond the scope of this essay but can be examined in Richard B. Brandt, *Ethical Theory: The Problems of Normative and Critical Ethics* (Englewood Cliffs, NJ: Prentice-Hall, 1959) and William K. Frankena, *Ethics*, 2nd ed. (Englewood Cliffs, NJ: Prentice-Hall, 1973). Other informative books are Fred Feldman, *Introductory Ethics* (Englewood Cliffs, NJ: Prentice-Hall, 1978) and Paul W. Taylor, *Principles of Ethics: An Introduction* (Encino, CA: Dickenson Publishing Company, 1975). All of these books are addressed to the general reader.

3. STC might develop mechanisms designed to reduce the price that individual technical communicators have to pay for acting ethically. For their part, individuals may have an obligation to work for the development and implementation of such mechanisms.

Index

Abstract words, 145–146
Abstracts, 343–345. *See also* Executive summaries
  descriptive, 344
  examples, 365–366, 401
  informative, 344–345
  purpose of, 343–344.
Acronyms, 452
Agreement
  pronoun, 443–444
  subject/verb, 446–447
APA reference style
  citations in text, 349–350
  reference list, preparation of, 350–355
Apostrophe, 447–448
Appendix, 347
  example, 369–370
Application letters. *See* Job application letters
Ascending or descending order of importance, organization of, 76–77
Audience. *See* Reader

Back matter, 346–347
Bar graphs, 102–104
Bibliography. *See* Reference list
Brackets, 452
Brochures, examples of, 122–126, 179–180, 247–248
Bulletins, examples of, 24–33, 151–156
Bullets, 120, 577–579

Capital letters, 453–454
Cause and effect, organization of, 77
  for definitions, 166
Changes in meaning
  for definitions, 167–168
Chronological organization. *See also* Instructions; Procedures; Process explanations
  for definitions, 77–78
Classification, organization of, 78
  for definitions, 166
Coherence in paragraphs, 73–75
Collaborative writing
  conflicts in, 19–22
  management review, 19–21
  on a team, 17–19, 602–604
  with a technical editor, 21–22
  with a partner, 16–17
Colon, 448
Comma, 448–449
Comparison or contrast, organization of, 79–80
  for definitions, 167
Computer documentation, 566–572
  online, 543–546, 585–589
Concrete words, 145–146
Conflicts among writers, handling, 19–22
Cutaway drawings, 109–110
  example of 112

Dangling modifiers, 441–442
Dash, 449

Data, interpreting, 312–313. *See also* Reports; Research
Definition
- examples of, 173–180
- expanded, 165–170
- formal sentence, 162–165
- informal, 162
- organization of, 80, 165–170
- placement in text, 170–171
- purpose of, 161

Description
- details needed, 186–189
- examples of, 193–197, 199–208
- for definitions, 167
- graphics for, 189–190
- language, 187–189
- model outline, 192
- organization of, 190–197
- purpose, 184–185
- subjects, 185–186

Descriptive abstract, 344
Document design. *See also* Document design elements; Graphics
- audience, 464–465
- case study, 558–562
- principles, 96–98, 557–558
- structuring, 466–470.

Document design elements. *See also* Document design; Graphics
- bullets, 120, 577–579
- headers and footers, 116–117
- headings, 113–116, 469
- jumplines, 117
- lists, 119–121, 573–579
- logos, 117
- typefaces, 118–119, 466–468, 596–601
- white space, 117–118, 597–598.

Documenting sources
- APA reference style, 349–355
- citations in text, 349–350, 355
- number-reference style, 355
- purpose of, 346–349

Editing, levels of, 501–510. *See also* Revising
Ellipsis, 454–455
Equations, style for, 457–458
Ethics, 482–484, 495–500, 605–612
- case analyses, 608–612
- duty-based principles, 606–607
- goal-based principles, 606
- rights-based principles, 607
- self-testing quizzes, 497, 500, 612

Etiquette, business, 524–527
Etymology, organization of
- for definitions, 168

Examples, organization of
- for definitions, 168–169

Exclamation point, 449
Executive summaries, 343–346
- purpose of, 343–344

Expanded definitions, 165–170. *See also* Definition, organization of
- examples of, 173–180

Exploded drawings, 110–111
- example of, 113

Fad words, 147–148. *See also* Word choice
Feasibility study, 378–381
- example of, 398–405

Flowcharts, 107
- examples of, 109–110, 230, 477–478

Front matter, 341–346

Gathering information, 13, 306–309. *See also* Reader; Research
Glossary, 346–347
- example of, 373–374

Gobbledygook, 145–147
Grammar, 441–447
- dangling modifiers, 441–442
- misplaced modifiers, 442
- parallelism, 443
- pronoun agreement, 443–444
- pronoun reference, 444
- reflexive pronouns, 444–445
- sentence faults, 445–446
- squinting modifiers, 442–443
- subject/verb agreement, 446–447

Graphics. *See also* Document design; Document design elements; Typefaces
- bar graphs, 102–103
- cutaway drawings, 109–110, 112
- exploded drawings, 110–111, 113
- flowcharts, 107, 109–110, 477–478
- in instructions, 213–214
- in oral presentations, 537–542
- in procedures, 213–214, 474–481
- in process explanations, 213–214
- line drawings, 107, 109, 111
- line graphs, 105–106
- maps, 111, 114
- organization charts, 106–108
- photographs, 111–112, 115
- pictographs, 103–105

pie graphs, 106–107
purpose of, 98–100, 546–548
tables, 100–101.

Headers and footers, 116–117
Headings, 113–116, 469
Hyphen, 449–450

Incident report, 381–383
example of, 406–407
Indexes. *See* Secondary sources
Information gathering. *See also* Interviewing; Reader; Research
primary sources, 308–309
secondary sources, 306–308
Informative abstract, 344–345
example of, 401
Initialisms, 452
Instructions. *See also* Procedures
definition of, 213
details, 220–222
examples of, 129–130, 219, 222, 233–238
language, 220–222
model outline, 216
organization, 215–220
readers, 214–215
troubleshooting, 222
warnings and cautions, 217.
International communication, 489–494
Interviewing, 43–45
Investigative report, 383–387
examples of, 302–304, 326–337, 408–413
Italics, 455

Jargon, 142–143
Job application letters, 270–271. *See also* Letters
example of, 288–289
Job interview, 485–488
Jumplines, 117

Language. *See also* Revising; Tone; Word choice; You-attitude
natural, 254
positive, 255
sexist, 143–144, 528–529.
Layout. *See* Document design; Graphics
Letters. *See also* Memorandums
direct organization, 256–259
examples of, 258, 260, 262–263, 276–281, 288–289
format, 267–270
indirect organization, 259–261
job application, 270–271
persuasive, 261–265
purpose, 253
tone, 253–256
transmittal, 342
Library sources, 307–308. *See also* Research
Line drawings, 107, 109, 111
examples of, 194, 200, 205, 208, 234–237
Line graphs, 105
example of, 106
List of figures, 343
example of, 363–364
List of symbols, 346–347
List of tables, 343
Lists
formatting, 573–579
punctuation of, 577
self-testing quiz, 574
using boxes, 120
using bullets, 120, 577–579
using numbers, 119–120, 577
using squares, 120
Logos, 117

Maps, 111
example of, 114
Measurements, style of, 455–456
Memorandums. *See also* Letters; Reports
direct organization, 256–259
examples of, 83–84, 177–178, 282–287
format, 270
indirect organization, 259–261
persuasive, 261–265
principles, 265–267
purpose, 253
tone, 253–256
transmittal, 342
Method-of-operation organization: for definitions, 169
Misplaced modifiers, 442
Modifiers
dangling, 441–442
misplaced, 442
squinting, 442–443
Multiple readers, 40–42. *See also* Reader

NASA tech brief, 199–200

Negation organization
  for definitions, 169
Newsletter article, examples of, 49–53, 89–91
Note-taking, 310–311. *See also* Research
Number-reference documentation style, 355. *See also* Documenting sources; Reference list
Numbers, style for, 456–457

On-the-job writing, types of, 550–552
Oral presentations
  advantages, 429–430
  anxiety, 563–565
  audience, 530–534
  delivery, 436–438, 517–521
  disadvantages, 430
  graphics for, 537–542
  model outline, 433–434
  organization, 431–434
  preparing for, 434–435, 511–517
  purpose of, 429
  self-checklist, 516
  teams for, 438
  types of, 430–431
  workshops, 521–523
Organization, 13–14, 66–67. *See also* Definition; Description; Instructions; Letters; Memorandums; Oral Presentations; Outlines; Paragraphs; Procedures; Process explanations; Reports, types of
Organization charts, 106–107
  example of, 108
Outlines. *See also* Oral presentations; and specific type of document
  examples of, 68, 69, 71, 72, 433–434
  formal, 69
  informal, 67–68
  sentence, 70
  topic, 70.

Paragraphs. *See also* Organization
  ascending or descending order of importance organization, 76–77
  cause and effect organization, 77
  chronological organization, 77–78
  classification organization, 78
  coherence in, 73–75
  comparison and contrast organization, 79–80
  definition organization, 80–81
  development of, 75–76
  partition organization, 78–79
  spatial organization, 81
  unity in, 72–73.
Parallel construction, 443
Paraphrasing, need for, 348
Parentheses, 450
Partition organization, 78–79
  for definitions, 169–170
Photographs, 111–112
  example of, 115
Pictographs, 103
  example of, 105
Pie graphs, 106–107
  example of, 107
Plagiarism, 348
Police report, example of, 200–203
Primary sources, 308–309. *See also* Gathering information; Reader; Reports; Research
Procedures
  decision logic tables, 478–479
  decision trees, 480
  examples of, 24–33, 153–156, 225, 239–244
  flowchart format, 477–478
  for equipment, 224–226
  for one employee, 223–224
  for several employees, 224
  narrative format, 474–475
  playscript format, 224, 476
  purpose, 213, 222–223
  question list format, 476–477
  step-by-step format, 223–224, 475
  topic format, 224–226
Process explanations
  details, 229
  examples of, 227, 230, 245–248
  language, 231
  model outline, 228
  organization, 227–229
  purpose, 213, 226–227
  readers, 226–227
  types of, 226
Product description
  example of, 196–197, 204–208
Progress report, 387–390
  example of,414–416
Pronouns
  agreement, 443–444
  reference, 444
  reflexive, 444–445
Proofreading, 137–138, 460–462
  self-testing quiz, 461–462

Proposal, 392–397
example of, 420–424
Punctuation
apostrophe, 447–448
colon, 448
comma, 448–449;
dash, 449
exclamation point, 449
hyphen, 449–450
parentheses, 450
question mark, 451
quotation marks, 450–451
semicolon, 451
slash, 451

Question mark, 451
Quotation marks, 450–451

Reader
analyzing, 4–5, 9, 36–42, 590–595
informal checking on, 42–43
interviewing, 43–45
multiple, 40–42
testing documents, 45–47, 590–595
Recommendation report, examples of, 85–87, 177–178, 317–322
Reference list, 346, 350–355. *See also* Documenting sources
example of, 368
Report supplements. *See* Back matter; Front matter
Reports. *See also* Reports, types of
analyzing problem, 305–306
examples of, 85–88, 177–178, 303–304, 317–337, 398–424
formal, 341–356
gathering information, 306–310
interpreting data, 312–313
long, 305–315
organization, 300–302, 305, 313–315
purpose, 299–300
short, 300–305
taking notes, 310–311.
Reports, types of
feasibility study, 378–381
incident report, 381–383
investigative, 383–387
progress, 387–390
proposal, 392–397
trip, 390–391
Research
gathering information, 306–309
interpreting data, 312–313
primary sources, 308–309
secondary sources, 306–308
taking notes, 310–311
Résumés, 271–274, 535–536. *See also* Job application letters
examples of, 290–293
Revising
checklist, 15–16
clipped sentences, 139–140
concrete and abstract words, 145–146
content, 135–136
coordination edit, 505
copy clarification edit, 506
copy editing, 501–504
format edit, 506–507
gobbledygook, 146–147
integrity edit, 505–506
jargon, 142–143
language edit, 507–508
lengthy sentences, 140
mechanical style edit, 506–507
organization, 136–137
overloaded sentences, 138–139
policy edit, 505
screening edit, 506
substantive edit, 508
techniques, 137–138
sexist language, 143–144
wordiness, 148–149

Secondary sources, 306–308
Self-testing quizzes
ethics, 497, 500, 612
proofreading, 461–462
writing lists, 574
Semicolons, 451
Sentence faults
comma splice, 445
fragment, 445–446
run-on, 446
Sentences
clipped, 139–140
lengthy, 140
overloaded, 138–139
Sexist language, 143–144, 528–529
Slash, 451
Social work court report, 323
Spatial organization, 81
Squinting modifiers, 442–443
Style. *See* Language; Tone; Word choice, You-attitude

Table of contents, 342–343
  example of, 361–362
Tables, 100–101. *See also* Graphics
Technical article, examples of 52–59, 88–91
Title page, 341–342
  examples of, 357–358, 400
Tone, 253–256, 580–584. *See also* Language; Word choice; You-attitude
Transmittal letters, 342
  examples of, 371–372, 399
Transmittal memorandums, 342
  example of, 359–360
Trip report, 390–391
  example of, 417–419
Typefaces, 118–119; 596–601. *See also* Document design elements

Unity, in paragraphs, 72–73

Visuals. *See* Graphics
Voice
  active, 140–142
  passive, 140–142

White space, 117–118, 597–598. *See also* Document design elements
Word choice, 142–149, 253–256, 580–584
Word processing, revising with, 134–135
Wordiness, 148–149
Workshops, 521–523. *See also* Oral presentations
Writing process. *See also* Gathering information, Organization; Reader
  analyzing purpose, 9–12
  analyzing readers, 9
  analyzing situation, 12–13
  Chicago engineers, 6–8
  drafting, 15
  gathering information, 13
  organizing, 13–14
  revising and editing, 15–16
Writing purpose, 4–5, 9–12
Writing situation, 4–5, 12–13

You-Attitude, 256. *See also* Language; Tone; Word choice

Credits *(Continued)*

Pages 60–61. Ko, H. C. and Brown, R. R., from "Enthalpy of Formation of $2CdO.CdSO_4$." Report of Investigations 8751. (Washington, D.C.: U.S. Bureau of Mines, U.S. Department of the Interior, 1983).

Pages 63–64. Druse, Ken. "Garden Guide," from *House Beautiful* 130 (August 1988). Reprinted by permission of *House Beautiful*, copyright © 1988, The Hearst Publications, Inc. All rights reserved.

Page 74. Schnable, R. R. and Pionke, H. B., from "Acid Rain," in *Meteorology Source Book*. Copyright © 1988 by McGraw-Hill Publishing Co. Reprinted with permission of the publisher.

Page 78. Puchstein, A. F., from "Motor," in *Encyclopedia of Energy*, 2d ed. Copyright © 1981 by McGraw-Hill Publishing Co. Reprinted with permission of the publisher.

Page 79. "The Central Processing Unit," from *The Prentice-Hall Encyclopedia of Information Technology* by Robert A. Edmunds. Copyright © 1987. Used by permission of the publisher, Prentice-Hall, Inc., Englewood Cliffs, N.J.

Page 81. Golden, J. H., from "Tornado," in *Meteorology Source Book*. Copyright © 1988 by McGraw-Hill Publishing Company. Reprinted with permission of the publisher.

Pages 89–91. Horner, B., "The News in Community Broadcasting," *LPTV Report* 4 (April 1989). Reprinted with permission of Kompas/Biel and Associates, Inc.

Page 101. "Agricultural Services, Number of Establishments, Gross Receipts and Payroll, by State, 1978." *Water-Related Technologies for Sustainable Agriculture in U.S. Arid/Semiarid Lands* (Washington, D.C.: U.S. Congress, Office of Technology Assessment, OTA-F-212, October 1983).

Page 102. "Selected Crops Harvested," from U.S. Bureau of the Census, U.S. Department of Commerce, *Census of Agriculture*, 1987. Geographic Area Series, vol. 1 (Washington, D.C.: U.S. Government Printing Office, 1987).

Page 103. "Additions to Lower 48 Natural Gas Proved Reserves: Revisions as Reported, 1966–1983," from *U.S. Natural Gas Availability* (Washington, D.C.: U.S. Congress, Office of Technology Assessment, 1985).

Page 104. From "Trends in U.S. Bureau of Mines R&D." U.S. Congress, Office of Technology Assessment, *Copper: Technology and Competitiveness*, OTA-E-367 (Washington, D.C.: U.S. Government Printing Office, September 1988).

Page 107. "World Titanium Pigment Manufacturing Capacity," from U.S. Congress, Office of Technology Assessment, *Marine Minerals: Exploring Our New Ocean Frontier*, OTA-0-342 (Washington, D.C.: U.S. Government Printing Office, July 1987).

Page 108. Organization Chart. From U.S. Bureau of the Census, U.S. Department of Commerce, *Census of Agriculture*, 1982. Subject Series, vol. 2. (4) (Washington, D.C.: U.S. Government Printing Office, July 1987).

Page 109. "Flow Sheets for Copper Production," from U.S. Congress, Office of Technology Assessment, *Copper: Technology and Competitiveness*, OTA-E-367 (Washington, D.C.: U.S. Government Printing Office, September 1988).

Page 110. "Technologies for Processing Offshore Mineral Ores," from U.S. Congress, Office of Technology Assessment, *Marine Minerals: Exploring Our New Ocean Frontier*, OTA-0-342 (Washington, D.C.: U.S. Government Printing Office, July 1987).

Page 111. "One-Leg-Stand Test," from U.S. Department of Transportation, National Highway Traffic Safety Administration, *Improved Sobriety Testing*, 1984 (Washington, D.C.: U.S. Government Printing Office, 1984).

Page 112. "Components of a Section Dredge," from U.S. Congress, Office of Technology Assessment, *Marine Minerals: Exploring Our New Ocean Frontier,* OTA-O-342 (Washington, D.C.: U.S. Government Printing Office, July 1987).

Page 114. "Hispanic as a Proportion of Total State Population, 1985," from U.S. Bureau of the Census, Current Population Reports, series p-25, no. 1040-RD-1. *Population Estimates by Race and Hispanic Origin for States, Metropolitan Areas, and Selected Counties, 1980 to 1985* (Washington D.C.: U.S. Government Printing Office, 1989).

Page 115. Photograph of 1062 Interior Lobby ATM. Reprinted with permission of Diebold, Inc.

Pages 123–126. From "Exercise and Your Heart." Reproduced with permission. © "Exercise and Your Heart," 1989. Copyright American Heart Association.

Page 128. From *1988 Annual Report.* Reprinted with permission of Chesapeake Corporation.

Page 130. From "Operating Instructions," for Panasonic VCR, Model PV2812. Reprinted with permission of Panasonic Company, a division of Matsushita Electric Corporation of America.

Page 167. Nowak, R. M. and Paradiso, J. L., "Bison," in *Walker's Mammals of the World,* 4th ed., copyright © Johns Hopkins Press, 1983. Reprinted with permission of the publisher.

Page 168. Cadle, R. D., from "Smog," in *Meteorology Source Book,* copyright © 1988 by McGraw-Hill Publishing Company. Reprinted with permission of the publisher.

Page 168. "Tulip," copyright © 1986 by Houghton Mifflin Company. Reprinted by permission from *Word Mysteries and Histories: From Quiche to Humble Pie.*

Pages 170, 173–174. Edmunds, R., "COBOL Programming Language" and "Hacking," in *The Prentice-Hall Encyclopedia of Information Technology.* Copyright © 1987 by Prentice-Hall, Inc. Reprinted with permission of the publisher.

Pages 175–176. Kaplan, L. D., "Greenhouse Effect," in *Meteorology Source Book,* copyright © 1988, McGraw-Hill Publishing Co. Reprinted with permission of the publisher.

Pages 179–180. From "Living with Multiple Sclerosis: A Practical Guide," prepared by Lynn Wasserman, copyright © 1978. Reprinted with permission of the Massachusetts Chapter, National Multiple Sclerosis Society.

Page 186. Hickok, R., "Squash Racquets," in *New Encyclopedia of Sports,* copyright © 1987 by McGraw-Hill Publishing Co. Reprinted with permission of the publisher.

Page 188. From "Supplier Side," *LPTV Report* 4 (October 1989). Reprinted with permission of Kompas/Biel and Associates, Inc.

Pages 194–195. "Cameras." Reprinted with permission of THE ENCYCLOPEDIA AMERICANA, 1988 edition, © Grolier Inc.

Page 200. From NASA *Tech Brief,* 87-1 (Washington, D.C.: National Aeronautics and Space Administration 1987).

Pages 204–208. From Diebold 1000 Modular Delivery System product description manual. Reprinted with permission of Diebold, Inc.

Page 219. "Changing the Ribbon Cassette," reprinted by permission from Actionwriter 1 Typewriter Operator Guide, copyright © 1985 by International Business Machine Corporation.

Page 230. From "Copper." Reprinted with permission of THE ENCYCLOPEDIA AMERICANA, 1988 edition, © Grolier Inc.

Pages 234–238. From Diebold 1060 Everywhere Teller Machine (Model II) Operating Guide. Reprinted with permission of Diebold, Inc.

Pages 242–245. Griffith, H. W., *Complete Guide to Medical Tests,* copyright © 1988 by Fisher Books. Reprinted with permission of the publisher.

Page 246. From "Football." Reprinted with permission of THE ENCYCLOPEDIA AMERICANA, 1988 edition, © Grolier Inc.

Pages 247–248. From "Smoke Detector Technology," National Fire Prevention and Control Administration, U.S. Department of Commerce. (Washington, D.C.: U.S. Government Printing Office, 1977.)

Page 274. Adapted from Harcourt, J. and Krizan, A. C., "A Comparison of Résumé Content Preference of Fortune 500 Personnel Administrators and Business Communication Instructors," *Journal of Business Communication* 26(2) (Spring 1989). Reprinted with permission of the authors.

Pages 326–337. Pirelli, Thomas E., Myron Glassman, Rebecca O. Barclay, and Walter E. Oliu. *Technical Communications in Aeronautics: Results of an Exploratory Study—An Analysis of Managers' and Nonprofit Managers' Responses,* August 1989. (Washington, D.C.: National Aeronautics and Space Administration. NASA TM-101625.) Available from NTIS, Springfield, VA.

Pages 399–405. Revised and adapted from Knapp, Janice M., "Feasibility Study of Computer Systems for Peppermint Stick Nursery School." Reprinted with permission of the author.

Pages 408–413. Gonzalez, G. M., "Early Onset of Drinking as a Predictor of Alcohol Consumption and Alcohol-Related Problems in College," *Journal of Drug Education* 19. Copyright © 1989, Baywood Publishing Co., Inc. Reprinted with permission from the author and publisher.

Pages 460–462. Bagin, C. B. and Van Doren, J., "How to Avoid Costly Proofreading Errors," from *Simply Stated* 65 (April 1986). Reprinted with permission of Document Design Center, American Institutes for Research.

Pages 463–473. Benson, P. J., "Writing Visually: Design Considerations in Technical Publications," from *Technical Communication* 32(4) (1985). Reprinted with permission of Society for Technical Communication.

Pages 474–481. Berry, E., "How to Get Users to Follow Procedures," *Journal of Systems Management* 32 (July 1981). Reprinted with permission of the Association for Systems Management.

Pages 482–485. Blank, S. J., "Greater Concern for Ethics and the 'Bigger Backyard.' " *Management Review* (July 1986). Reprinted by permission of publisher, from *Management Review,* July/1986 © 1986. American Management Association, New York. All rights reserved.

Pages 485–488. Boe, E., "The Art of the Interview," *Microwaves and RF* 24 (December 1985). Reprinted with permission of Penton Publishing Co. Inc.

Pages 489–494. Bowman, J. P. and Okuda, T., "Japanese-American Communication: Mysteries, Enigmas, and Possibilities," *The Bulletin of the Association for Business Communication* 48 (December 1985). Reprinted with permission of the authors.

Pages 495–500. "Is Ethics Good Business?" by Abby Brown. Reprinted with permission from *Personnel Administrator* (February 1987) published by the Society for Human Resource Management, Alexandria, VA.

Pages 501–510. Buehler, M. F., "Defining Terms in Technical Editing: The Levels of Edit as a Model," *Technical Communication* 28(4) (1981). Reprinted with permission of the Society for Technical Communication.

Pages 511–523. Calabrese, R., "Designing and Delivering Presentations and Workshops," *The Bulletin of the Association for Business Communication* 52 (June 1989). Reprinted with permission of the author.

Pages 524–527. Davidson, Jeffrey P., "Astute DP Professionals Pay Attention to Business Etiquette," *Data Management* 24 (October 1986). Reprinted with permission of the author.

Pages 528–529. "Eliminating Gender Bias in Language," *Simply Stated* 28 (August 1982). Reprinted with permission of Document Design Center, American Institutes for Research.

Pages 530–534. Elsea, J. G., "Strategies for Effective Presentations," *Personnel Journal* 64 (September 1985). Reprinted with permission from *Personnel Journal,* Costa Mesa, California. Copyright September 1985. All rights reserved.

Pages 535–536. Erdlen, J. D., "A Good Résumé Counts Most," *Machine Design* 58 (October 23, 1986). Reprinted with permission of Penton Publishing, Inc.

Pages 537–542. Holcombe, M. W. and Stein, J. K., "How to Deliver Dynamic Presentations: Use of Visuals for Impact," *Business Marketing* 71 (June 1986). Reprinted with permission from *Business Marketing*. Copyright Crain Communications Inc.

Pages 543–546. Horton, W., "Writing Online Documentation," *Technical Communication* 35 (November 1988). Reprinted with permission of Society for Technical Communication.

Pages 547–549. "Is It Worth a Thousand Words?" *Simply Stated* 67 (July–August 1986). Reprinted with permission of the Document Design Center, American Institutes for Research.

Pages 550–552. Adapted from Kirtz, M. K. and Reep, D. C., "A Survey of the Frequency, Types and Importance of Writing Tasks in Four Career Areas," *The Bulletin of the Association for Business Communication* (in press). Reprinted with permission of the authors.

Pages 553–556. Kleim, R. L., "Writing Technical Procedures," *Journal of Systems Management* 35 (October 1984). Reprinted with permission of the Association for Systems Management.

Pages 557–562. Marra, J. L., "For Writers: Understanding the Art of Layout," *Technical Communication* 28 (3) (1981). Reprinted with permission of the Society for Technical Communication.

Pages 563–565. Martel, M., "Combating Speech Anxiety." Reprinted with permission from the July 1981 issue of *Public Relations Journal*. Copyright 1981 by the Public Relations Society of America.

Pages 566–572. Perryman, P., "Technical Writing for Computer Software," *Journal of Systems Management* 35 (February 1984). Reprinted with permission of the Association for Systems Management.

Pages 573–579. Plunka, G. A., "The Editor's Nightmare: Formatting Lists Within the Text," *Technical Communication* 35 (February 1988). Reprinted with permission of the Society for Technical Communication.

Pages 580–584. "The Practical Writer," *The Royal Bank Letter* 62 (January/February 1981). Reprinted with permission of The Royal Bank of Canada.

Pages 585–589. Redish, J., "Creating Computer Menus that People Understand Easily," *Simply Stated* 61 (November/December 1985). Reprinted with permission of Document Design Center, American Institutes for Research.

Pages 590–595. Ridgway, L., "The Writer as Market Researcher," *Technical Communication* 32(1) (1985). Reprinted with permission of the Society for Technical Communication.

Pages 596–601. "Six Graphics Guidelines," *Simply Stated* 30 (October 1982). Reprinted with permission of the Document Design Center, American Institutes for Research.

Pages 602–604. "When You're Not the Whole Staff," *Simply Stated* 68 (September 1986). Reprinted with permission of the Document Design Center, American Institutes for Research.

Pages 605–612. Wicclair, M. R. and Farkas, D. K., "Ethical Reasoning in Technical Communication: A Practical Framework," *Technical Communication* 31(2) (1984). Reprinted with permission of the Society for Technical Communication.